ANIDIS
italian national association for earthquake engineering

AFPS
french association for earthquake engineering

PROCEEDINGS OF
THE FRENCH-ITALIAN CONFERENCE
ON

SLOPE STABILITY IN SEISMIC AREAS

May 14-15, 1992
Bordighera (Imperia), Italy

Edited by
E. FACCIOLI & A. PECKER

OUEST ÉDITIONS
Presses Académiques

■

Printed in France by VANDEN, Vigneux-de-Bretagne (Loire-Atlantique)
ISBN 2-908261-52-9

Phone (33) 40.14.34.34 — Fax (33) 40.14.36.36

CONTENTS

FOREWORD

The papers in this volume were presented at a symposium which inaugurated a series of periodic meetings to be jointly held with the Italian Association for Earthquake Engineering (ANIDIS) and the French Association for Earthquake Engineering (AFPS).

The choice of the subject "Seismic Stability of Slopes" was dictated by the need to review and develop methods of analysis in a field where practical design tools are lacking. In that spirit, communications were accepted on related subjects such as earthdams, as well as some papers not directly relevant to the subject, but results of which appear promising and could probably be extended to the seismic analysis of slopes.

As it is readily apparent from the table of contents, the symposium centered around three main themes: case histories and monitoring, microzonation and vulnerability to landslide hazard, modelization and computational aspects.

It is unfortunate that relatively little attention was devoted to the determination of the mechanical characteristics of the constitutive slope materials since it became obvious from the presentations that parameters such as the initial stress conditions, or the evolution in time or with strain of the shear resistance along the assumed failure surfaces, strongly affect the computed results.

Of particular interest for the first theme, appear the studies on historical landslides reactivated by the 1980 Southern Italy earthquake, and characterized by concentrated deformations on well-defined lines of fracture. These point to a subdivision of the landslide into a set of rigid blocks, and thus lend support to the applicability of a class of popular models for simplified dynamic analyses. The case histories in question are also among the rare instances which report quantitative observations of permanent displacements after the earthquake and in successive time intervals. While the pattern of separately moving blocks is observed also in the much larger Puriscal (Costarica) landslide, the displacement field in the latter appears more complex and the causal relation between specific earthquakes and ground deformation less evident.

All participants seemed to agree on the need for more data on the magnitude of coseismic slope movements; the instrumental observation system illustrated in one of the contributions could certainly stimulate similar initiatives on other existing landslides, but different means of investigation should also be explored, e.g. based on satellite images or aerophotographs, to obtain a picture of the displacement field at different times.

In seismic microzoning and in mapping vulnerability to landslide hazard, while the adoption of simple models for stability analysis seems hardly escapable, the use of the sophisticated Geographical Information Systems (GIS) looks in theory ideal for exploring and identifiying critical combinations of triggering factors. The discussion pointed out that the effort required for producing and handling all the relevant cartographic information should not be underestimated, and that GIS pilot studies are critically needed both on a regional (1:50 000) and local (1:5 000) scale, especially in areas characterized by strong variation of seismic ground shaking hazard in the presence of roughly comparable geological conditions. There is a justified expectation in the efforts recently initiated in this field in at least two different research projects sponsored by the EC programme EPOCH on natural hazards.

The presently used methods of analysis can be classified into three categories: the pseudo-static approach, the end result of which is a single parameter, the safety factor; methods based on a simplified evaluation of permanent displacements which all resort to a Newmark type of analysis; the finite element method based on more sophisticated modelizations of the material constitutive behavior. From the presentations and discussions, it appeared that the choice of the relevant method must be made in accordance with the objective: for seismic codes (like Eurocode 8), small structures or structures of secondary importance, and microzonation studies which encompass large areas, the only practical and economical method remains the pseudo-static approach. However, the acceptable "safety factor" has still to be specified. Clear evidence was given that a large static safety factor would result in negligible permanent displacements under dynamic conditions and results of this type could be a basis for a rational choice of an acceptable "dynamic safety factor".

For important structures, the designer has the choice between Newmark types of analyses, and finite element analyses. It seems that the latter should, at the present time, be restricted to man-made structures for which they are capable of giving some qualitative and quantitative results concerning the seismic response of a dam. The input data required by non-linear finite element models seem to be too sophisticated for natural deposits; furthermore these methods appear to be very sensitive to the choice of the constitutive soil parameters, which may govern the numerical stability of the solution. Few oral presentations illustrated failures of non-linear finite element analyses in the prediction of the dynamic stability of natural slopes.

From the modelization session, it was therefore concluded that Newmark types of analyses are the most appropriate for natural slopes: they are easy to use, amenable to parametric studies and cost effective. Moreover, they have been, in some cases, validated by comparison with actual recorded case histories.

In conclusion, based on the quality of the presentations and the wealth of discussions, we believe that the symposium was a success. Andrea del Grosso's participation in all the main aspects of its practical organization was certainly an important contributing factor.

The progress achieved in the field of the seismic stability of natural slopes needs to be assessed in the next two to three years, and this undoubtedly corresponds to the wishes of both Associations.

Ezzio FACCIOLI **A. PECKER**

Thème 1

CASE HISTORIES AND MONITORING

LANDSLIDE OR EARTHQUAKE ? CASE HISTORIES FROM HISTORICAL SOURCES

Paola ALBINI * and Jean VOGT **

Abstract: Many sources, ancient and modern, use the word "earthquake" in a wide sense, including landslides, rockfalls, etc. This use however lead to misinterpretations by uncritical seismic catalogues, listing not only geological processes triggered by earthquakes but also those occurring independently. Examples are either developed or briefly mentioned.

Riassunto: Molte fonti, antiche e moderne, usano il termine "terremoto" in un ampio senso, includendo frane, cadute di rocce, etc. Legata a questo uso è la presenza in cataloghi sismologici non critici non solo di processi geologici innescati da terremoti, ma anche di alcuni avvenuti indipendentemente. Alcuni esempi vengono sviluppati, altri citati in breve.

1 Introduction

While most papers of this workshop are devoted to rockfalls and landslides clearly triggered by earthquakes, this contribution is trying to explain the way such geological processes have been falsely linked with earthquakes, mostly in the past. In a general way recent studies of historical seismology used, among others, the tool of the critical re-appraisal of "catalogues" of events, considered as the last mediation, in a chronological perspective, between us, the sources and the phenomena to be studied. Retrieval and direct reading of the same sources show, among many others, the problem of the very identification of earthquakes. This is specially the case of landslides, rockfalls and so on, thought to be associated with earthquakes. Actually, many of such processes included in catalogues or listings are not at all triggered by earthquakes. Such misinterpretations are linked either to an original "confusion" of the very meaning of the word earthquake, especially "terrae motus", or to a use in a wide sense of the word including whatsoever movement of the earth and genuine earthquake. Moreover, such misinterpretations may have important implications for the discussion of the seismotectonic frame, the appraisal of maximum intensity and estimate of seismic risk. From a methodological point of view, some cases will be discussed at some length, even if misinterpretations have already been corrected to some extent by more critical catalogues, that is even if people agree now that such events are not to be associated with earthquakes. Further, some random examples, gathered during casual reading in the last years, without any systematic research, of more or less ambiguous uses of the word "earthquake" or similar words are mentioned briefly in the frame of a grid, showing on one hand the kind of geological processes and on the other hand a typology of misinterpretations. Only some of these examples will be commented from a methodological point of view: these examples should of course be discussed later in a deeper way, with a larger set of sources and more argumented discussion.

---

* Historian, Istituto per la Geofisica della Litosfera - CNR, Milano, Italy
** Rué du Docteur Woehrlin, 1, 67000 Strasbourg- Robertsau, France

2 Landslides described as earthquakes or misinterpreted as such

2.1 ANNO 1248: "LES ABYMES DE MYANS"

The formation of "les Abymes de Myans", near Chambery in Savoia [Fig. 1], is today interpreted as a consequence of a big landslide fallen from Mont Granier, in the Chartreuse massif [1].

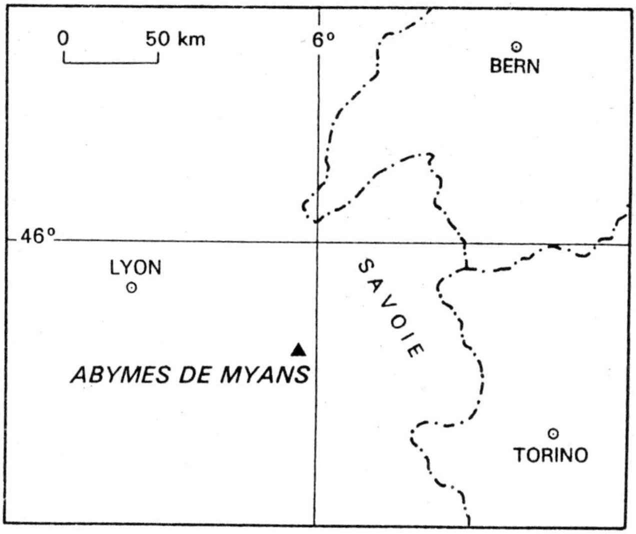

Fig. 1: Geographical location of "les Abymes de Myans", near Chambery.

But this phenomenon has sometimes been considered during the centuries as a consequence of an earthquake. Fig. 2 shows the principal relations among the main sources of information used by some descriptive seismological catalogues of 19th and early 20th century [2, 3, 4, 5, 6] and those quoted by one recent contribution on the earthquakes in Western Europe between 394 and 1259 [7]. Among these authors, some describe the phenomenon merely as a landslide [plain square], some describe it as a landslide triggered by an earthquake [grey square], some raise doubtful questions on the nature of the event [square with a black shadow].

In Tab. 1 three original texts are presented, chosen as pertaining to the range of "primary" sources of information and because they are representative of the more common descriptions of this event.

Both Salimbene [8] and Mattheus Paris [9] are considered reliable chroniclers by current historiography, but none of them was witness of the event. Salimbene seems to be more precise in his description: he is very accurate on geographical location ("in valley Morienne, between the town of Grenoble and the castle of Chambery, one league from Chambery"), on the mechanism of the event ("a mountain, which fell during the night and filled all that valley") and even in referring his source of information ("some years later, 1269, I asked monk Guglielmus, minister of Bourgogne, [...] about the case of that mountain, and what I knew from his talk, I exactly and veraciously wrote").

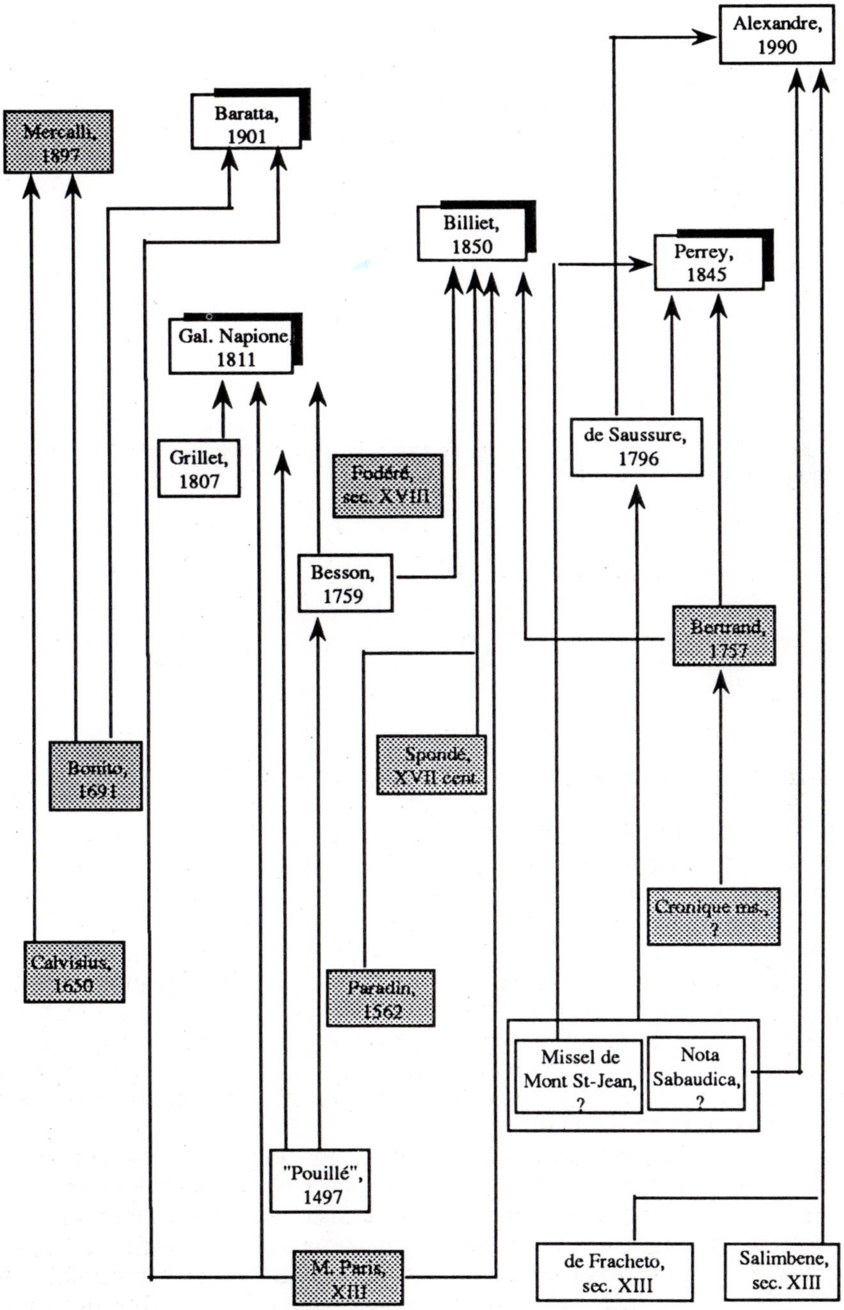

Fig. 2:. Relations among the main sources of information on the 1248 event. Arrows indicate the direction of information.

Salimbene da Parma, sec. XIII	Mattheus Paris, sec. XIII	Pouillé of the dioceses of Grenoble, 1497
"Item in predicto millesimo [1248] in valle Morienne, [...] inter Gracinopolim civitatem et Cambariacum castrum, prope Cambariacum ad unam leucam [...] super quam mons altissimus eminebat, qui de nocte cadens replevit totam vallem illam. Et durat **ruina** in longitudine per unam leucam et in latitudine per unam et dimidiam"	"Eodem tempore [1248] in partibus Sabaudie, in vallibus scilicet Moriani, quaedam villae, scilicet quinque, cum suis bovilibus, caulis, et molendinis adiacentibus, obrutae et absorptae montibus et scopulis, qui vicini erant, de loco creationi suae, **facto horribili terrae motu** [...] "	"Ecclesia sancti Mauricii de Marchiis, etc. Infra dictam parrochiam est venerabilis conventum Fr. Minorum de observantia Beate Mariae de Myana: ibi terminata fuit **ruina** quae de anno Domini 1248, septimo calendas decembris [November 25, cecidit de aspero monte, et quinque parochias destruxit, et signanter villam ecclesiamque sancti Andreae, quae erat titulus dicti Decanatus Sabaudiae"

Tab. 1: Three texts describing the 1248 event, in original language (Latin).

Paris describes the phenomenon as a landslide, but then attributes it to "horribili terrae motu", to introduce, in the second part of his description, his remarks on the sinful attitudes of the people living in the valley, the reason of God's punishment by means of an earthquake. The third testimony [10], not yet retrieved in original, but only in the quotation given by Besson [11], is particularly interesting for being inserted in one official administrative document on the state of the dioceses of Grenoble in the 15th century:. The phenomenon is described with the same expression used by Salimbene ("ruina ... de monte"), and also the day in which the landslide fell is given: "septimo calendas decembris", which corresponds to November 25 according to our calendary. The same day, reported as "St. Catharine's day", which is worshipped on November 25, is given also by a text quoted by de Saussure [12]. This text is reported by Perrey [3] and Alexandre [7] under two different definitions: note from a "Missel de Mont St-Jean" and "Nota Sabaudica", evidenced in Fig. 1 in a bigger square including the two sources.

Serious doubts on the phenomenon being the effect of an earthquake, or an earthquake "tout court", were already been expressed in the 19th century, especially by Billiet [4]: "Sans doute la chute d'une masse aussi énorme a dû produire un grave ébranlement dans tout le décanat de Savoie; mais a-t-elle été produite elle-même par un véritable tremblement de terre?". Unluckily his critical talk has not been used properly by later seismological catalogues, as this event appears in unprinted catalogue by Rothé [13], who gives only part of the text by Paris, taking it from Billiet, but not considering his critical observations. Such event, the nature of which apparently seems clear when one reads the "primary" sources, has originated a sort of "seismic mythology", with duplications and different location of "earthquakes" according to following centuries authors. This aspect has been partly explained by Alexandre [7], to whom authors address those who want to deepen this problem.

2.2 ANNO 1513: THE "BUZZA DI BIASCA" (BELLINZONA)

Two more examples will now be analysed in some details, both concerning phenomena in Central Alps area. Fig. 3 shows the epicentral location and intensity of two earthquakes as recorded by the Italian seismic catalogue [14]: the first in 1513, Bellinzona, the second in 1618, Chiuro.

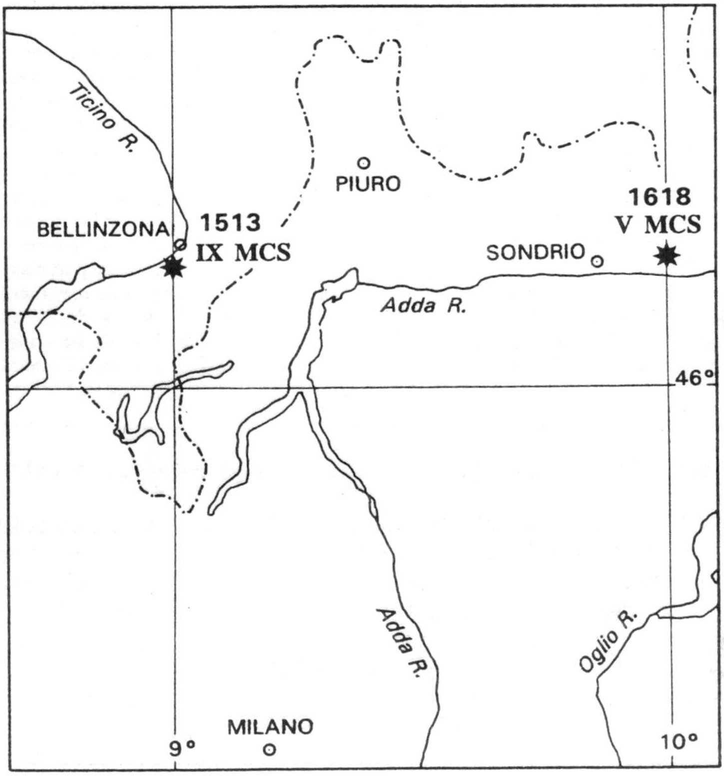

Fig. 3: Epicentral location and intensity of the 1513 and 1618 "earthquakes" according to the catalogue of Italian earthquakes [14].

The 1513 event is recorded in the Italian catalogue under the date of February 10, with epicentral intensity IX MCS, while in the Swiss catalogue [15] the same event is referred as occurred in the year 1512, with epicentral intensity VII MSK. The location of this event near the Insubric lineament makes it a key event for the knowledge of the seismicity of this area.

A recent study [16] has proposed a re-appraisal of the event using both Italian and Swiss sources. The main features of the obtained results will be briefly recalled here.

The research started with the aim of deepening the knowledge of an earthquake, the 1513 February 10, included in the Italian catalogue. The first step consisted in retrieving the sources used by Italian and Swiss catalogues, as they disagreed both on the date and on the maximum intensity of the event.

The scheme of the principal relations among the sources of the two catalogues is presented in Fig. 4.

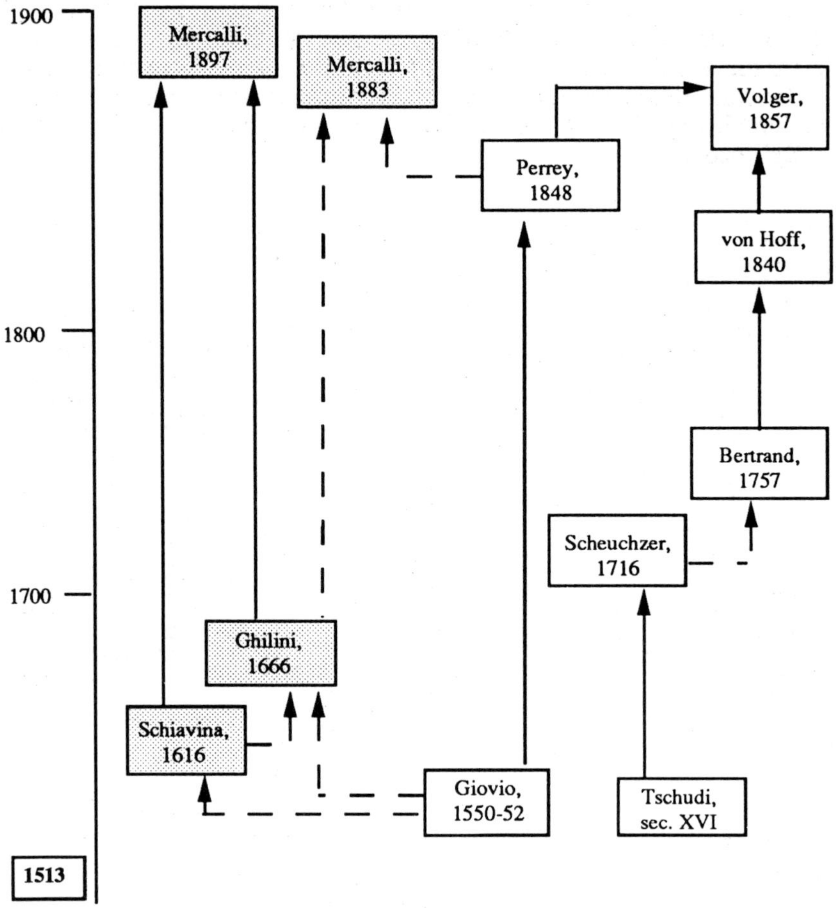

Fig. 4: Principal relations among sources used by Italian [grey square] and Swiss [plain square] catalogues. Arrows indicate the direction of information, dashed line a possible link.

Source of the Italian catalogue is Mercalli [5] who simply quotes the text of Schiavina [17]. The latter author seems to establish no links between the perception of an earthquake in Alessandria and the event near Bellinzona. Mercalli, on the contrary, seems to be sure of such a link, as in one precedent work [18] he lists an event near Bellinzona in the year 1513. Ghilini [19] also quoted by Mercalli, does not add much, but that the earthquake was felt "in other places", without indicating them nor the sources from which he took the information. Looking around in time, searching for other earthquakes occurred in Northern Italy in the same period, we obtain the following list:

1510 February	Alessandria (Piemonte), great part of Italy [19]
1511 February	Alessandria (Piemonte), great part of Italy [17]
1511 March 26	Friuli, Northern Italy, from many sources
1511 February 8	Verona and Venice (Veneto) [6]
1513 February 10	Alessandria [17, 19].

But going back to Mercalli and his sources, this is the information one gets for the year 1513: a landslide north of Bellinzona, an earthquake felt in Alessandria. So it seems clear that the link between the two pieces of information has been created by Mercalli. The compiler of the Italian catalogue adopted uncritically the information given by Mercalli, attributing a IX MCS maximum intensity to the event, in the locality of Bellinzona.

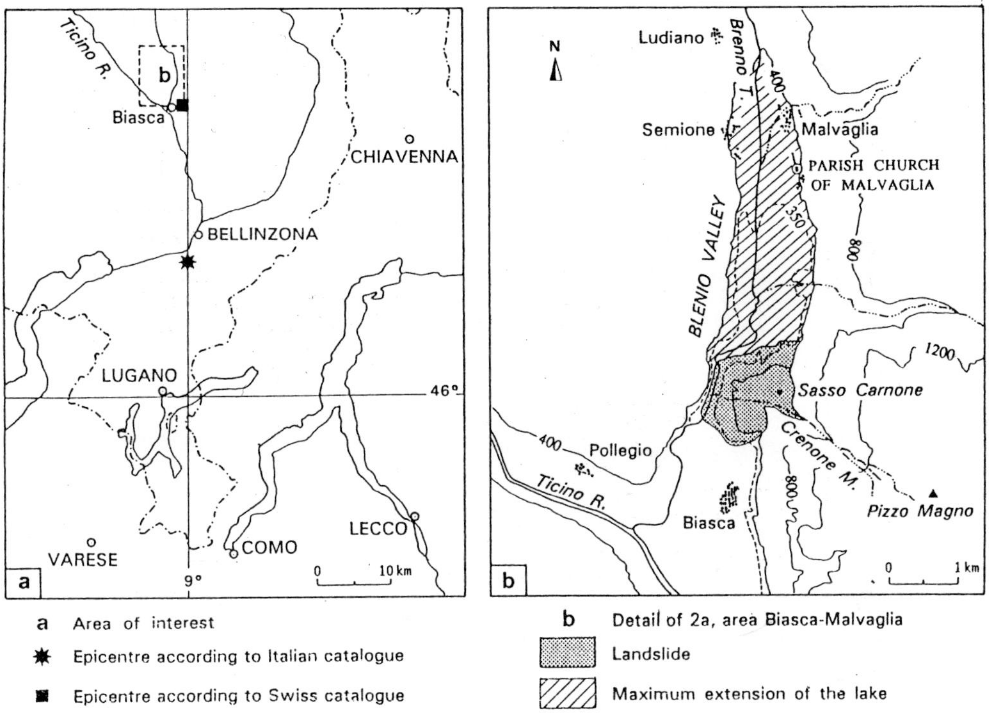

Fig. 5: The "Buzza di Biasca", 1513

As already recalled, the Swiss catalogue [15] lists the event in the year 1512, without indication of month and day, locating it in Biasca, and with an epicentral intensity of VII MSK. Source of the Swiss catalogue [Fig 4] is Volger [20], whose information have a complex rooting, to which is not unrelated Giovio [21], an Italian author coeval to the event, who probably was known also by the other Italian sources quoted above. This is the description he gives of the event, in English translation, while the original is in Latin:

"In the same time [1513] at the roots of the Alps, across which one goes to Switzerland, over Bellinzona a very big mountain, shaken by an earthquake [terraemotu concussus] fell with a big noise ..." [21].

The main interpretative problem seems here to be how has to be read the expression "terraemotu concussus": does he intend, substantially, "landslide" or "earthquake"?

Anyhow, apart from the date problem, both Italian and Swiss sources agree in describing a landslide. The landslide, of big dimension (Fig. 5) detached from mount Crenone, at the mouth of the Blenio Valley, about one kilometre north of the today locality of Biasca. It is well visible also today, and has also a particular name, given by the inhabitants of the valley : "Buzza di Biasca".

The date of the landslide can be fixed, with sufficient reliability, to 1513, September 30, after a document from a notary of Bellinzona, Nicolino Rusca [22], retrieved with a specific research on this event. He describes with accuracy the collapse of part of the mountain and the formation of a lake, after the landslide blocked up the Brenno torrent. The lake overflew about two years after, causing great damage even to Bellinzona. Date of the landslide and of the overflowing are confirmed by an inscription, that may be of the year 1555, like the frescoes surrounding it, and that is still visible today in St. Apollinare chapel of the parish church of Malvaglia, which stayed submerged under the waters for about two years.

Confirmation of the possible seismic triggering of the landslide has been searched in Northern Italy chronicles coeval to the event. No information has been retrieved on any earthquake occurred in the year 1513, even in sources sensible to natural phenomena, particularly to "real" earthquakes occurred in the same period in Northern Italy, as the Friuli earthquake of 1511, March 26. The ensemble of these considerations permit to infer that the landslide was not triggered by an earthquake, particularly not of such proportion as the IX MCS epicentral intensity suggests.

2.3 ANNO 1618: "PLURSIUM TERRAE MOTU OBRUTUM"

The second case concerns an event occurred in Val Chiavenna, in 1618 (Fig. 3). In that year, the locality of Piuro was a wealthy, prosperous locality, in the Swiss Canton of Graubünden, halt on the caravan route going through the Alpine passes of Spluga and Maloia from Italy to the Rhine Valley. On August 25, heavy rains fell in that area, also producing small landslips [23]. Some peasants felt "the earth trembling under the feet" [23]. In the afternoon of September 4, the inhabitants of Piuro attended a religious function, intended to keep away possible catastrophes. But between 21 and 24 hour, according to the Italian style, corresponding to the sunset, a big landslide from mount Conto buried more or less entirely the little town. The flow of the river Mera was stopped for about three hours; 930 on about 1000 inhabitants died.

Reports, letters and also drawings [Fig. 6, 24] recorded the event: whatever expressions they use to indicate the phenomenon, all the evidences tend towards an interpretation of the geological process as not linked with a seismic event; moreover they report that it was possible to foresee such an event. In the reports written by people staying in localities near Piuro in the same days, no information of the perception of an earthquake is reported. Fortunato Sprecher, general delegate of the Swiss government, living in Chiavenna, two kilometres from Piuro, does not use the word earthquake in his description. It has to be stressed that at the time this event occurred Valtellina and Val Chiavenna were under Swiss dominion.

How it happens that such event is interpreted as an earthquake and how could one trace back and suggest the reasons of its being listed in the Italian catalogue?

The main source of Baratta [6], from whom the Italian catalogue takes the information, is a report of the same year [25]. The anonymous compiler defined the event in Piuro "terremoto". A brief review of the main sources of some descriptive seismological catalogues for the Italian peninsula [26, 18, 27] has been carried out.

Fig. 6: The locality of Piuro after the landslide of September 4, 1618 [24].

Here follows some preliminary results: Bonito's sources are lists of natural catastrophes compiled in 17th century. Coeval Swiss authors, like Anhorn [28] are quoted by Bertrand [29] to describe the Piuro event. But then Bertrand adds the information on an earthquake felt in Valtellina in the same year, without quoting specific sources. The integral texts of Bertrand, Perrey, Mercalli and Baratta have been put side by side in Tab. 2.

Bertrand, 1757	Perrey, 1848	Mercalli, 1883	Baratta, 1901
"En mille-six-cent-dix-huit, Pleurs fut enseveli par la chute de la montagne de Conto. Cet accident funeste arriva le 25 aout, pendant la nuit. Les habitants avec leurs maisons furent ensevelis. Il perit plus de douze-cent personnes.	"1618 - 25 aout. Secousses en Suisse; la ville de Pleurs (Grisons) fut ensevelie par la chute d'une montagne.		"1618 settembre 4 - Piuro e Valtellina A 24 ore circa del 4 settembre (25 agosto secondo il Perrey) spiccò in seguito ad un terremoto la montagna di Ciliano e rovinò addosso al paese di Piuro ... insieme a Piuro subissava il villaggio di Schillano"
On essuya aussi le meme tremblement dans la Valteline. On en sentit les secousses dans la plupart des villes du Pays de Vaud, et à Neufchatel"	Ces secousses ayant ebranlé la Valteline doivent figurer dans ce catalogue"	"1618. Agosto 25 in Valtellina e Canton Grigione"	

Tab. 2: Texts of some descriptive seismological catalogues describing 1618 event.

So it seems that Bertrand is responsible for the interpretation of the phenomenon as linked to an earthquake occurred in Valtellina, in 1618, and felt also at Neuchatel and in the Vaud region. But in reality is Perrey who, referring to Bertrand information, amalgamated it. Then Mercalli synthesized the information given by Perrey. Baratta, even if interrogating himself on the nature of the event, is at the end responsible for the insertion of such event in the catalogue of Italian earthquakes. In fact he located the event in "Piuro and Valtellina". Actually, Piuro is in Val Chiavenna, but there is another locality with a similar name, Chiuro, in Valtellina [Fig. 3]: the compiler of the catalogue has been deceived and this is why the epicentre has been located in the latter locality, which is 50 kilometres far from Piuro. Even if one wants to attribute the phenomenon to a low intensity earthquake (V MCS, as suggested by the catalogue), such a landslide could not be originated by an earthquake of epicentral intensity V MCS occurred at a distance of 50 kilometres.

Moreover, some preliminary cross-checks have been made on earthquakes felt in the same area in the same period, studying some coeval sources from Valtellina: they do not refer earthquakes in the year 1618, even if they seem to be interested by natural phenomena, as they clearly refer to other earthquakes in 1621 and 1623.

3 Random examples

After three outstanding, well known cases, some random examples, without references, will now be presented, with a short methodological discussion of some of them. These cases are inserted in the following grid [Fig. 7] intended to show on one hand the kind of geological processes described by the sources (vertical column) and on the other hand a tentative typology of possible misinterpretations of the phenomena (horizontal column). The cases will be commented following the horizontal direction, that has been built like a "crescendo" movement: first, misinterpretations linked to a wide or not precise use of words (terrae motus, earthquake, tremblement de terre, Erdbeben, and so on)today used in a strictly "technical" sense; then "spatial and chronological amalgams" of different phenomena not clearly understood, from a "scientific" point of view, by testimonies; at the end cases referring to a more critic attitude towards a "definition" of the phenomena.

The two examples for the Maghreb area among others are linked to the wide sense of the Arab word "zelzel", which in these regions is indifferently used to indicate earthquakes and whatsoever movements of the earth. A coeval source referring on an indisputable landslide occurred in 1869 in the region of Dahra (north-western Algeria) comments that "Les Arabes des environs ont cru d'abord à un tremblement de terre", as the official letter from the local caid begins with the statement "There has been an earthquake in our town" and goes on describing the landslide [30]. In Kroumirie (north-western Tunisia) a landslide in 1928 left "... impression durable de frayeur ressentie par les habitants qui parlent encore [in 1960] de “tremblement de terre“"[31]. These modern examples can help understanding better the ancient confusions, especially when one has to deepen the knowledge of ancient phenomena in scarcely populated regions.

While geological processes are in most cases easily understood from sources, the different uses of the word earthquake lead to different interpretations, often difficult to be correctly balanced. An extreme example is the use of the word earthquake in titles of newspapers describing normal geological processes. For instance, a paper describing a landslide at Saint-Rabier (Périgord) in 1856, is entitled "Affreux

	PROBABLE USE IN A WIDE SENSE OF THE WORD TERRAE MOTUS, ETC.	FALSE USE OF THE WORD EARTHQUAKE	SPATIAL AMALGAM (WITH AN EARTHQUAKE OCCURRING AT SOME DISTANCE)	CHRONOLOGICAL AMALGAM (WITH AN EARTHQUAKE OCCURRING MORE OR LESS IN THE SAME PERIOD)	INTERROGATION ABOUT AN ASSOCIATION WITH AN EARTHQUAKE	COMPARISON WITH AN EARTHQUAKE	OBSERVATIONS OF PHENOMENA POSSIBLY LEADING TO A SEISMIC INTERPRETATION (NOISE, IMPRESSION OF SMALL SHOCKS LINKED TO THE GEOLOGICAL PROCESS ITSELF)
NOT EASILY IDENTIFIED PROCESSES (DESCRIPTION OF CREVASSES)	1178, Oxenhale (region of Durham, Scotland) 1607, region of Wurzburg 1821, Ligier	1654, Senones (Vosges) *(by a modern author)*					
LANDSLIDES (IN A STRICT SENSE)	1560, Klingenberg (near Dresden) 1593-94, Glarus 1682, Trübenhausen (Hessen) 1726, Peterwardein 1733, Pardines (Auvergne) 1772, Honfleur (Normandie) 1869, Dahra (N-W Algeria) 1927, Kroumirie (N-W Tunisia)	1856, Saint Rabier (Perigord) 1885, Boltana (Aragon)	1356, Rothenburg an der Tauber 1755, Vivarais (Rhine valley) 1776, Gerri de la Sal (Catalunia) 1824, Krinau (Toggenburg) 1828, Oudenaarde (Belgium) *(by popular opinion)* 1865, Spain	1356, Rothenburg an der Tauber 1571, Hertfordshire 1755, Vivarais (Rhine Valley) 1776, Gerri de la Sal (Catalunia) 1824, Krinau (Toggenburg) 1828, Oudenaarde (Belgium) *(by popular opinion)* 1865, Spain	1583, near Dorchester 1790, Monte Pace (near Messina) 1828, Oudenaarde *(by a later author)*	1625, region of Bamberg 1740, Heidelberg 1789, Laver ?, Mosella valley 1806, Goldau (canton Schwyz) 1881, Paptelke (Hungary)	1625, region of Bamberg 1776, Gerri de la Sal (Catalunia)
ROCKFALLS (ALSO IN QUARRIES)	1839, Roussillon	1845, Gapennes (Picardie) 1875, Ronsdorf (Wupperthal)		1828, Sieg valley *(by popular opinion)*			1755, Yorkshire
SOIL SETTLEMENT		1571, Homburg auf der Ohm (Hessen) 1605, region of Wurzburg	1918, Draguignan		1696, region of Portland	1758, Muttenz 1853, Finland	1758, Muttenz
FLOOD							1868, Ticino

Fig. 7: Grid containing date and location of random examples: vertical column indicate geological processes described by the sources, the horizontal one a tentative typology of possible misinterpretations of the phenomena.

tremblement de terre...." [32]. It seems that such a kind of title ("Terremotos en Boltaña", Aragon) is responsible of the presence in the Spanish catalogue of a landslide occurred in 1885.

Sources or authors sometimes connect geological processes with more or less distant earthquakes occurring more or less in the same time-window. Among the cases of such amalgam, a fine example is of course the Lisbon earthquake of 1755, considered responsible for geological processes in some parts of France, for instance Vivarais (Périgord), actually linked to a period of heavy rains. Sometimes, the amalgam seems to be less arbitrary.This is the case for instance of the 1828 landslide in Oudenaarde (Belgium): a landslide in March is linked by contemporary sources to an earthquake felt in the area on February 1828, actually lightly felt in Oudenaarde. A later source raise doubtful questions on the connection with the February earthquake. To the same event a rockfall is connected, occurred two months later in the Sieg valley, where the earthquake was not felt.

Interrogation about a possible link between a landslide having been the effect of a previous big earthquake occurred in the area is raised about a landslide on March 3 , 1790 Monte Pace (Messina): "[...] Peut-être les secousses de 1783 [the big earthquake of Calabria and Sicily, Italy] avaient-elles formé des crevasses dans cette montagne et la mer en minant sa base depuis ce temps lui a-t-elle fait perdre son equilibre" [33].

It also seems that noises or small shocks due to the very process of a landslide may be enough to lead to a misinterpretation, by means of a comparison This happens for the famous landslide in Goldau (canton Schwyz), in 1806: the trembling of the soil, windows and roofs making some noise lead people to compare such effects to those caused by an earthquake.

These case studies and random examples should not be attributed to a general approach by scientists to such a problem. There have always been enlightened people able to distinguish a genuine earthquake through centuries. One fine example is a report on Hoentwiel (Hegau) landslide in 1588, which says:"Es ist aber ganz und gar kein Miraculum nach Erdbidem darau zu verspüren, sondern es ist ein natürlicher Fall" [It is not at all a miracle or an earthquake, but a natural slide] [34]. Another example can be taken from a 17th century Englishman's diary: "1696, 2nd February. [...] An earthquake in Dorsetshire by Portland, or rather a sinking of the ground suddenly for a large space, near the quarries of stone [...]" [35].

General statements are made, also by ancient sources, to avoid confusing reports on phenomena. A particularly "enlightened" man seems to have been Castillon [36], a French author writing in 1771: "On confond très-souvent les frémissemens de la terre avec les commotions des tremblements, et c'est d'après cette erreur qu'on a cru pouvoir mettre dans la classe des phénomènes de ce genre le malheureux événement qui ravagea le territoire de Chieti, capitale de l'Abruzze [Italy], le 22 du mois de Juin 1764: mais il suffit de rapporter les faits tels qu'ils sont passés alors, pour connoître combien se sont tromés ceux qui ont pris les circonstances de ce triste accident, pour autant d'effets marquées d'un tremblement de terre".

4 An open conclusion

Many other famous and enlightening case histories have surely been disregarded in this contribution, that was, in a general sense, intended also to show how the methodology developed in researches of historical seismology [see 37 for other references] can contribute in deepening the knowledge of other natural phenomena, particularly those, truly or falsely, often linked with seismic events.

As historical sources offer a really wealth information on past natural phenomena, the proceeding of the research will lead to discover other cases showing other kinds of possible misinterpretations. This is the reason why authors prefer to conclude without proposing a strict classification of such misinterpretations, that could be misused by superficial compilers of catalogues, as it has sometimes happened in the past. The more critic way to proceed is in fact to consider singularly each case, and trace back to the "original" confusion.

Moreover and despite their utterly critical discussion, the authors of course show keen interest for genuine earthquake-linked geological processes. Beside famous cases, hundreds and thousands, less known, deserve inventories and discussions, by means of ad hoc research. Untangling such processes is one of the most interesting task of historical seismology with its interdisciplinary mind and open perspective.

Acknowledgements

The authors would like to thank M. Stucchi for his help and useful suggestions, and E. Faccioli and D. Slejko for their critical remarks.

Bibliography

[1] Goguel J. and Pachoud A. (1972). Géologie et dynamique de l'écroulement du Mont Granier, dans le Massif de Chartreuse, en novembre 1248. *Bulletin du BRGM*, II série, Seccion III, n° 1, p. 29-38.

[2] Galeani Napione G. (1811). Ricerche intorno ai terremoti del Piemonte. *Mémoires de l'Academie impériale des sciences, littérature et beaux-arts de Turin pour les années 1809-1810*, p. II, t. XIX, p. 143-159.

[3] Perrey A. (1845). Mémoire sur les tremblements de terre ressentis dans le bassin du Rhone, *Annales des sciences physiques et naturelles, d'agriculture et d'industrie puliées par la Société royale d'Agriculture de Lyon*, t. VIII, p. 265-370.

[4] Billiet A. (1850). Mémoire sur les tremblements de terre ressentis en Savoie. *Mémoires de l'Académie Royale de Savoie*, XIII, p. 245-288.

[5] Mercalli G. (1897). *I terremoti della Liguria e del Piemonte*. Napoli.

[6] Baratta M. (1901). *I terremoti d'Italia*. Torino.

[7] Alexandre P. (1990). *Les séismes en Europe occidentale de 394 à 1259. Nouveau catalogue critique*. Bruxelles.

[8] Salimbene da Parma (XIII cent.). *Chronica*. O. Holder-Egger ed., 1905-1913, M.G.H., SS., 32, Hannover and Lipsia.

[9] Mattheus Paris (XIII cent.). *Chronica majora*. H.R. Luard ed., 1880, vol. 5.

[10] *"Pouillé" of the dioceses of Grenoble* (1497). Besson ed., 1759.

[11] Besson (1759). *Mémoires pour l'histoire Ecclesiastique des Diocèses de Genéve, Tarantaise, Aoste et Mauriane, et du Décanat de Savoye*, Nancy.

[12] de Saussure H.B. (1796). *Voyage dans les Alpes*, t. III, Neuchâtel.

[13] Rothé E. (n.d.). *Unprinted seismic catalogue of France*.

[14] Postpischl D., ed. (1985). *Catalogo dei terremoti italiani dall'anno 1000 al 1980*, Quad. Ric. Scient., 114 2B, Bologna.

[15] Mayer-Rosa D. (1988). *Catalogue of the Swiss Seismological Service* (print-out).

[16] Albini P., Bellani A., Stucchi M. (1990). *Terremoti e frane nelle Alpi centrali.* Atti 7° Convegno Annuale Gruppo Nazionale Geofisica Terra Solida, pp. 129-146.

[17] Schiavina G. (1616) *Annales Alexandrini.* V.F. Ponzilioni ed., 1863, HPM, Scriptores, t. IV, Torino.

[18] Mercalli G. (1883). *Vulcani e fenomeni vulcanici in Italia*, Milano.

[19] Ghilini G. (1666). *Annali di Alessandria.* A. Bossola ed., 1903, Alessandria.

[20] Volger O. (1857). *Untersuchungen über das Phänomen der Erdbeben in der Schweiz*, Gotha.

[21] Giovio P. (1550-52). *Historiarum sui temporis.* D. Visconti ed., 1957, t. I, Roma.

[22] Rusca N. (1515). *Documents of May 6 and 25, 1515.* L. Brentani ed., 1963, Antichi maestri d'arte e di scuola delle terre ticinesi, vol. VII, Lugano.

[23] Crollalanza G .B. (1870). *Storia del Contado di Chiavenna.* Milano.

[24] *Theatrum Europaeum* (1662). t. I, 1618-1628, Frankfurt.

[25] Anonymous (1618). *Vera Relatione del grandissimo terremoto e compassionevole infortunio successo à dì 4 di settembre 1618 alla infelice terra di Piur nel stato de' Grisoni.* Viterbo.

[26] Bonito M. (1691). *Terra Tremante.* Napoli

[27] Perrey A. (1848). Mémoire sur les tremblements de terre de la peninsule italique, *Mémoires couronnés et mémoires des savant étrangers*, t. XXII, Bruxelles.

[28] Anhorn B. j. (1665). *Zornzeichen Gottes*, Basel.

[29] Bertrand M.E. (1757). *Mémoires historiques et physiques sur les tremblements de terre*, La Haye.

[30] Bourdon G. (1871). Etude geographique sur le Dahra. *Bulletin de la Société de Géographie de Paris*, 6a s., t. I.

[31] Marthelot P. (1960). *Note sur un décollement de versant dans la vallée des Attafa.* Actes 84e Congrès des Sociétés Savantes, Dijon, 1959, Section de Geographie.

[32] *Courier des Alpes* (1856). May 28.

[33] Anonymous (XVIII cent.]. *Manuscript*, Biblioteque National de Paris, NAF 3135.

[34] Anonymous [XVI cent.]. *Manuscript*, Würtembergische Landesarchiv.

[35] Evelyn J. (XVII cent.). *Diary.* London 1906.

[36] Castillon L. (1771). *Der dernières Révolutions du Globe ....* Bouillon.

[37] Stucchi M. and Albini A. (1991). *New developments in macroseismic investigation.* Proceedings of the Mexico-EC Workshop "Seismology and Earthquake Engineering", Mexico City, April 22-26, 1991 (in press).

DYNAMIC ASPECTS OF A LANDSLIDE REACTIVATED BY THE NOVEMBER 23, 1980 IRPINIA EARTHQUAKE (SOUTHERN ITALY)

Beniamino D'ELIA*

Abstract: The landslide (1.3 km long, 500 m wide, and about 50 m thick) has displaced a gentle slope formed mostly of hard clays. The movement was a multiple rotational slide. By the comparison between displacements as observed 12 hours after the seismic event and displacements as calculated with the Newmark method, the triggering factors and the failure mechanism are considered and discussed.

Riassunto: La frana (lunghezza 1,3 km; larghezza 500 m, e spessore ca. 50 m) ha interessato un pendio a debole inclinazione costituito principalmente da argille consistenti. Il movimento fu uno scorrimento rotazionale multiplo. Per mezzo di un confronto tra spostamenti calcolati col metodo di Newmark e spostamenti osservati 12 ore dopo il sisma si discute sui fattori che hanno scatenato il movimento e sul meccanismo di rottura.

1. Introduction

The 1980 Irpinia earthquake (M_L = 6.5; Duration: 50 s with acceleration peaks > 0.05 g) reactivated many large and thick quiescent landslides in o.c. clay and structurally complex formations [1]. The "Andretta" landslide is one of these landslides. It is located 1.5 km SW of the village of Andretta (near Avellino) and is a 1.3 km long multiple rotational slide averaging a width of 500 m and a maximum thickness of 50 m. The slide displaced a gentle (≈ 8°) slope formed of o.c. hard clay and conglomerate deposits, which had been stable —at least in its middle-upper part— for the latest 200 years (Fig. 1).

The groundwater conditions at the time of the earthquake and the geometric and kinematic characters of the slide have been defined on the basis of previous studies [2], which have also allowed hypotheses on the slide governing and triggering factors and on the failure mechanism.

The aim of the paper is the evaluation of the informations on the failure mechanism and triggering factors, which can be obtained from an *a posteriori* analysis of the phenomenon. Thus, an analysis of permanent displacements adopting the traditional Newmark method was carried out.

The critical acceleration has been obtained with parametric stability analyses taking into account inertia forces and the possible seismically-induced pore water pressure increase as variables.

Calculated displacements have been compared to the displacements that were observed 12 hours after the seismic event.

---

* Professor of Geotechnics, "La Sapienza" University, Rome, Italy

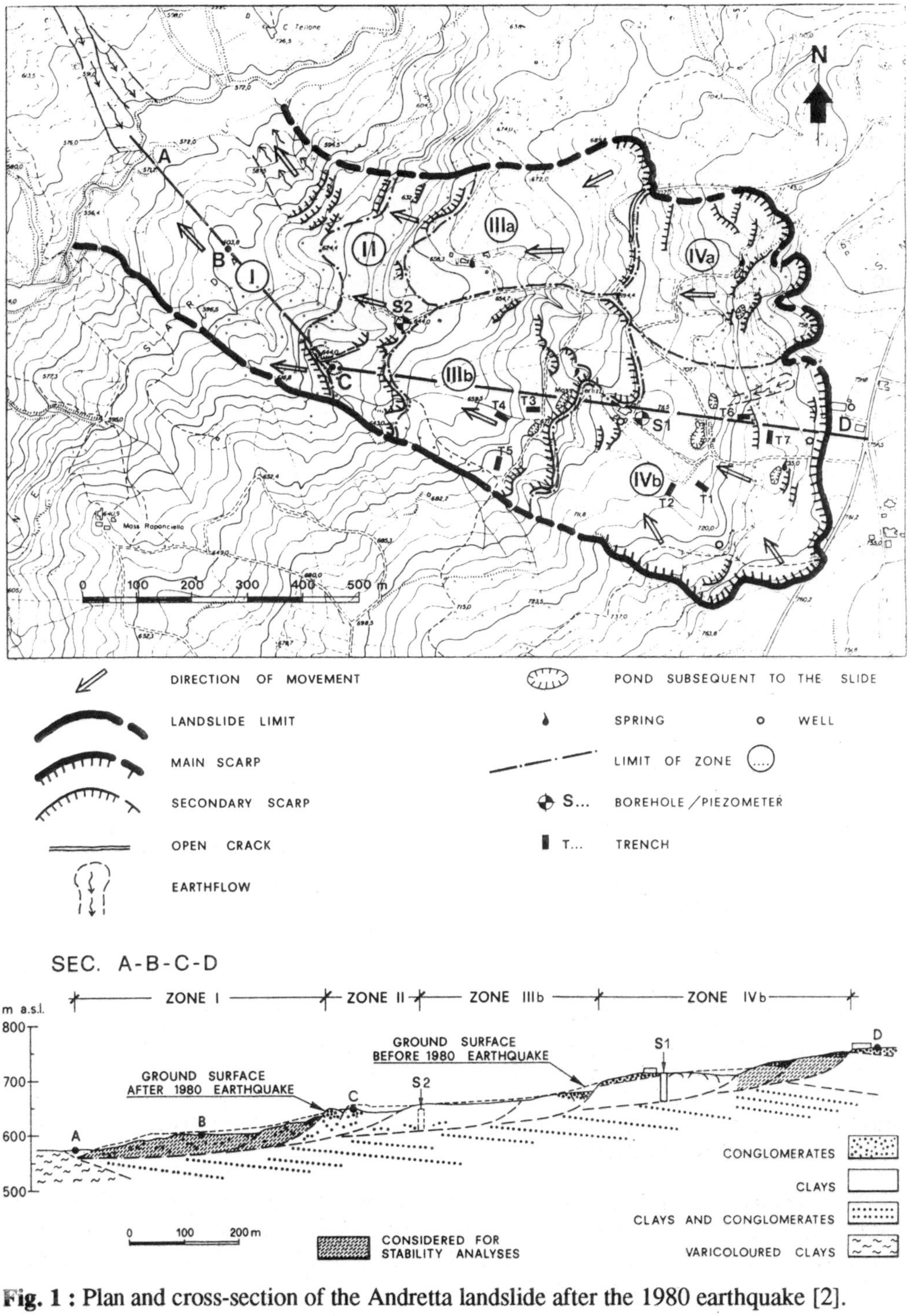

Fig. 1 : Plan and cross-section of the Andretta landslide after the 1980 earthquake [2].

2. Slope stability analyses: evaluation of the critical acceleration

The analyses refer to two parts of the landslide that are different for their geomorphological and hydraulic features (Fig. 1):

a) the lower part (Zone I), where the movements occurred along pre-existing slip surfaces and have involved softened and reworked clayey materials ($\gamma \approx 20$ kN/m^3);

b) the upper part (Zone IVb) —which is delimited by the pre-existing main scarp downhill, and the crown that formed after the earthquake uphill—, where the movements occurred along a newly-formed slip surface and have mostly involved unweathered and not softened clay ($\gamma \approx 22$ kN/m^3).

The stability analysis of the two parts of the landslide body have been carried out independently of one another because of the slide structure consisting in large blocks, each of which controlled by its own equilibrium conditions.

Stability analyses were carried out with the limit equilibrium method in terms of effective stresses.

The seismically-induced pore pressure increase (Δu), which is one of the variable in the analyses, has been given in percent of the average pre-earthquake pore pressure.

Pre-earthquake pore pressure distribution refers to conditions that were possible to assume on the basis of available data, i.e.: water table at the ground level (lower part of the landslide) and at depths varying from 0 to 8 m below ground level (upper part of the landslide).

With taking into consideration the results obtained with laboratory tests on the landslide body clay soils and their structural features, the shear strength parameters in terms of effective stresses have been evaluated as follows.

a) Lower part of the landslide (residual parameters along the pre-existing slip surface): $c'_r = 0$; $\varphi'_r = 16° \div 18°$. The φ'_r range as obtained from tests on softened clay soils ($\varphi'_r = 13° \div 15°$) was increased because the landslide body is composed also of gravel and sand soils.

b) Upper part of the landslide (peak parameters along the newly-formed slip surface): $c' = 0 \div 20$ kPa; $\varphi' = 31°$. The c' value as obtained by laboratory tests on intact samples (c' = 40 kPa) was reduced so as to take into account the clay jointing and fissuring.

The stability analysis of the upper part of the landslide takes into account the constraints imposed by the part of the landslide immediately downhill (Zone IIIb) (Fig. 1). Displacements in such zone have been greater than those in the the upper part of the slide. This fact was taken into account by assuming that the horizontal stabilizing force exerted by Zone IIIb varied from the at-rest state before the earthquake to the active state during and after the earthquake.

The results of stability analyses are shown by the plots of Fig. 2 where the factors of safety F of the upper and lower parts of the landslide are shown as a function of the horizontal seismic coefficient K, of the increase in pore pressure Δu, and of the shear strength parameters.

As to the lower part of the landslide, the factor of safety before the earthquake is in the order of 1.2. If $\Delta u = 0$, the condition $F = 1$ is obtained when $K = 0.02 \div 0.04$. If $K = 0$, the condition $F = 1$ is obtained when $\Delta u \approx 20\%$. The movement onset can thus be the effect either of a low inertia forces, or of a limited increase in pore pressure.

As to the upper part of the landslide, the factor of safety before the earthquake is in the order of 1.4. If $\Delta u = 0$, the condition $F = 1$ is obtained when $K = 0.08 \div 0.10$. If $K = 0$, the condition $F = 1$ is obtained when $\Delta u \approx 40\%$.

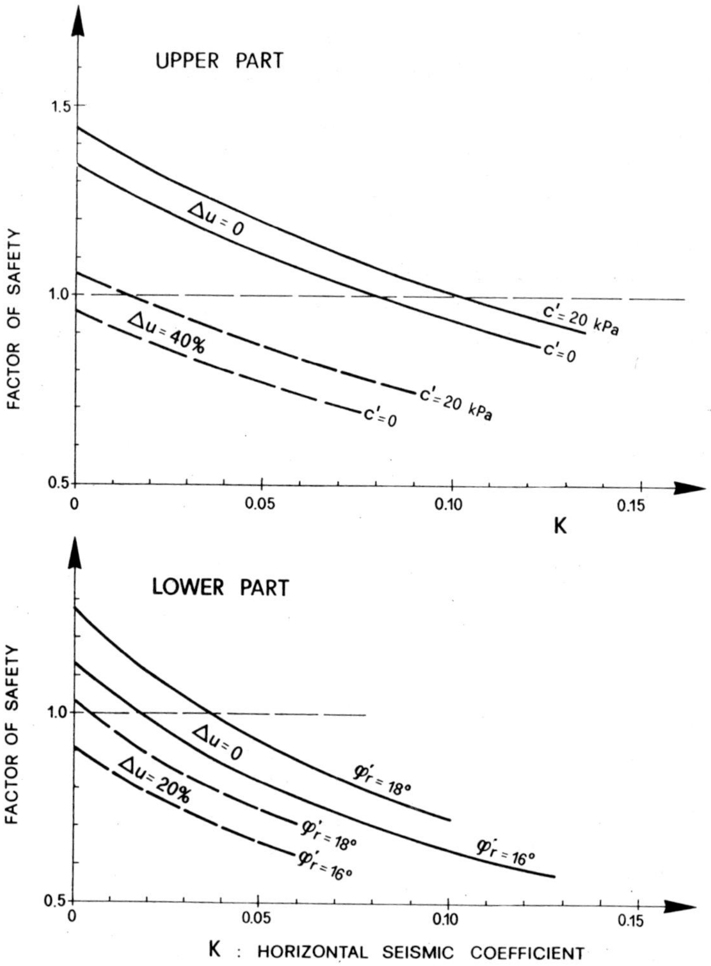

Fig. 2 : Factor of safety vs. horizontal seismic coefficient for different values of the pore pressure increase Δu and shear strength parameters in the considered two parts of the landslide.

The movement onset in this part of the slope can not be attributable only to the increase in pore pressure. As a matter of fact, the stiffness of clay leads to exclude that the earthquake could have caused an increase in pore pressure in the order of the calculated values.

This conclusion is supported by the results of cyclic triaxial tests carried out on o.c. clay [3], showing that the first cycles give rise to a negative pore pressure whereas subsequent cycles lead to a gradual increase in pore pressure, which becomes positive. However, positive pore pressures of samples having OCR > 2 are negligible up to 10^4 cycles.

Thus, the bearing of inertia forces would have been preponderant in governing the limit equilibrium condition of the landslide upper part at the time of the earthquake.

3. Seismic input and displacement analysis

Because ground motion records of the Nov. 23, 1980 earthquake in the landslide zone were not available, the accelerograms recorded by the 10-km-away Calitri accelerometric station were used in the analyses.

To use these records is justified because:

— the Calitri station and the landslide area are at the same distance (about 17 km) from the NW-SE trending seismogenic structure;

— the Calitri station is located on the top of a hill (el. 650 m a.s.l.) formed by a sequence of terrains that is similar to that forming the hill that has been affected by the landslide;

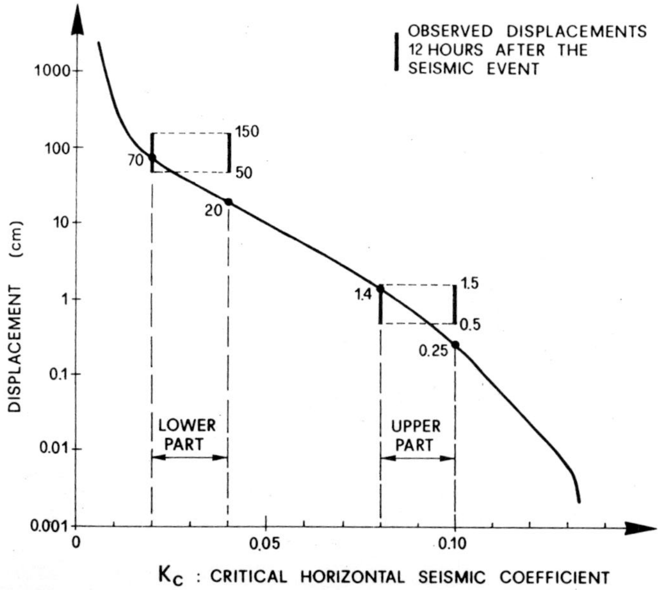

Fig. 3 : Calculated displacements vs critical horizontal seismic coefficient. The observed displacements within the K_C ranges of the lower and upper parts of the slide are also shown.

— following the Nov. 23, 1980 earthquake, the reactivation of a large rotational slide occurred also in Calitri [4].

With keeping into account the orientation of the slid-down slope, the E-W accelerogram was used in the analysis and reference was made to the horizontal seismic accelerations only [5].

By assuming the critical acceleration to be constant during the earthquake, displacements were calculated in the hypothesis of a rigid block sliding on a plane [6]. Displacements have been obtained with the double integration of a digitized corrected accelerogram.

Figure 3 shows the displacements as calculated in the range of values of the horizontal critical acceleration between zero and the maximum corrected acceleration (0.134 g).

For a comparison between the results of the Newmark analysis and the landslide actual displacements, the two parts of the landslide body considered in the stability analyses have been examined.

Based on the critical accelerations or the horizontal critical seismic coefficients (K_c) of the two parts of the landslide as obtained from the aforesaid analyses, the calculated displacements are (Figures 2 and 3):

a) Lower part of the landslide: being $\Delta u = 0$ and $\varphi'_r = 16° \div 18°$, K_c varies from 0.02 to 0.04 corresponding to displacements in the range 70-20 cm.

b) Upper part of the landslide: being $\Delta u = 0$ and $c' = 0 \div 20$ kPa, K_c varies from 0.08 to 0.10 corresponding to displacements in the range 1.4-0.25 cm.

If such displacements are compared to the ones that were observed the morning after the earthquake (50-150 cm in the lower part of the landslide, and 0.5-1.5 cm in the upper one), we see that calculated displacements are of the same order of magnitude as the observed ones. This is particularly true if we consider that in the 12 hours following the main shock, displacements had increased as an effect of aftershocks and of the phenomenon evolution [2].

The slide stopped in about 20 days after the earthquake, resulting displacements being of about 25 m in the lower part of the landslide and about 2 m in its upper part.

4. Triggering factors and failure mechanism

Ground evidences and analytical results are in a rather good agreement, thus suggesting that it is reasonable to think that the chosen theoretical model based on inertia forces is adequate to describe the failure phenomenon, and that these forces may be considered as the triggering factor of the slide.

The reduction of the soils shear strength owing to an increase in pore pressure might have contributed to the development of the landslide only at the slope toe.

Soon after the earthquake, general failure conditions set in within the slope and different displacements occurred in the various parts of the landslide body. In the upper part of it a new slip surface formed at the very time of the earthquake.

5. Discussion on the possibility of improving analyses

The analyses as carried out include assumptions on the geotechnical properties of the slope forming soils, on the seismic input and on the role played by the pore pressure increase. On this respect we can make a few considerations to see whether it is possible to improve the reliability of the obtained results.

Soil shear strength - In the analyses, it has been neglected that the soils shear strength depends on seismic loading conditions.

When we considered the upper part of the landslide, where the failure had occurred along a newly-formed slip surface, it was not taken into account either the possible increase in shear strength owing to dynamic loading and its probable decrease under cyclic loading [7], or the decrease of the shear strength parameters as displacements progressed. On the other hand, when considering the lower part of the landslide where the failure had occurred along a pre-existing slip surface, it was not taken into account that the shear strength along this surface depends on the loading rate. Recent studies have shown that the residual shear strength of cohesive soils strictly depends upon the displacement rate and the soil plasticity [8, 9]. In the case of soils with low clay fraction such as the soils forming the landslide body, it was observed a drop of strength during fast loading to values lower than those obtained with slow loading tests.

For a more detailed analysis, these features —if proved by specific laboratory tests on Andretta soil samples— can be taken into account and would give a better evaluation of displacements at the end of the earthquake.

Accelerogram adopted in the displacements computation - If the distance of the accelerometric station from the site under study is taken into account, the adopted accelerogram may not be representative of the actual ground motion in the landslide area.

Moreover, the shaking intensity in the landslide area might be lower than the actual one due to the topographic effect given by the position of the Calitri accelerometric station on the top of a hill with rather steep slopes.

These two aspects may somehow be taken into account, but even in this case the results could be affected by the same degree of uncertainty as the results obtained with the analyses as carried out.

Seismically-induced pore pressure increase - According to the analysis results, increase in pore pressure would have not played a prominent role in triggering the failure process. However, first displacements might have been somewhat delayed with respect to the ground shaking, which is a fact that cannot be excluded because of a lack of observations in the 12 hours following the earthquake. If a delay occurred, inertia forces are not the triggering factor of the landslide, and the delay can be justified by the increase in pore pressure and its subsequent redistribution within the slope.

Laboratory tests are a tool to study the ground response to cyclic loading in undrained conditions in terms of relationships between shear deformations and increase in pore pressure. Numerical analyses considering such relationships can help in evaluating the slope response to the ground motion history in terms of seismically induced excess pore pressure. The results so obtained are undoubtedly useful for a first evaluation. However, they are still affected by uncertainties that cannot be eliminated because depend, for instance, on the dubious definition of the state of stress within the slope before the earthquake.

The above considerations show that a better evaluation of the landslide process is possible by improving the quality of data used in the analyses and pondering the influence of the increase in pore pressure factor. Unfortunately, uncertainties attributable to the lack of information about the local seismic input and to the complex geomorphologic and geotechnical features of the phenomenon — particularly to the unknown pre-earthquake slope conditions— cannot be eliminated.

6. Conclusions

The occurrence of the Andretta landslide is to be seen in the long duration of the 1980 Irpinia earthquake and in the presence of a large mass of ground in conditions close to the limit equilibrium.

Even if uncertainties are allowed as far as the actual geotechnical parameters, the local variations of the ground shaking intensity and the true displacements of the landslide immediately after the earthquake are concerned, the Newmark model seems to represent the studied phenomenon with a rather good approximation.

This conclusion is a point in favour of the hypotheses previously made, that is:
i) inertia forces are the triggering factor, and
ii) soon after the earthquake, general failure conditions set in within the slope.

References

[1] D'Elia B., Esu F., Pellegrino A., and Pescatore T. (1985). *Some effects on natural slope stability induced by the 1980 Italian earthquake*. Proc. XI ICSMFE, San Francisco, Vol. 4, p. 1934-1943.

[2] D'Elia B., Federico G., Pescatore T., and Rippa F. (1986) - *Occurence and development of a large landslide (Andretta-Italy) reactivated by the Nov. 23, 1980 earthquake*. Proc. Int. Symp. on "Engineering Geology Problems in Seismic Areas", Bari, Vol. 2, p. 365-381.

[3] Matsui T., Ohara H., and Ito T. (1980). Cyclic stress-strain hystory and shear characteristic of clay. *Journ. of Geotech. Eng. Div., Proc. ASCE*, Vol. 106, n. GT10.

[4] Hutchinson J. N., and Del Prete M. (1985). Landslides at Calitri, Southern Appennines, Reactivated by the Earthquake of 23rd November 1980. *Geologia Applicata ed Idrogeologia*, Vol.20, Parte I, p. 9-27.

[5] Sarma S. K. (1975). Seismic stability of earth dams and embankments. *Géotechnique*, Vol. 25, n. 4, p. 743-761.

[6] Sarma S. K. (1981). Seismic displacement analysis of earth dams. *Journ. of Geotech. Eng. Div., Proc. ASCE*, Vol. 107, n. GT12, p. 1735-1739.

[7] Crespellani T., Ghinelli A., Madiai C., and Vannucchi G. (1990). Analisi di stabilità dei pendii naturali in condizioni sismiche. *Rivista Italiana di Geotecnica*, Vol. 26, n.2, p. 49-74.

[8] Lemos L., Skempton A. W., and Vaughan P. R. (1985). *Earthquake loading of shear surfaces on slopes*. Proc. XI ICSMFE, San Francisco, Vol. 4, p. 1955-1958.

[9] Lemos L. (1991). *Shear strength of shear surfaces under fast loading*. Proc. X ECSMFE, Firenze, Vol. 1, p. 137-141.

REACTIVATION OF MUDSLIDES AFTER A LONG QUIESCENT PERIOD : THE CASE OF BUONINVENTRE IN SOUTHERN APENNINES

Mario DEL PRETE* and Giuliana TRISORIO LIUZZI**

Abstract: The present paper draws attention to some aspects concerned with the reactivation of large mudslides that have been quiescent for a long period in seismic areas. It is presented the case of Buoninventre mudslide in Southern Apennines (Italy) reactivated after a quiescent period of over 30 years by the Irpinian earthquake of 1980.

Riassunto: Lo scopo del lavoro e' quello di presentare alcuni aspetti connessi alla riattivazione da terremoto di grandi colate quiescenti per lunghi periodi di tempo.
Viene presentato il caso della frana di Buoninventre nell'Appennino Meridionale (Italia), riattivata dal terremoto Irpino del 1980.

1. Introduction

The landslides triggered by earthquakes are a topic of great importance in the evalutation of geological hazards.

When they are triggered at same time of the earthquake they are called direct, and if they start later, they are called indirect.

During the Southern Italy earthquake in 1980 there was a small number of direct landslides, mainly rockfalls in the calcareous-dolomitic slope [1].

Indirect landslides were mostly large rotational slips and above all mudslides and mudflows [2],[3]. The sites of the main mudslides in varicoloured clays reactivated or not by the earthquake, are shown in Fig. 1, whereas the main geometrical characteristics and the distance from the epicentre are listed in Tab. 1.

The Buoninventre mudslide, quite close to Caposele town in Irpinia, has been chosen as an example of reactivated mudslide.

This paper draws attention to the reactivation mechanism of a large lanslide body after a long quiescent period.

* Professor of Engineering Geology - University of Basilicata.
** Associate Professor of Soil Conservation - University of Bari.

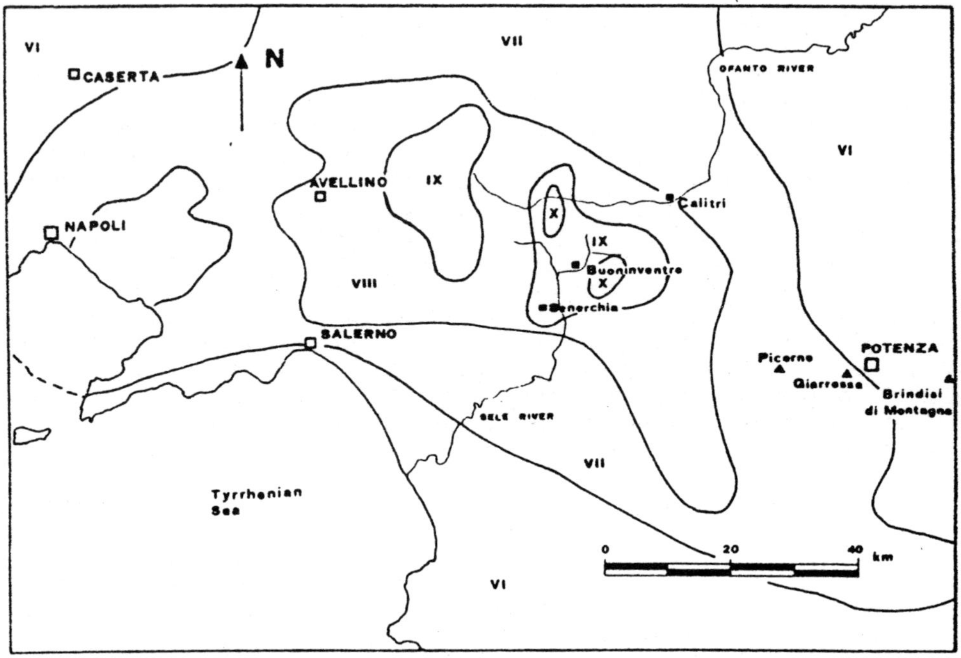

Fig. 1 : Location of the main mudslides in varicoloured clays. ■ reactivated by the 1980 earthquake; ▲ not reactivated by earthquake; ● active before earthquake.

Tab. 1 : Main geometrical characteristics of the studied mudslides bodies.

Mudslide	d (km)	L (m)	B (m)	h (m)	β (°)	α (°)	Orientaz.
Buoninventre	10	2800	230	20	7.5	3.5	N - S
Senerchia	14	2400	330	30	7.5	3.5	ONO - ESE
Calitri	19	480	215	10	8		N - S
Picerno	28	2000	300	20	10		N - S
Giarrossa	35	2600	150	25	9	6.5	N - S
Brind. di M.	54	600	80	5	10	5.5	N - S

L = Length of mudslide
B = Mean width of mudslide channel
h = Mean depth of shear surface
d = distance fram the epicentre
β = mean slope of the channel
α = mean slope of accumulation zone

2. Geological and Geotechnical aspects

The Buoninventre mudslide zone is made up mainly by a pelitic succession of deep sea deposits, known as varicolored clay [4] being part of orogenically transported Units of Appenine Chain. This formation has been subject to severe macro and micro structural strains that strongly affect their mechanical behaviour. Main and secondary structural discontinuities, bedding planes, contacts between lithologically different elements, fractures and scaly planes detect macro and micro clay-elements, affected by softening and chemical and mineralogical weathering phenomena leading to the genesis of a detritic cover that can be over 10 m deep. These covering deposits are accumulated in the valley depressions and they supply mudslides prone to periodic movements and variable quiescent periods.

Mudslide bodies have a trend to a stable morphological shape. According to their morphological evolution they can be reactivated as a consequence of metereological and seismic events. It may happen that some of them can be reactivated seasonally. The oldest among them can be mobilized after a quiescent period of more than 20 years, as we saw after the Irpinia earthquake of 23 November 1980, for the Senerchia, Calitri and Buoninventre mudslides.

The geotechnical characterization of mudslide materials is difficult because of their heterogeneity, the confuse assembly of different litotypes, the frequency and extension of discontinuities, the dimensions of scaly elements, the influence of the weathering process on the mineralogical composition of the pelitic fraction.

Tab. 2.: Main geological and geotechnical characteristics of the studied mudslide bodies.

Mudslide	WL(%)	IP(%)	CF(%)	Ø'r(°)
Buoninventre	45-70	25-35	25-60	9-10
Senerchia	40-90	15-50	25-60	9-13
Calitri	60-70	30-40	40-60	10-12
Picerno	40-65	20-35	25-50	10-12
Giarrossa	35-55	18-35	20-50	
Brind. di M.	60-110	30-70	25-40	10-12

WL = Liquid Limit
IP = Plasticity Index
CF = Clay fraction
Ø'r = residual friction angle

It is really difficult to define their mechanical behaviour in the laboratory, also because it is not always possible to operate on sufficiently rappresentative volume elements.

Actually even the simplest geotechnical characteristics, such as Atterberg limits, and particle size distribution, are strongly affected by the used techniques [5] [6].

In tab. 2 there is a summary of the main geotechnical characteristics of varicoulored clays outcropping in Campania-Basilicata areas affected by mudslides. The data are referred only to the clay fraction that should affect the mechanical behaviour of the formation more than the lithorelicts existing in mudslides bodies.

The studied soils are mainly made up of varicoulored clays, which are different for the mineralogical components. They are silty clays and clayey silts with high plasticity and mean clay fraction actitivity , characterized by small residual friction angles.

For clay specimens coming from the Buoninventre and Senerchia mudslides Bellino e Maugeri [7] have obtained residual friction angles between 5° and 7°. These values are small and not consistent with the back-analysis results. This discrepancy is probabily due to tests carried out only on completely remoulded clay samples where scaly clay elements have been destroyed. The obtained minimal values are a lower limit of the residual friction angle.

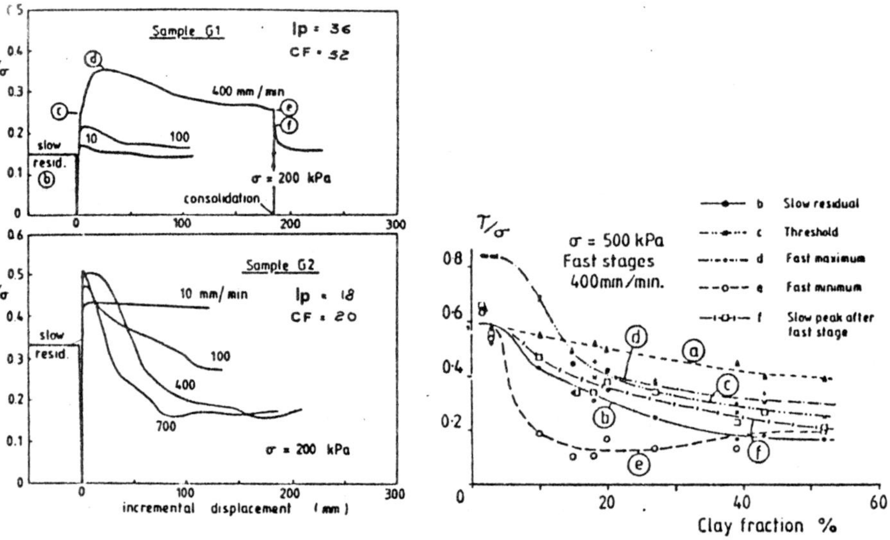

Fig. 2 : Typical strength in fast shear (a); Variation of strength components with clay fraction (b)(after Vaughan et al.[9]).

In situ the presence of rigid not disgregated scaly clay elements probably gives an increment of shear strength along the slip surface [8].

Research on the influence of the shear velocity on the residual shear strength are under way following [9]. They used velocities much higher than those measured for the studied mudslides during their active phase, even if it could be reached during seismic events. In fact for Buoninventre mudslide it has been calculated that it was possible to reach an istantaneous velocity of 0.85 m/s [10]. Probably it is an overstimated value, because it is referred to a strong-motion accelerogram recorded on calcareous-dolomitic rocks.

The greatest shear velocities used at Imperial College were between 400 and 700 mm/min. The data show that for a sudden increase of shear velocity there is an increment of shear strength, and that beyond a threshold value the shear strength decreases to a typical value. This value is higher than the residual strength at slow velocity when the clayey fraction is more than 40% (Fig. 2).

The tests carried out in a range of velocities between 0.0015 and 1.2 mm/min on Senerchia varicoloured clay do not show severe differences in the residual strength, even if there are little increments when the velocity increases (Fig. 3).

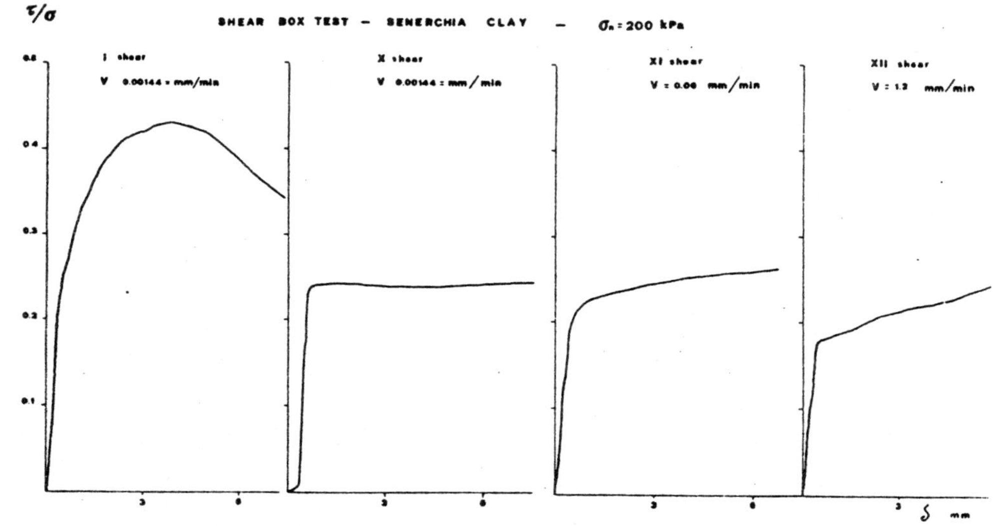

Fig. 3 : Shear-Box tests at different velocity on varicoloured clay sample.

It is important to consider that seismic acceleration and the associated hypotetical simultaneus displacements are subject to a continuous inversion of the direction. For this reason Ambraseys [11] remarks that the most dangerous situation occurs when the landslide length is egual or lower than the earthquake wave half length. It is important to underline that as regards the disturbance on soils caused by earthquake, the possible displacement along the shear surfaces during earthquake must be summed, even if they have opposite direction. This aspect is of great importance when the resistance along slip surfaces is evaluatated after long quiescent period, as in the case of the landslides mobilized by the 1980 earthquake. A recovery of strength for tixotropical effects can be considered, as it has been recently proposed [12] [13]. This little increment of strength can be effectiv in the case of heavy rainfalls, while it vanishes with a brittle failure after small displacements during dynamic loading (Fig. 4).

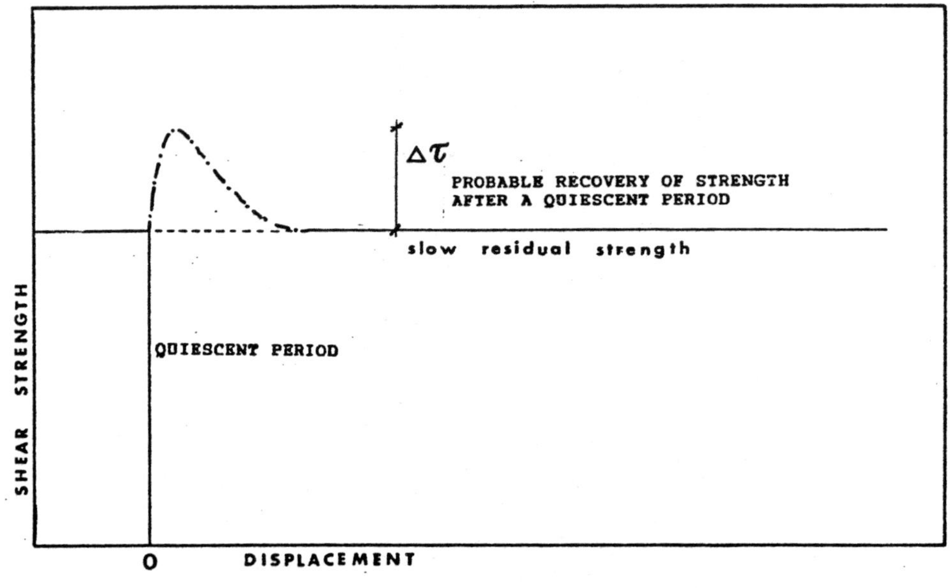

Fig. 4 : Recovery of strength after a quiescent period.

3. Buoninventre mudslide reactivation mechanisms

The Buoninventre mudslide (Fig. 5), as the Senerchia one, has been reactivated after a quiescent period of more than 30 years. This is demonstrated by historical research and by old, small country houses built on the mudslide body [14]. After the 23 November main shock of 1980 many observations have been carried out by aerial photographs and field surveys. In Fig. 6 are reported the main phases of the reactivation phenomena.

The phenomena started with an upstream rotational slip having an extension of 40.000 m^2 On the downstream edge of this area, tension cracks took place and a part of the slipped mass was disgregated flowing as a viscous fluid (Fig. 7).

During the following ten days the phenomenon spread towards the valley, bringing out a viscous flow over the mudslide body. After this a complete mudslide body mobilization took place along the basal slip surface (Fig. 6b; 8a).

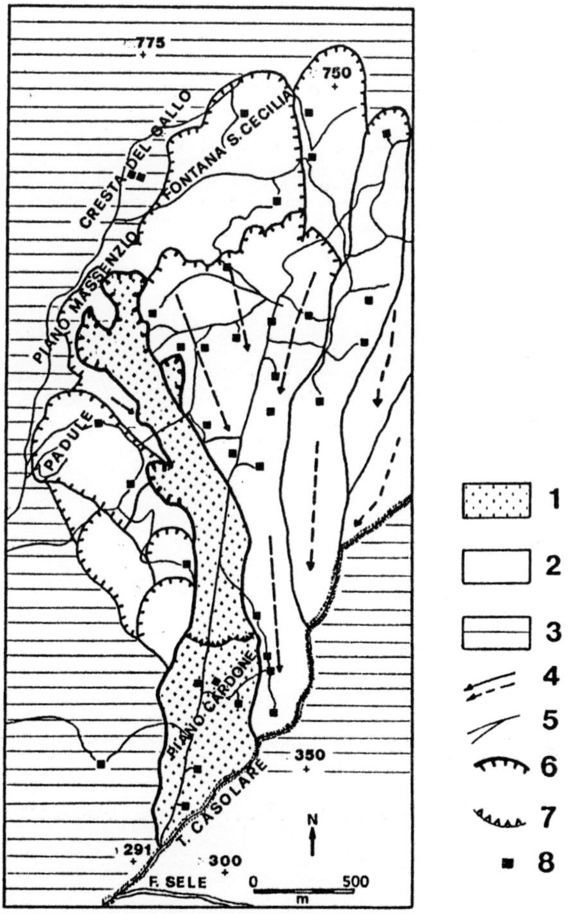

Fig.5 : Geological and geomorphological map of Buoniventre area: 1) mudslide reactivated by the 23/11/80 earthquake; 2) old mudslide not reactivated by the earthquake; 3) Varicoloured clay of Sicilids Units; 4) main direction of mudslides; 5) roads; 6) scarps; 7) passive failures; 8)houses.

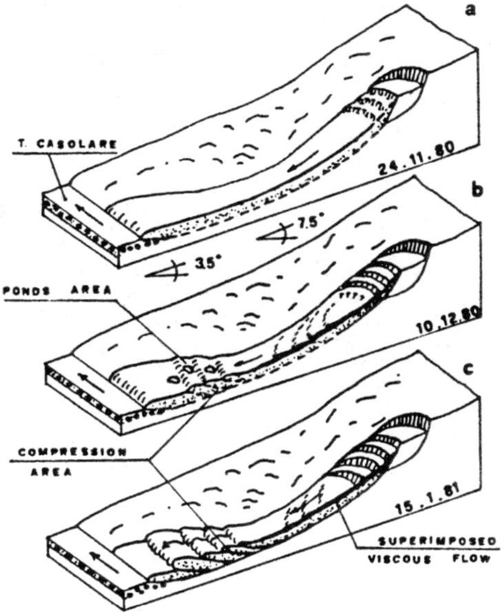

Fig.6 : Different phases of Buoninventre mudslide reactivation.

Fig. 7 : View of the source area and of the channel of Buoninventre mudslide.

a

b

Fig. 8 : Aerial photographs of different reactivation stages of Buoninventremudslide: (a) 7/12/1980; (b) after a complete reactivation.

Fig. 9 : Passive failures area

Fig. 10 : Slickensides at the left side of the Buoninventre mudslide channel

Starting from the 10th of December, in the zone where mudslide channel ends, compression strains took place, with the formation of passive failures, pointed out by topographic undulations (Fig. 6c; 9) and slickensides along lateral shear boundaries. (Fig. 10).

As a consequence of compression, small ponds developed following the increase of pore pressure (Fig. 11).

Fig. 11 : Ponds in the compression zone of Buoninventre mudslide.

At the beginnig of January 1981 the mobilization of the accumulation zone took place(Fig. 6c).

The complete reactivation process, very similar to that of the Senerchia mudslide, can be assumed as a typical example of seismic reactivation mechanism of huge mudslides.

The main aspect regards the propagation of movement from the upstream to the downslope area due to a series of interconnected causes and effects.

It is necessary to consider separately the phases of reactivation of the feeding area, the channel area and the accumulation area. In the first one there was a rotational slip more or less simultaneous to the main shock. The slipped mass was in a limit equilibrium condition due to high pore pressure connected with the presence of springs. This condition was verified assuming groundwater level coincident with slope surface, and $\phi_r' = 12^\circ$ and $c' = 0$.

The value of strength parameters was assumed considering the effect of small displacement simultaneous to the shocks.

The channel mobilization can be explained by loss of strength along the basal slip surface and by increase of pore pressure level. According to Matsui et al. [15] it is possible to obtain a pore pressure increase after seismic shocks if a threshold of strain is exceeded (Fig. 12). This threshold of strain increases with overconsolidation ratio and plasticity index of soil.

In the case of Buoninventre, the increase of the pore pressure must be connected not only with earthquake schoks but also with the undrained loading of superimposed mudflow coming from the alimentation area.

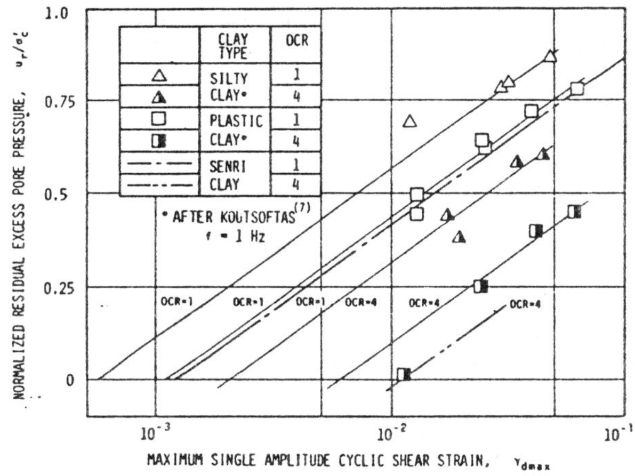

Fig. 12 : Effects of overconsolidation ratio and clay type on normalized residual pore pressure after cyclic loading (after Matsui et al. [15]).

The channel was mobilized after about 20 days as a rigid body moving along a basal slipsurface (ß=7.5°).This loaded the resisting accumulation zone (ß=3.5°) creating passive failures.

The movement of the accumulation area was strongly delayed in comparison with the channel. This implies a direct effect of the channel load.

A model considering a rigid body motion with a load of the channel on the accumulation zone, gives, for the last one, a factor of safety more than one.

To explain such delayed movement of the downslope zone it is necessary to think of a variation of pore pressure as an effect of undrained loading due to the pushing action of the channel.

The results obtained by the back-analysis confirm this hypotesis.

References

[1] D' Elia B., Esu F., Pellegrino A., Pescatore T. (1985). *Some effects on natural slope stability induced by the 1980 Italian earthquake* Proc. XI I.C.S.M.F.E., San Francisco (4),1943- 1949.

[2] Cotecchia V., Del Prete M. (1984). *The reactivation of large flows in the parts of Southern Italy affected by the earthquake of November 1980, with references to the evolutive mechanism* IV Int. Symp. on Landslide, Toronto, vol. 2, 33-38.

[3] Hutchinson J.N., Del Prete M. (1985). *Landslides at Calitri, Southern Apennines, reactivated by the earthquake of 23rd November 1980*.Geol. Appl. ed Idrog., vol. XX, p. I, 9-38.

[4] Ogniben L. (1969). *Schema introduttivo alla geologia del confine calabro-lucano* Mem. Soc. Geol. It., 8, 453-763.

[5] Cotecchia V., Del Prete M. (1970). *Geologia dei dintorni di Vietri di Potenza e particolari caratteri d' instabilita' dei versanti in flysch ed argille varicolori* Geol Appl. ed Idrog., vol. V, 57-134.

[6] Rippa F., Picarelli L. (1977). *Some consideration on index properties of southern Italy Shales*. Proc. Int. Symp. "The Geotechnics of Structurally Complex Formations", Capri, vol. 1, 401-406.

[7] Bellino S., Maugeri M. (1985). *Confronto fra valori di resistenza residua ottenuti con diverse apparecchiature anulari di taglio*. R.I.G., n.2, 101-111.

[8] Del Prete M., Guadagno F.M. (1988). *Observations on landslides in typical flysch sequences of southern Apennines (Italy)* V Int. Symp. on Landslide, Losanna, vol. 1, 109-114.

[9] Vaughan P.R., Lemos L., Tika T. (1985). *Strength Loss on Shear Sufaces Due To Rapid Loading* Proc. XI I.C.S.M.F.E., San Francisco, Discussion Session 7B, 30-36.

[10] Maugeri M., Motta E. (1985). *A note on the Residual Strength in a Landslide Induced by the 1980 Italian Earthquake* Proc. XI I.C.S.M.F.E, Fan Francisco, Discussion Session 7B, 37-41.

[11]Ambraseys N.N. (1977). *On the response of structures to travelling waves* Proc. Cento seminar on Earthquake Hazard Minimization, Teheran, 410-414.

[12] Simeone V., Melidoro N.L. (1991). *Recuperi tixotropici lungo superfici di taglio in condizioni di resistenza residua* VIII Riunione Ric. Grup. Naz. Coor. Studi Ing. Geot. C.N.R. Roma.

[13] Skempton A.W., Northey R.D. (1952). *The Sensitivity of clay* Geotechnique 3, n.1, 30-53.

[14] Del Prete M. (1990). *Examples of mudslide hazard in Southern Apennine (Italy)* Atti Convegno Irpinia dieci anni dopo, in press.

[15] Matsui T., Ohara H., Ito T. (1980). *Cycilc stress-strain history and shear characterisc of clay* Jour. Geot. Eng. Div., Proc. A.S.C.E, vol. 106, GT 10, 1101-1120.

DEFORMATION PROCESS OF A HILL SHAKEN BY THE IRPINIA EARTHQUAKE IN 1980

Giovanni Battista FENELLI*, Luciano PICARELLI*, Francesco SILVESTRI*

Abstract: The paper describes the deformation phenomena observed in the town of Bisaccia, still elapsing after the Irpinia earthquake in 1980. The overall experimental data collected so far seem to merge towards an unique hypothesis, which implies the dissipation of pore pressures induced by the earthquake.

Sommario: L'articolo descrive i fenomeni deformativi osservati nell'abitato di Bisaccia, tuttora in corso dopo il terremoto in Irpinia del 1980. L'insieme dei dati sperimentali ottenuti finora sembra confermare l'ipotesi che gli spostamenti registrati siano dovuti alla dissipazione di sovrappressioni interstiziali indotte dall'evento sismico.

1. Foreword

Earthquakes are considered to be a triggering cause of landslides in many geological materials. The real causes of these phenomena are not thoroughly understood; according to the italian experience they appear to be different in the different geological materials [1,2,3,4,5].

Inertial forces due to rapid loading may cause falls and toppling in rock masses, whereas a number of documented cases of landslides due to liquefaction of saturated loose sands highlights the role of pore pressures induced by the seismic shaking.

In clay deposits, several factors can interact in a complex fashion, so producing variable effects. Sometimes it is possible to recognize long-term displacements as the delayed consequence of pore pressure buildup induced by cyclic loading. The settlements still occurring in Bisaccia, a town severely shaken by the Irpinia earthquake twelve years ago, appear to belong to such a kind of phenomena.

A joint investigation on various aspects regarding the stability of the Bisaccia hill has been carried out by the Universities of Napoli and Roma; the paper reviews and furtherly develops some previously published experimental data and observations on the effects of Irpinia earthquake [4,6,7,8,9,10].

* Istituto di Tecnica delle Fondazioni e Costruzioni in Terra (Geotecnica), Università di Napoli "Federico II"

2. Geomorphological framework

The town of Bisaccia is located in the Apennine chain, about 100 km far from Naples. It stands on an elongated hill delimited by two narrow valleys (fig. 1).

The hill is constituted by a thick slab of marine conglomerates consisting of large rounded pebbles and blocks (maximum size of about 0.5 m) included in a slightly cemented sandy matrix. It rests on a very thick formation of clay shales, called "scaly clays" due to their typical structure.

A fault is supposed to cross transversely the hill, lowering its northern part, where the most ancient portion of the town is built [4]. Furthermore, several vertical fractures divide the slab into rather regular blocks.

A number of boreholes, drilled by the local administration after the 1980 earthquake, allowed to obtain more information about the real size of the conglomerate slab. The measured overall thickness is more than 100 m; a further borehole drilled care of the University of Napoli in site D (fig. 1) reached a depth of 151 m, showing that locally the conglomerates have a thickness of 115 m.

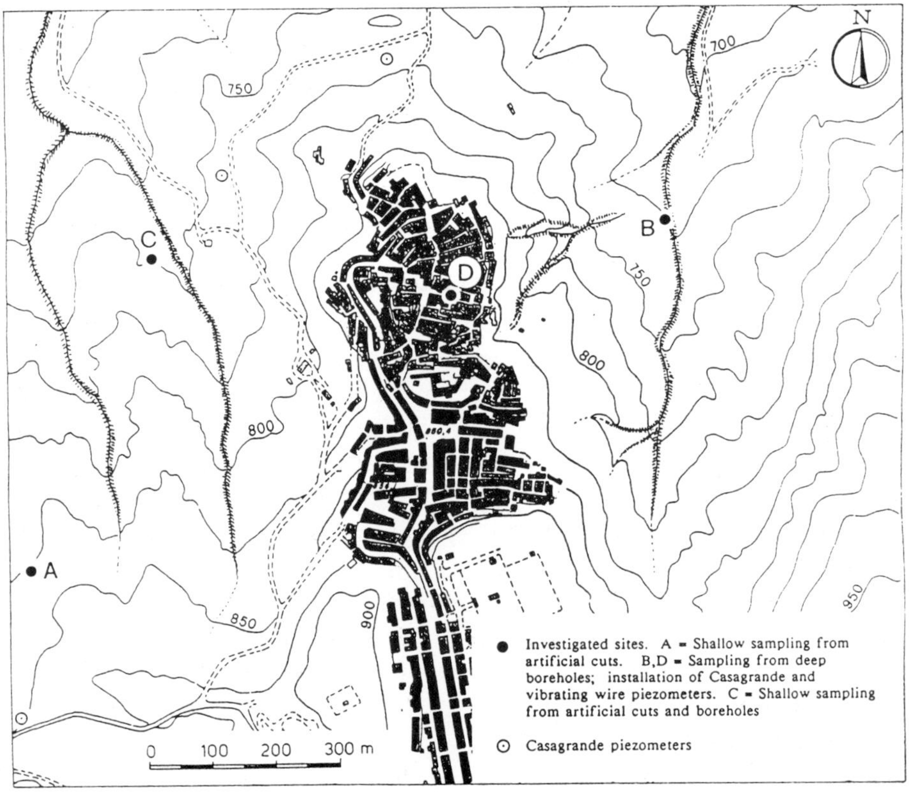

Fig. 1: Plan of the Bisaccia hill and investigated sites.

The thickness of the clay shale formation is supposed to be considerable [11]; as a matter of fact, the boreholes drilled in site B have crossed such materials down to a maximum depth of 65 m [8].

According to the geologists opinion [11], the formation of the narrow valleys bordering laterally the hill (fig. 1) is probably due to erosion processes started maybe 300.000 years ago along two parallel faults caused by tectonic disturbances. The erosion has brought about a dismantling of the conglomerate cover and a successive indenting of the underlying clay shales.

At present, the erosion is certainly still active; furthermore, the lateral slopes of Bisaccia hill are subjected to landslides, whose foot involves the underlying clay shales; within the same area, numerous shallow slides or flows are active on gentle slopes where the conglomerates are absent.

3. Groundwater regime

The area has been investigated by means of a number of boreholes (sites B, C and D), the installation of piezometers (sites B, D and other sites indicated in fig. 1) and shallow sampling (sites A and C).

The groundwater regime has been investigated by means of Casagrande and vibrating wire piezometers. The piezometers alongside the contour of the hill show that, all over the investigated area, the groundwater level practically coincides with the top of the clayey formation.

In site B (i.e. at the center of the eastern valley) seven Casagrande and vibrating wire piezometers have been installed down to a maximum depth of 56.8 m. These instruments currently reveal deficient pore pressures (fig. 2-a). According with the previously mentioned hypothesis about the mechanism of valleys formation, such a phenomenon is caused by the total stress reduction due to the erosion, which is rapid compared to the time needed for the pore pressure equalization [8].

On the basis of the same geological hypotheses, Urciuoli [10] has justified these results by a FEM analysis adopting the computer code CRISP [12]. The erosion phenomenon has been simulated as a 2-D constant rate excavation, being the water table at the top of the clay deposit; the behaviour of conglomerates has been considered linear elastic, whereas the clay shales have been modelled following the modified Cam-clay constitutive law.

The main results of this analysis have been reported in fig. 2-a, and show quite a good agreement with the measurements. The differences concerning the shallowest piezometers depend on the presence of a saturated debris cover (4 m thick), which has not been taken into account in the analysis. Within this layer the pore pressures are practically hydrostatic.

In site D (i.e. beneath the center of the town) two piezometers have been recently installed: the first one is located inside the conglomerates at a depth of 94.3 m; the deepest one, in the argillaceous formation, has been installed at a depth of 149.5 m. At present, the recorded pore pressures are maybe not yet equalized, but stabilizing around the upper bound of the range reported in fig. 2-b, together with the results of the above mentioned numerical analyses. The pore pressures lie around a hypothetical steady state condition based on the current position of the groundwater level under the town and the lateral valleys, and are significantly higher than the predicted ones.

According to the erosion analysis, the Bisaccia hill should be subjected to a very slow sinking phenomenon, whereas the floor of lateral valleys should be raising due to the associated swelling process.

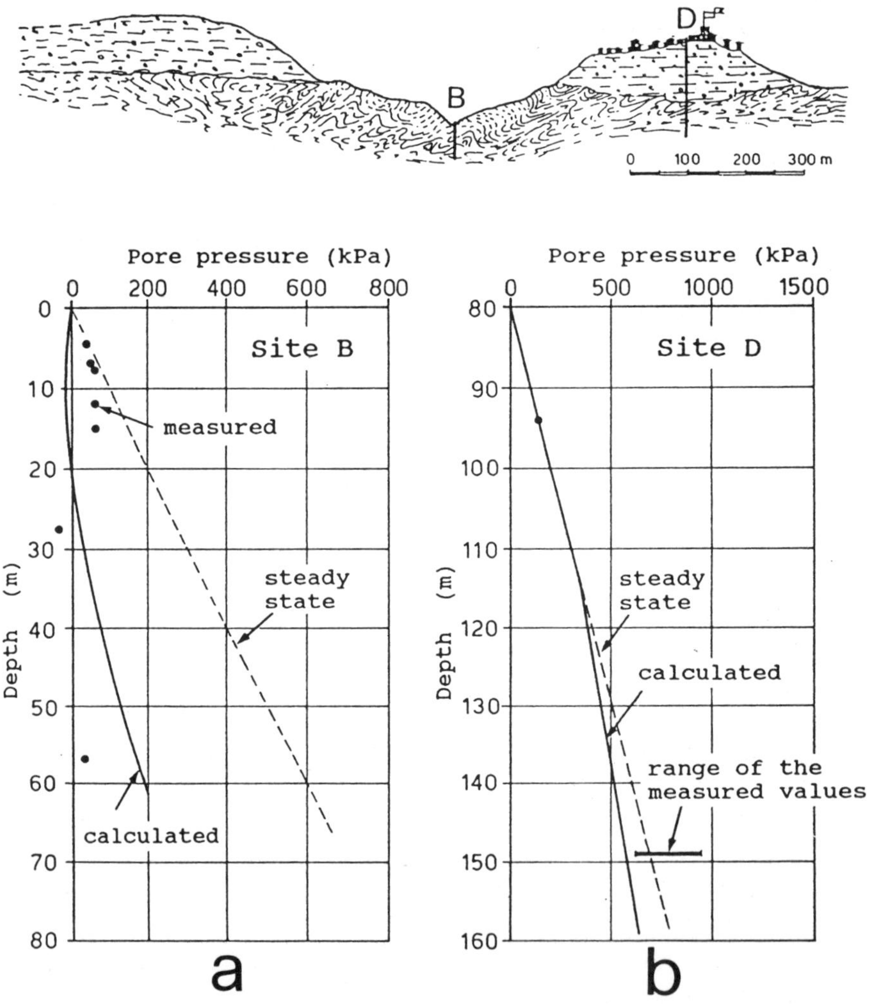

Fig. 2: Measured pore pressures in sites B and D.

4. Physico-mechanical properties of the clay shales

Due to the size of single elements of the conglomerate formation, it has not ben possible to carry out any experimental program on their mechanical properties.

Conversely, a very wide investigation has been accomplished on clay shales. Laboratory tests have been performed on samples obtained from four different sites: sites A and C, where shallow samples have been drawn from artificial cuts executed on natural slopes, and sites B, C and D, where both short and very deep boreholes have been drilled inside the argillaceous formation.

Due to the aim of this paper and the limited available space, only some significant data will be synthesized in the following pages.

4.1 INDEX PROPERTIES

The overall obtained data, which concern some hundred samples taken either close to the ground surface either down to a maximum depth of about 148 m (site D), reflect the considerable homogeneity of the argillaceous formation.

This consists of intensely fissured clay shales crossed by an intricate network of principal shears, having variable azimuth and dip, and being spaced a few centimeters. The single "chunks" are characterized by an agglomeration of millimetric shear lenses.

These materials may be defined as silty clays whose clay fraction ranges between 50 and 70 %; the plasticity is very high, being represented by a liquidity limit varying between 80 and 160 %, with a maximum value of the order of 190 %. The average plasticity index, PI, ranges between 85 and 90 %.

In site D, the natural water content is slightly decreasing down to the maximum investigated depth (148 m); its average value is 0.20, and the corresponding liquidity index is about -0.25.

Finally, the data obtained in sites A, B and C show that the average porosity varies between 35 and 38 %, and the degree of saturation is close to one.

4.2 STATIC STRESS-STRAIN BEHAVIOUR

A large number of samples (taken in sites A, B, and C) has been subjected to laboratory static test in order to obtain statistically significant data.

One-dimensional and triaxial compression tests have been carried out on samples of different size (maximum diameter 97 mm).

Oedometer tests have been performed attaining a maximum vertical stress of 23 MPa, being the preconsolidation pressure of the order of 2 MPa. For stresses ranging between 2 and 5 MPa, the constrained modulus of samples taken in site B varies between 35 and 72 MPa; for the deepest samples, a maximum modulus of 125 MPa has been obtained. In the same tests, the consolidation coefficient, c_v, resulted of $6{\cdot}10^{-10} \div 2{\cdot}10^{-9}$ m²/s.

The CID tests have shown a stress-strain behaviour which is typical of the "scaly clay shales" [6]: according to the their high OCR values, within the low stress range a well pronounced dilatancy following an initial volume decrease has been observed. However, the stress-strain curves don't exhibit any peak, showing a ductile behaviour. This is probably due to the structure of the material: although its low porosity brings about a global volume increase under the deviatoric stresses, the presence of shear discontinuities forces the failure surface to follow the pre-existing discontinuities.

The confining pressures have reached 4 MPa; the Mohr's failure envelope is strongly non linear: at least for the specimens coming from site A, it is characterized by $\phi' = 14°$, $c' = 14$ kPa for $\sigma' < 1$ MPa, and $\phi' = 3°$, $c' = 200$ kPa for higher stresses. Such a low friction angle agrees quite well with the residual value.

For $\sigma_3 < 200$ kPa, the pore pressures developped at failure in CIU tests are negative, being characterized by a Skempton's coefficient A_f ranging between -0.1 and -0.2; in any case, for axial strains ε_a less than 5 %, the induced pore pressures are positive (fig. 3).

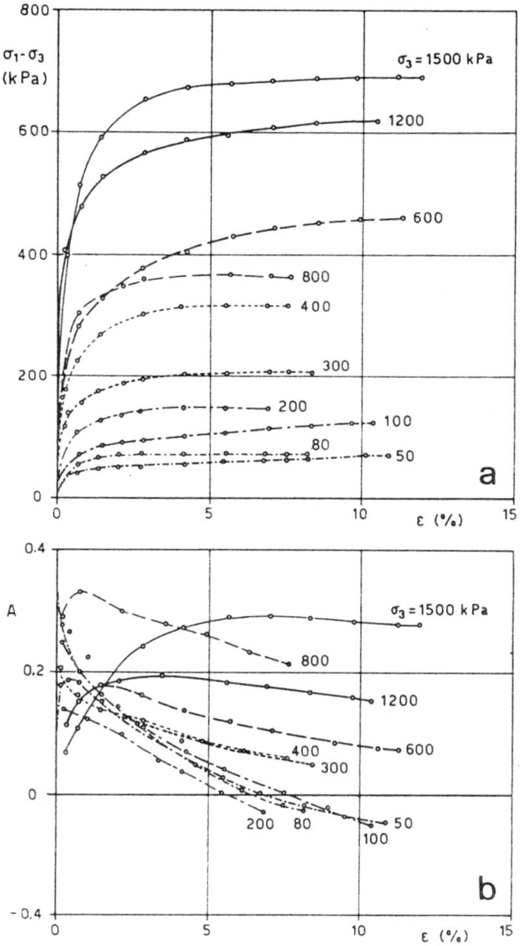

Fig. 3: Stress-strain behaviour of the clay shales: (a) deviatoric stress and (b) pore pressure coefficient A plotted versus axial strain.

The same tests have shown that the deformability modulus in the low strain range ($\varepsilon_a < 1$ %) may be as high as some hundred MPa, with a significant reduction for higher strains.

4.3 CYCLIC AND DYNAMIC BEHAVIOUR

The response of Bisaccia clay to cyclic and dynamic loading has been investigated by means of different laboratory tests, carried out on eleven specimens carved from undisturbed shallow samples taken at site A.

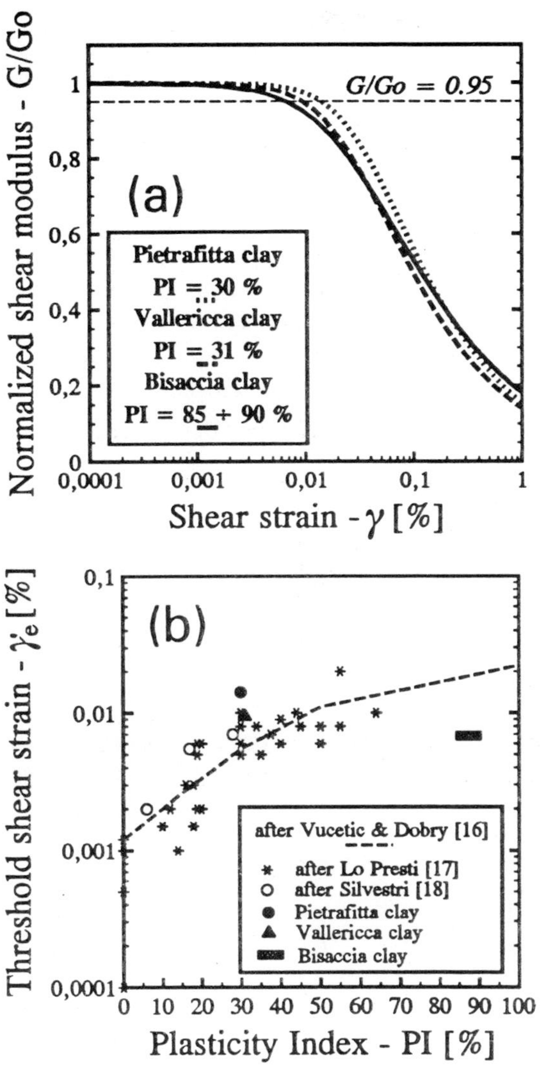

Fig. 4: Stiffness-strain behaviour of Bisaccia clay compared to data from literature: (a) G/G_0-γ relationship, (b) correlation between linear threshold strain and plasticity index.

Since the clay shale is believed to have been shaken beyond the linear range, but well below the limit state, a characterization of its stress-strain behaviour and dissipative properties has been pursued, together with an evaluation of pore water pressure buildup due to the application of cyclic loads.

The testing program has included cyclic triaxial, direct and torsional shear, axial and torsional resonant column tests. During cyclic tests, steps of 20 or 25 cycles with frequencies ranging between 0.5 and 4 Hz have been adopted; the testing criteria and some results have been already published [4, 13].

The initial values of shear modulus, G_0, obtained by means of low amplitude resonant tests, show a relatively slight variation with the mean confining pressure, increasing from 140 to 252 *MPa* for p'_0 ranging between 200 and 1380 *kPa*. Such a result is typical for heavily overconsolidated soils.

The normalized modulus-strain relationships, G/G_0-γ, are negligibly affected by the mean confining stress, p'_0, and the overconsolidation ratio, *OCR*. They can be curve-fitted by the Ramberg-Osgood model [14], giving out the average curve plotted in fig. 4-a. It is interesting to compare it to the G/G_0-γ curve of two intact natural plio-pleistocenic clays, featured by lower plasticity indexes [15].

A conventional "engineering" value of the linear threshold strain level, γ_e, can be analytically obtained as the value corresponding to $G/G_0 = 0.95$ in the Ramberg-Osgood model (see fig. 4-a).

The extent of the linear behaviour for fine-grained soils is conventionally associated to their mineralogic and microstructural properties, by means of a correlation between γ_e and *PI* which generally shows an increase of γ_e with *PI* (fig. 4-b). A physico-mechanical explanation of such a relationship has been proposed by Vucetic and Dobry [16], on the basis of the influence of electro-chemical interparticle bonds on the stress-strain behaviour of clays[1]. The logical framework holds for homogeneous soils, unless significant effects of structure appear; this might be the case of cementation due to ageing effects, producing an increase of γ_e at a given *PI*, as perhaps it happens for Pietrafitta and Vallericca clays (see fig. 4). Conversely, a macrostructure featured by the presence of shear lenses, like Bisaccia clay, can be affected by precoceous sliding between particles, which for the homogeneous material should occur at higher strain levels.

As a matter of fact, for Bisaccia clay γ_e is about 0.07 %, significantly lower than what should be expected from materials with comparable values of *PI*. It is striking to observe that soils featured by significantly lower plasticity indexes (like Pietrafitta and Vallericca clays, having *PI* ≈ 30 %), exhibit a threshold strain even higher than Bisaccia clay.

Following analogous considerations, the effect of the scaly structure should be relevant also on the threshold level, γ_v, where sensible coupling of shear-volumetric effects occurs [18].

During cyclic tests, even with the highest back pressures it has been difficult to achieve values of B coefficient high enough to rely upon the complete saturation of the specimens. Nevertheless, significant pore water pressure buildups have been found to occur since strain levels of the order of 0.05 %. Similar values for γ_v have been obtained during a recent experimental work on the above mentioned plio-pleistocenic clays [18].

While the evaluation of the amplitude of the pore pressure buildup associated to the shear load level represents what is known as the "short-term" effect of the cyclic load, the so-called "long-term" effect is related to their delayed dissipation with time.

Regarding this point, some useful considerations may be borrowed from the researches carried out by Yasuhara and Andersen [19], in order to analyze the behaviour of offshore gravity platforms founded on overconsolidated clays subjected to intermittent cyclic loads. Both the short and long term effects due to sequences of cyclic loads on Drammen clay have been investigated in the laboratory by means of cyclic simple shear apparatus. It has been shown that a generation of positive pore pressure is induced in the overconsolidated clay, and that its amplitude increases within subsequent cyclic loading stages (short-term effect). After each serie of cyclic loads, the soil consolidates due to dissipation of

---

[1] As the plasticity index increases, this influence is supposed to be much more relevant if compared to the effect of gravitational loads, which dominate the frictional phenomena, hence the plastic straining of soils. In terms of shear strains, this should paradoxically result in a wider elastic domain pertaining to higher plasticity soils [18].

induced pore pressures (long-term effect), leading to larger volume changes than what could be caused by equal changes of effective stresses due to compression-swelling-recompression cycles.

5. The main historical earthquakes and their consequences

The southern Apennine chain is seismically very active; in this mountainous area the town of Bisaccia has been frequently shocked by earthquakes having intensities up to X-XI MSK scale. The best known earthquakes occurred in 1694, 1732, 1851, 1910, 1930, 1980; in particular the earthquakes in 1694, 1732, 1930 and 1980 have certainly been very strong and destructive.

Very useful information about the earthquake effects have been reported in the old and recent chronicles.

The seism of 1694 stroke all the region (Irpinia): in Bisaccia 200 houses were

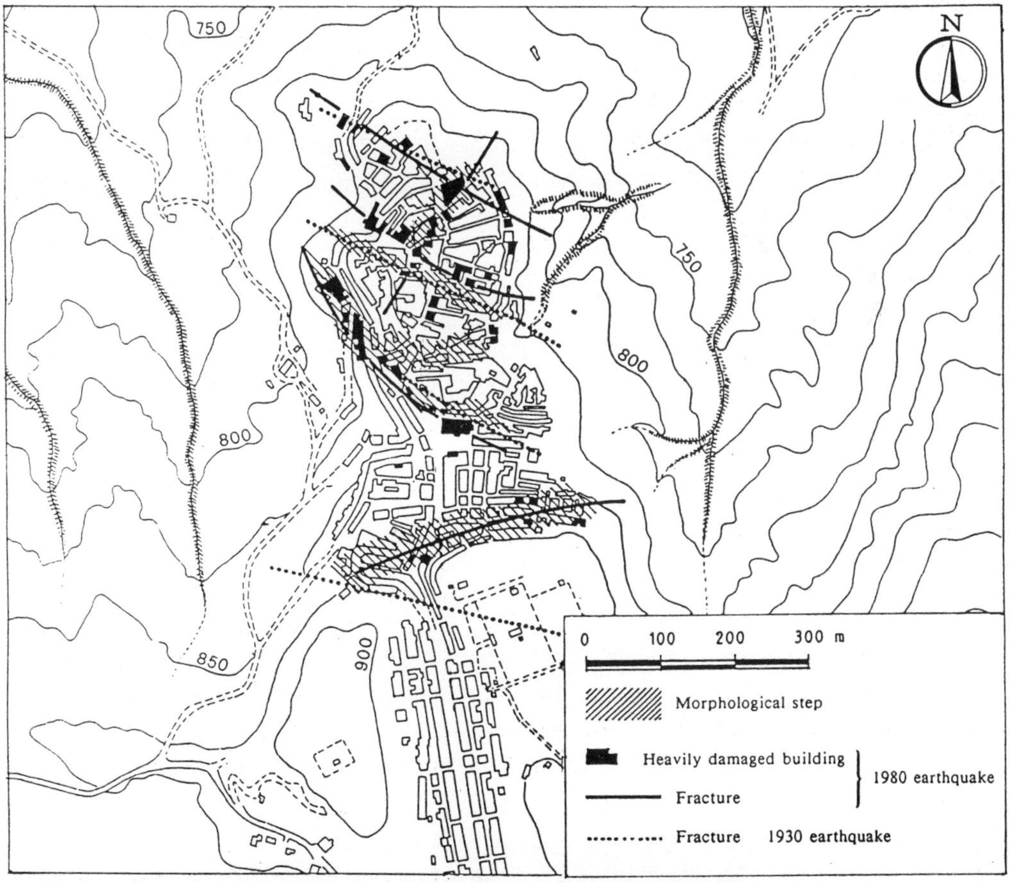

Fig. 5: Main consequences of the earthquake in 1980.

razed to the ground; 8 people died [20]. In 1732, 15 houses were destroyed and a man was killed [20]. The earthquake of 1930 was evaluated as strong as IX-X MSK, and 18 people died [21,22,23].

All the descriptions of the seismic effects are surprisingly similar and congruent. The mentioned earthquakes have generally determined a limited number of victims and injured people, but severe damages on the buildings, which had to be demolished. Fractures and continuous cracks in the road pavements have always been reported by the chronicles; hence, it is likely that the damages suffered by the buildings are not due to the direct action of the seismic forces, but the result of induced ground displacements.

This latter hypothesis has been confirmed on the occasion of the earthquake in 1980 [4]. The main shock of the seism lasted about 50 s for accelerations > 0.05 g and 80 s for accelerations > 0.01 g, and was followed by numerous aftershocks. A peak ground acceleration as high as 0.10 g has been recorded in the area.

The visual observation of the main consequences of the earthquake, and an analysis of the topographic measurements carried out all over the town within the period 1981-1988, altogether give quite a clear framework of the deformation mechanism determined by the seismic action.

Fig. 5 provides a complete mapping of the main effects induced by the earthquake in 1980. In the figure the main continuous cracks on the pavements due to the seismic event in 1930 are also reported, together with some pre-existing morphological steps [7].

All the information collected appear to merge, giving a clear picture of the overall deformation mechanism: the quakes have been producing finite displacements of single blocks that divide the conglomerate slab, determining fractures of the pavements and damaging the neighbour buildings; obviously, the most damaged constructions are located just along the fractures (see fig. 5). These phenomena replicate with successive seisms bringing about the formation of steps separating single blocks.

Some sketches of typical damages are reported in fig. 6: these lively pictures confirm that the most significant phenomena are due to concentrated movements along some well defined lines.

High precision levellings have been carried out on behalf of the University of Ancona inside the town in the period 1981-82 and lastly, in 1988, with care of the University of Napoli. Some results are reported in fig. 7.

The obtained data show that the settlements are rather high, with a maximum of about 15 cm (not indicated in fig. 7). Taking into account that the measurements started in February, 1981, i.e. about three months after the earthquake, and considering that there is a clear evidence of displacements immediately caused by the seism (see fig. 6), such phenomena appear relevant.

A longitudinal section S-N (fig. 7-a,b) shows rather clearly that the post-earthquake settlements may be interpreted as due to rigid movements (translation

A)

B)

C)

Fig. 6: Typical effects on buildings and pavements.

A. The earthquake has produced the cracks (a); the door frame has suffered a differential settlement of about 2 cm (b); just in front of it, the pavement shows a continuous step of about 15 cm (c). The house is certainly more recent than the neighbour buildings (d): has it been reconstructed after being destroyed by previous seisms ?

B. The step (a) (about 20 cm high) existed before the earthquake in 1980. The cracks (b) are due to the seism: they reveal a settlement of about 2.5 cm and a lateral displacement of the same value. They are localized just where previously repaired cracks have been detected.

C. The stairs (a) cover a morphological step of about 1 m (probably due to accumulated differential settlements). Before 1980 the stairs were already considerably deformed; after the seism further deformations and cracks (b) have been observed.

and tilting) of single blocks. As a matter of fact, very sharp discontinuities of the settlement plot can generally be observed just in the zones crossed by recognized continuous fractures on pavements; similar observations hold for all the investigated sections. The same diagrams allow to hypothesize the presence of further undetected cracks.

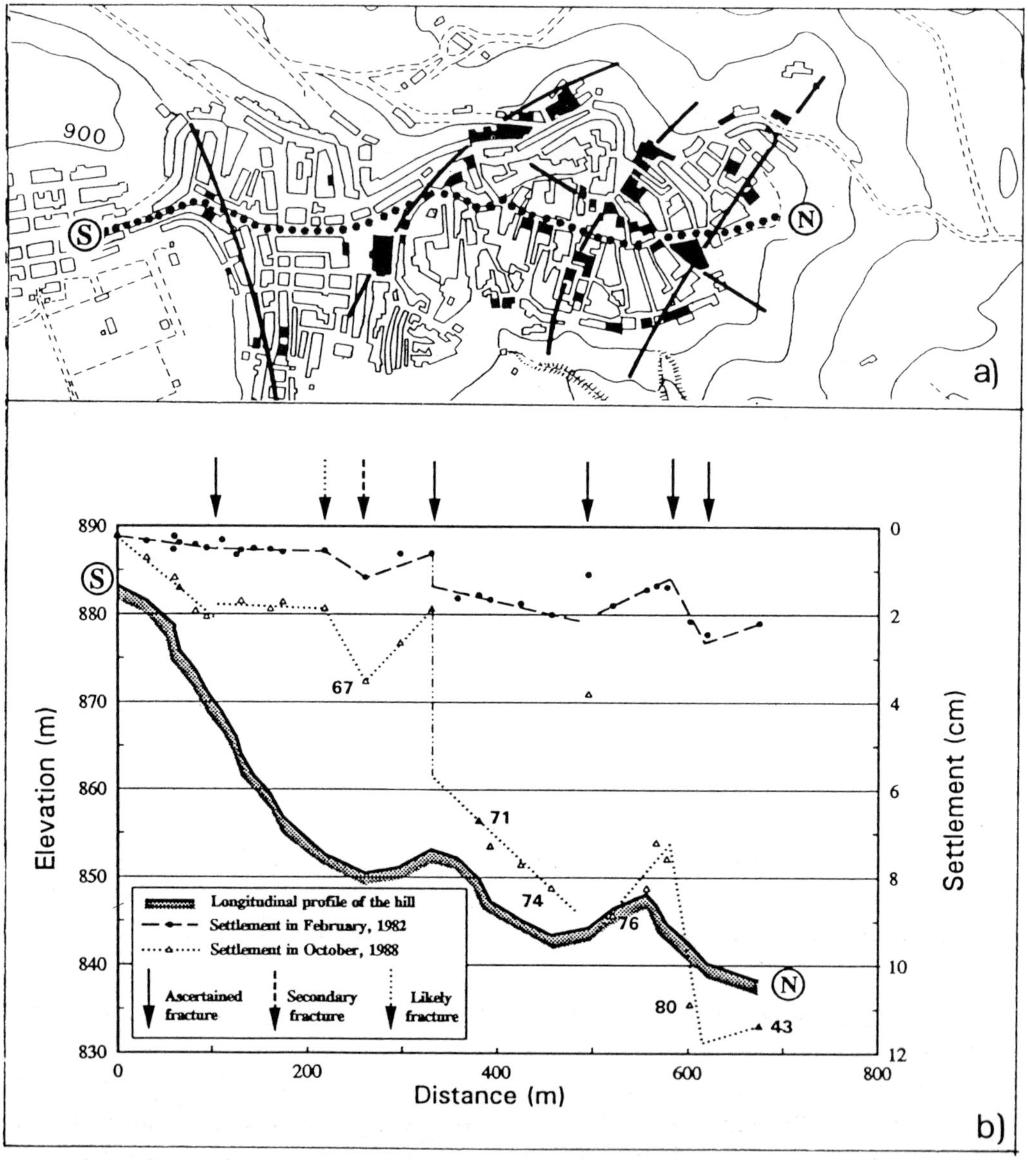

Fig. 7: Effects on the ground observed after 1981: (a) plan of the town and contours of the main blocks probably subdividing the slab; (b) settlements observed along a S-N section.

In fig. 7-b a topographic profile of the hill has also been reported; it is very interesting to observe that the segments representing the rigid settlements of single blocks frequently look like the corresponding local topographic profile. This seems to confirm that the present morphology of the hill might be the result of the geological phenomenon of sinking, featured by periodical accelerations due to earthquakes. The overall mechanism of displacements induced by the seisms appears to replicate following the same general fashion which has been affecting the hill through the centuries.

In fig. 8 the evolution of settlements with time is plotted for some significant benchmarks, belonging to the same S-N section in fig. 7; the settlements increase continuously, with a decreasing rate: at the moment they don't appear to have stopped. This is confirmed by a visual examination of numerous cracks today reopened although they were repaired after the seism. The slow rate of such a phenomenon could be justified by the very low consolidation coefficient of the underlying clay shales.

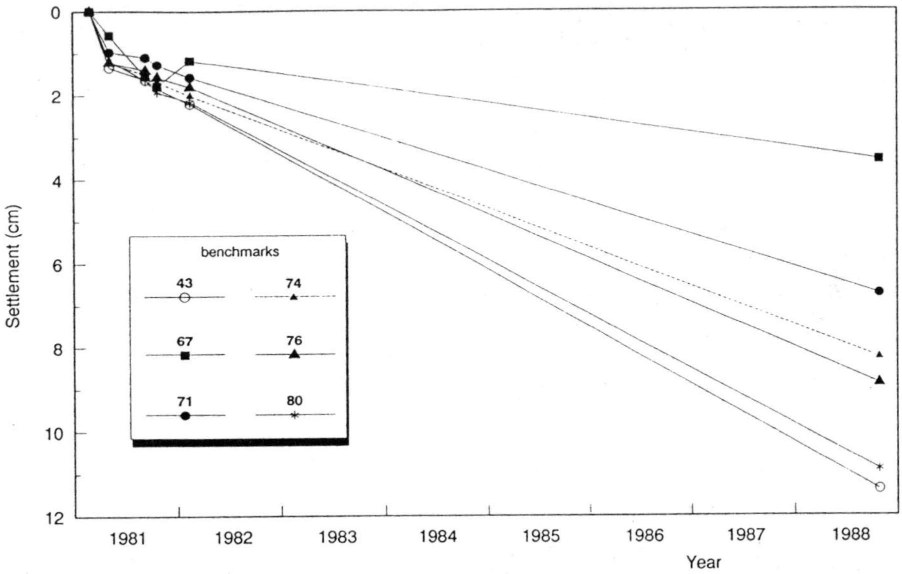

Fig. 8: Displacements vs. time for some significant benchmarks, referred in fig. 7.

6. A reliable hypothesis on the causes of observed phenomena

A careful processing of the data collected up today about the properties and behaviour of the Bisaccia hill allows to hypothesize that the displacements observed after the earthquake in 1980 have been induced by post-cyclic recompression phenomena triggered by the reiterated seismic actions.

A suitable testing program, indeed, has not yet been carried out on the Bisaccia clay; however, the available overall experiences on the clayey material, together with cyclic and dynamic measurements performed on other heavily overconsolidated soils [18] seem to show that shear strain levels of the order of 0.05 % are enough to be associated to sensible positive pore pressure buildup. A rough estimate of the strain amplitudes which might have been induced by the latest earthquake in 1980 suggests that the threshold strain level γ_v is likely to have been overcomed during the strongest seismic shocks.

Just to check such a hypothesis, it has been planned to install a number of piezometers underneath the conglomerate slab. Due to both technical and economic constraints, only two piezometers have been eventually installed in the subsoil of the town (see fig. 2-b). Nevertheless, the mentioned results of the measurements seem to give a further significant clue of the aforesaid assumption, being the measured pore pressures well beyond those calculated by Urciuoli [10], and currently drifting towards values higher than the steady state conditions.

Finally, both the trend and the rate of the observed settlements appear to be due to consolidation determined by induced pore pressures. On the other hand, the slow rate of pore pressure dissipation ought to be ascribed to the low permeability of clay shales along with the thickness of the deposit. These factors should concur in allowing accumulation of the pore water pressure buildup due to subsequent earthquakes, featured by a return period of some tens of years in this area. Therefore, although the single seismic events by themselves might have induced pore water pressure amplitudes relatively poor, the historical succession of strong earthquakes might be responsible for a superposition of effects, to be added to the erosion (and maybe creep strains [9]) in producing relevant settlements of the hill. In conclusion, considering the relative shortness of the return period of earthquakes compared to the long consolidation times, the observed sinking is likely to be continuous in time with periodical peaks determined by the most significant quakes.

Such arguments appear conceptually convincing, but they need to be quantitatively confirmed. Due to the lack of knowledge of the modifications in the pore pressure regime actually induced by the latest seism, numerical analysis might provide at least an estimate of the amplitudes of Δu, partly responsible for the observed sinking.

A reliable numerical approach to the complexity of the boundary problem should be based upon a non-linear, multi-directional constitutive law for the clayey soil, also accounting for pore water pressure buildup induced by deviatoric cyclic loading.

So far, a first 2-D computational analysis by FEM has already been performed by Esu et al. [24], in order to simulate the short-term effects due to the earthquake of 1980. The calculations have been based on a linear elastic behaviour for both the conglomerate and the clay, this latter featured by a bilinear correlation between pore pressure and octahedral shear stress, based on the results of the cyclic laboratory tests referred in § 4.3.

The results of this very first analysis, which probably underestimates the real situation, show that the seismic actions induce valuable pore pressures, especially under the border of the slab.

7. Final remarks

Researches as that one carried out after the 1980 earthquake are very cumbersome, due to different reasons, like: the difficulties in establishing the role of the geological factors, the wideness of the shaken area, the complex nature and peculiar properties of the involved soils, and the unavoidable troubles in measuring small displacements in an inhabited area, as well as pore pressures down to a relevant depth. Furthermore, such an investigation requires a noticeable financial support.

Nevertheless, a reliable and charming hypothesis has been drawn, supported by congruent experimental data. Such hypothesis appears also very significant for comparable situations observed in the same geological environment.

Further studies are obviously needed, in order to highlight the role of the different concurring factors. The next developments should include a better understanding of the static and cyclic behaviour of the soils, as well as a more evolved interpretation of the boundary problem.

Acknowledgements

The research has been carried out with the financial support of MPI (40 %) and CNR (GNDCI).

References

[1] Cotecchia V., Travaglini G., Melidoro G. (1969). *I movimenti franosi e gli sconvolgimenti della rete idrografica prodotti in Calabria dal terremoto del 1783*. Geologia Applicata e Idrogeologia, vol. 4, pp. 1-24.

[2] Govi M., Sorzana F. (1977). *Effetti geologici del terremoto: frane*. Rivista Italiana di Paleontologia e Stratigrafia, vol. 83, n. 2.

[3] Cotecchia V., Del Prete M. (1984). *The reactivation of large flows in the parts of Southern Italy affected by the earthquake of november 1980, with reference to the evolutive mechanism*. IV International Symposium on Landslides, Toronto.

[4] D'Elia B., Esu F., Pellegrino A., Pescatore T.S. (1985). *Some effects on natural slope stability induced by the 1980 italian earthquake*. Proc. XI ICSMFE, S. Francisco, 1985.

[5] D'Elia B., Federico G., Pescatore T.S., Rippa F. (1986). *Occurrence and development of a large landslide (Andretta, Italy) reactivated by the nov. 23, 1980 earthquake*. International Symposium on Engineering Geology problems in seismic areas, Bari.

[6] Esu F., Bilotta E., D'Elia B., Federico G., Fenelli G.B., Pellegrino A., Pescatore T.S., Picarelli L., Tancredi G. (1985). *Geotechnical properties and slope stability in structurally complex clay soils.* In "Geotechnical Engineering in Italy: An overview". A.G.I., ISSMFE Golden Jubilee Volume.

[7] Fenelli G.B. (1988). *Misure di spostamenti e pressioni neutre nell'abitato di Bisaccia.* Convegno Gruppo Nazionale Difesa Catastrofi Idrogeologiche, Bologna.

[8] Fenelli G.B., Picarelli L. (1990). *The pore pressure field built up in a rapidly eroded soil mass.* Canadian Geotechnical Journal, vol. 27, n. 3.

[9] D'Elia B. (1991). *Deformation problems in the italian structurally complex clay soils.* Invited lecture, X ECSMFE, Firenze.

[10] Urciuoli G. (1992). *Rigonfiamento di un'argilla di alta plasticità e modellazione dei fenomeni erosivi del colle di Bisaccia.* Tesi di Dottorato in Ingegneria Geotecnica, Università di Napoli Federico II.

[11] Di Nocera S., Iaccarino G. (1990). Private communication.

[12] Britto A.M., Gunn M.J. (1987). *Critical state soil mechanics via finite elements.* Ellis Horwood Ltd., Chichester.

[13] Papa V., Silvestri F. (1987). *Su un'attrezzatura per prove di colonna risonante e torsione ciclica.* Atti della Riunione del Gruppo Nazionale di Coordinamento per gli Studi di Ingegneria Geotecnica, CNR, Roma.

[14] Ramberg W., Osgood W.R. (1943). *Description of stress-strain curves by three parameters.* Technical Note 902, National Advisory Committee for Aeronautics, Washington, D.C.

[15] Silvestri F. (1991). *Analisi del comportamento dei terreni naturali in prove cicliche e dinamiche di taglio torsionale.* Tesi di Dottorato in Ingegneria Geotecnica, Università di Napoli Federico II.

[16] Vucetic M., Dobry R. (1991). *Effect of soil plasticity on cyclic response.* Journal of the Geotechnical Engineering Division, ASCE, Vol. 117, No. GT1.

[17] Lo Presti D.C.F. (1989). *Proprietà dinamiche dei terreni.* Atti delle conferenze di Geotecnica di Torino, XIV ciclo "Comportamento dei Terreni e delle Fondazioni in Campo Dinamico".

[18] Silvestri F. (1991). *Stress-strain behaviour of natural soils by means of cyclic/dynamic torsional shear tests.* Workshop on "Experimental Characterization and Modelling of Soils and Soft Rocks", Napoli.

[19] Yasuhara K., Andersen K.H. (1985). *Effect of cyclic loading on recompression of overconsolidated clay.* Proc. XI ICSMFE, S. Francisco.

[20] Baratta M. (1901). *I terremoti d'Italia.* Fratelli Bocca ed. Torino.

[21] Alfano G.B. (1931). *Il terremoto irpino del 1930.* Osservatorio di Pompei.

[22] Oddone E. (1931). *Il terremoto dell'Irpinia del 23 luglio 1930.* L'Universo.

[23] Chierici G. (1932). *I monumenti dell'Alta Irpinia ed il terremoto del 1930.* Avellino.

[24] Esu F., Nuti C., Palazzo S. (1987). *Comportamento dinamico del rilievo di Bisaccia durante il terremoto irpino del 23 novembre 1980.* III Convegno Nazionale "L'Ingegneria Sismica in Italia", Roma.

EARTHQUAKE SENSITIVITY OF THE PURISCAL LANDSLIDE, COSTA RICA

S. MORA*, E. LEROI**, F. DURAND**, J.P. ASTE**, P. MOUROUX**

Abstract : Having briefly reminded evaluation methods of slope instability, under dynamic conditions, we present a summary of the current knowledges about Santiago de Puriscal, Costa Rica landslide. A study of the slide mass displacements led to determine the effects of an earthquake on landslide instability.

Résumé : Après un bref rappel des méthodes d'évaluation de la stabilité des pentes, en conditions dynamiques, nous présentons une synthèse des connaissances sur le glissement de terrain de Santiago de Puriscal, Costa Rica. Une étude des déplacements du massif a permis de déterminer l'influence d'un séisme sur la stabilité du glissement.

1 Introduction

In certain slope sediments, such as clay and granular, well-drained soil, where a reference earthquake will induce only slight increase of the pore pressure and a low shear strength loss (less than 20%, [1]), three types of approach can be envisaged for evaluating the stability under dynamic conditions. These are :

i) so-called pseudostatic rupture methods, where the seismic acceleration is represented by a resultant force. The failure planes are either circular (Bishop, Sarma) or non-circular (Spencer, perturbations method). These methods give :

- either the safety factor F, which corresponds to given seismic coefficient K;
- or the critical seismic coefficient K_c that corresponds to $F = 1$.

ii) failure displacement methods ([2] and [3]), which enable to calculate a critical acceleration K_c, also called N. This is then used to determine a maximum cumulative displacement, taking into account the fact that, during an earthquake, the safety factor can several times drop below 1 for short moments.

iii) general stress/strain methods, which provide knowledges on stress, deformation and local safety coefficient in any part of the mass, and potential failure lines. From these results and from the stress distribution along these failure lines, it is possible to more accurately evaluate total safety coefficient.

However, if these pre-conditions are not fulfilled, in particular when the pore pressure strongly increases as a result of a seismic shock, liquefaction may occur. The solution of this problem require quite different models.

* Departamento de Geologia, Instituto Costarricense de Electricidad, San José, Costa Rica.
** Bureau de Recherches Géologiques et Minières, Marseille, France.

2 The Puriscal landslide, Costa Rica

2.1 REGIONAL SETTING

The town of Santiago de Puriscal, 10,500 inhabitants, is entirely located on an enormous landslide mass. The activity of this mass is known since at least 1918, but aerial photographs of 1945 show that this landslide could well be as old as two centuries, although no historical records of such movement have been preserved.

The slide mass consists of two sub-masses, covering about 4.5 km^2, and with a total estimated volume of 195 hm^3 (million cubic metres). This makes it the largest known slide mass in Central America and the Caribbean, and one of biggest in the world ([4] and [5]).

This landslide, like most other such features, has mainly resulted from the conjunction of two 'permanent' factors (geomorphology and geology) and two 'triggering' factors (precipitation and seismicity).

2.1.1 Geographical location and geomorphology

Santiago de Puriscal is located in the western foothills of the north part of the cordillera de Talamanca, about 25 km southwest of San José, Costa Rica's capital (Fig. 1). The surrounding topography shows that the geomorphological cycle is in a young stage. This is indicated by 'V'-shaped valley cross sections, and steep slopes that can be as much as 45°.

This kind of relief is conducive to slope-deterioration phenomena, like erosion, and mass movement, in particular where they are aggravated by human action.

2.1.2 Structural and neotectonic setting

Puriscal is located in a mountainous area underlain by plutonic and volcanic rocks, which are the expression of subduction of the Cocos plate beneath the Caribbean plate. Aerial photographs made by the Costa Rican Instituto Geographico Nacional and satellite images (SPOT, LANDSAT) show that, around Puriscal, the bedrock is affected by three active sets of faults :

- an E-W set south of Puriscal;
- a NW-SE set to the northeast and north of Puriscal;
- a NE-SW set to the northwest of Puriscal.

A fourth fault system, again NW striking, is found near Quivel, in the southwest part of the area investigated.

The intense seismic activity that has affected the region since March 1990, has reactivated many of these faults and fractures. In fact, the ground is strewn with open fractures, whose average density was calculated to be about 500 m/ha.

Their length varies from 5 to 80 m, their width ranges between 0.5 and 10 cm, and the measured relative displacements are at least 10 cm. The fractures severely disturb the existing infrastructure and communication networks as well as many buildings. For instance, Puriscal church is particularly affected in spite of its recent rebuilding.

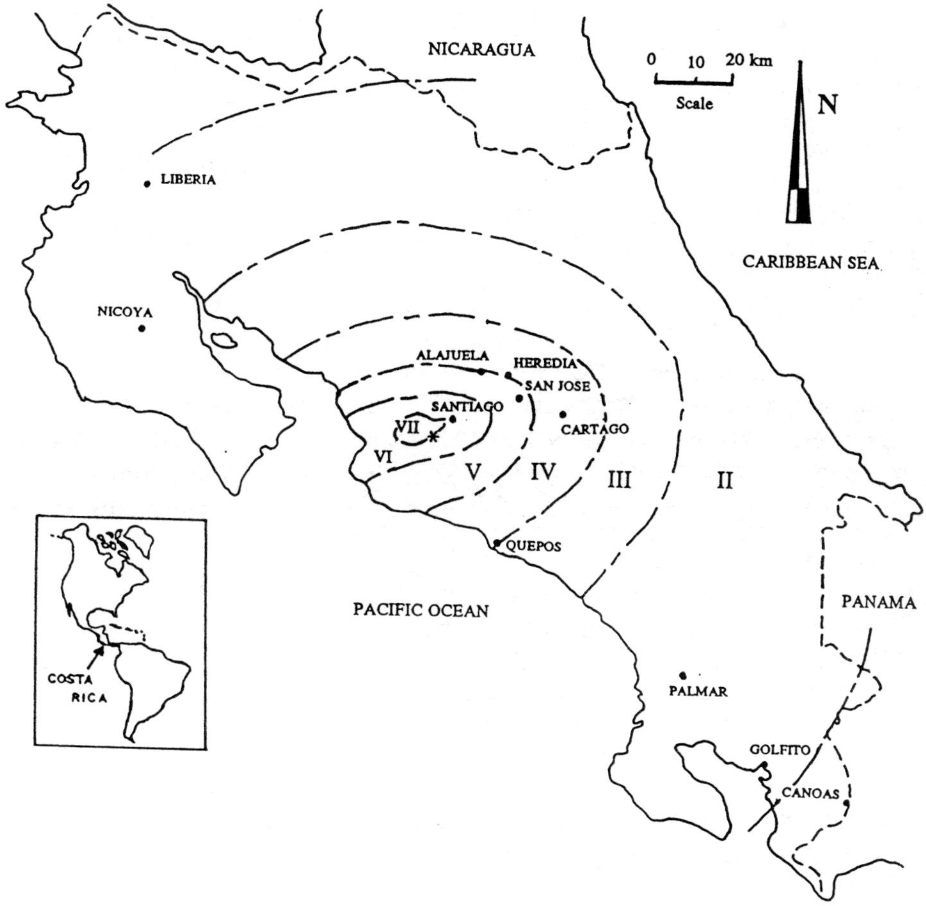

Figure 1 : Geographical location of Santiago de Puriscal and intensity map of earthquake of 30/06/90 (Epicentre : *). (From S. Mora [4]).

2.1.3 Local geology

Puriscal is built on volcanic rocks that overlie upon Oligocene to Early Miocene sedimentary rocks. This substratum has been strongly affected by weathering in the humid tropical climate of the region, as well as by hydrothermal activity related to the late magmatic phase. The result of this is a residual, regolithic soil that is 60 m thick on average (Fig. 2).

The first 40 metres, or A horizon, are composed of highly plastic silt and clayey silt, reddish to yellowish brown, with local white, greyish or purple kaolinite and montmorillonite that are the results of hydrothermal alteration.

The next 20 metres, or B horizon, are composed of strongly weathered bedrock, or regolith, with bedrock blocks surrounded by clayey and silty decomposition products.

Under this, lies the bedrock itself, at about a 60 m depth, composed of volcanic breccia, conglomerate and tuff, with intercalations of andesitic basalt.

2.1.4 Climate

Puriscal lies in the Pacific foothills of the cordillera, and has a wet tropical climate, with a precipitation of about 2541 mm/year. The rainy period is from May to November, with recorded maxima of 130 mm/ 12 hrs, 225 mm/ 24 hrs, and 300 mm/ 36 hrs for the months of September and October. These data were computed from records starting in 1940. The most intense rainfall, which is over 140 mm/hr in 15 min, or 70 mm/hr in 60 min, is estimated to have a recurence period of 10 years [6].

This precipitation causes severe erosion, characterized by a loss of soil that is estimated to be about 25 kg/m^2.

2.1.5 Seismicity

The historical seismicity in the Puriscal area, covering the 20th Century, shows a first cycle of intense activity that started around 1905 and had its paroxysmal event in the 1924 Orotina earthquake (Ms=6.5 ; MM=VII$^+$ in Puriscal). From 1927 to 1961, only few strong earthquakes were recorded, which were moreover widely spaced in time and of only moderate intensity (MM=IV to V). In 1961, another cycle has started with earthquake intensities of MM>V. Until April 1991, the two major events were the Cobano earthquake of 25/03/90 (Md=6.5 ; MM=VI in Puriscal) and the earthquake of 22/12/90 (Ml=5.8 ; MM=VII in Puriscal).

The Puriscal seismic swarm is the name given to a period of intense seismic activity that started in May 1990 and was still going on in April 1991, when the last data were collected for this paper. From May to December 1990, over 30,000 tremors were recorded, of which at least 1000 were felt by people [7]. It is suggested that this swarm is the result of the tectonic destabilization caused by the Cobano earthquake.

—

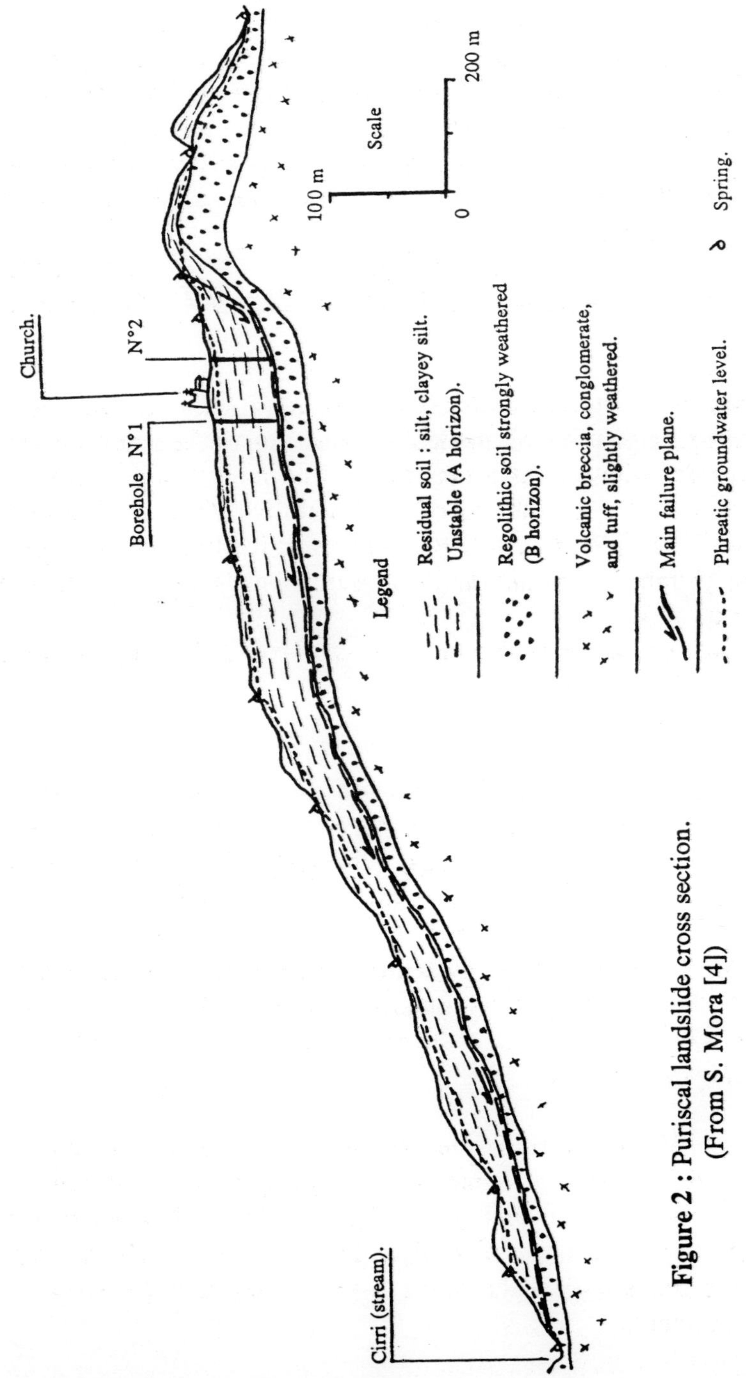

Figure 2 : Puriscal landslide cross section.
(From S. Mora [4])

The main event on 22/12/90, whose epicentre was 7 km northeast of Puriscal, generated strong ground accelerations over 10% g during 2 to 3 seconds. The second strongest event occured on 30/06/90, 7 km southwest of Puriscal (Ml=5.0 ; MM=VI$^+$ in Puriscal ; Fig. 1), and generated maximum ground accelerations of 16% g near the landslide during 3 to 4 seconds. The data used for the Puriscal landslide study were collected soon afterwards.

2.2 THE LANDSLIDE

2.2.1 Description

The landslide is irregularly shaped in plan and cross section, and it consists of two masses, called Santiago and Cirri-Carit. The first is 2.25 km^2 in size, has the town built on top of it and moves northwest, whereas the second is about 1.75 km^2 and moves north (Fig. 3). Since 1950, the average creep velocity has been estimated at 5 to 15 cm/year. Given these figures and an average thickness of the slide mass of 30 to 60 m, the respective volumes of moving ground for the two masses are 80 and 115 hm^3.

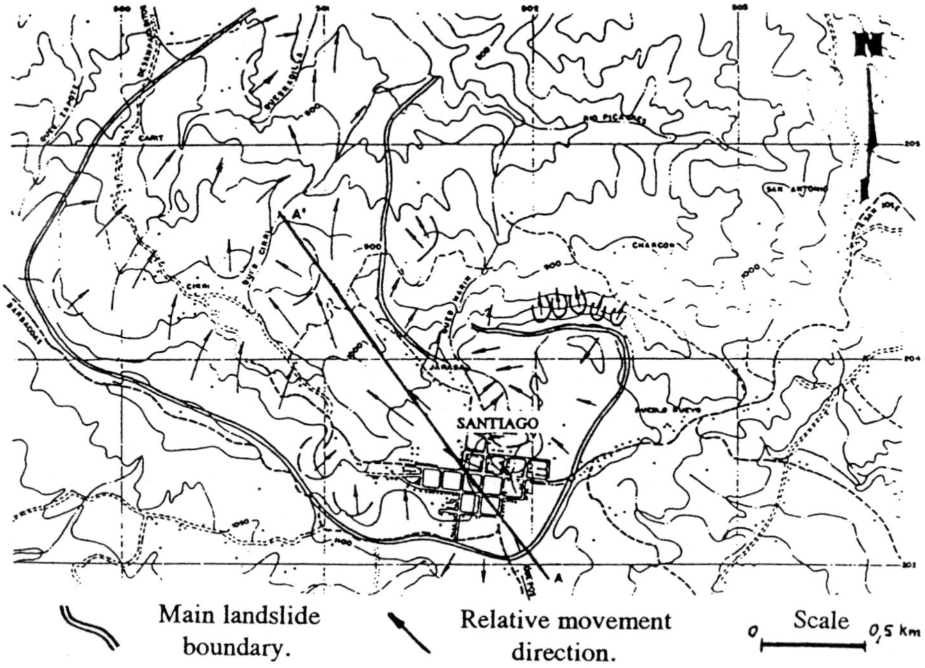

Figure 3 : Santiago de Puriscal landslide, interpreted from airphotos 1989 (I.G.N.); (From S. Mora [4]).

The failure plane is roughly parallel to the surface (Fig. 2) and irregularities in the ground, such as small scarps and cracks, show the existence of separately moving blocks. Electric resistivity profiles showed the presence of a 20 to 40 m thick layer, with internal secondary sliding surfaces. Clayey silt inclusions with their very high resistivity define shallow sliding planes, that are the main cause of the surface disturbances. The main failure plane, which in fact encompasses a set of parallel and interlocking discontinuities, lies at a depth of between 30 and 60 m, near the boundary between A and B horizons.

Table 1 shows the actual size of the areas affected by movement for different years. The estimates of these areas was made from aerial photographs.

Years	Active areas size (ha)
1945	86.7
1956	70.8
1961	112.0
1965	145.8
1970	156.0
1974	161.1
1981	165.3
1984	169.7
1989	194.3

Table 1 : Active area of Puriscal landslide for various years.

Figure 4 shows the correlation between the highly active areas, seismicity and strong rainfall above 50 mm/ 24 hrs. From 1945 to 1960, the landslide activity decreased, because both rainfall and seismicity were at low levels. From 1960 onwards, moderate seismicity but strong and frequent rainfall cause an increase in landslide activity of +7.6 ha/y until 1965. After this, 'normal' rainfall and seismicity led to a stabilization of the landslide activity (+1.7 ha/y). However, since 1985, seismicity has increased dramatically and this, coupled with heavy rainfall, led to an increase of activity (+4.5 ha/y) until the 1990 seismic swarm.

2.2.2 Geotechnical data

Geotechnical characteristics (Table 2) only concern the uppermost A horizon (Fig. 2). They were obtained from samples of N° 1 borehole that were taken at various depths. Shear-resistance tests were done under undrained, unconsolidated conditions (U.U. tests), which are the most representative of dynamic conditions.

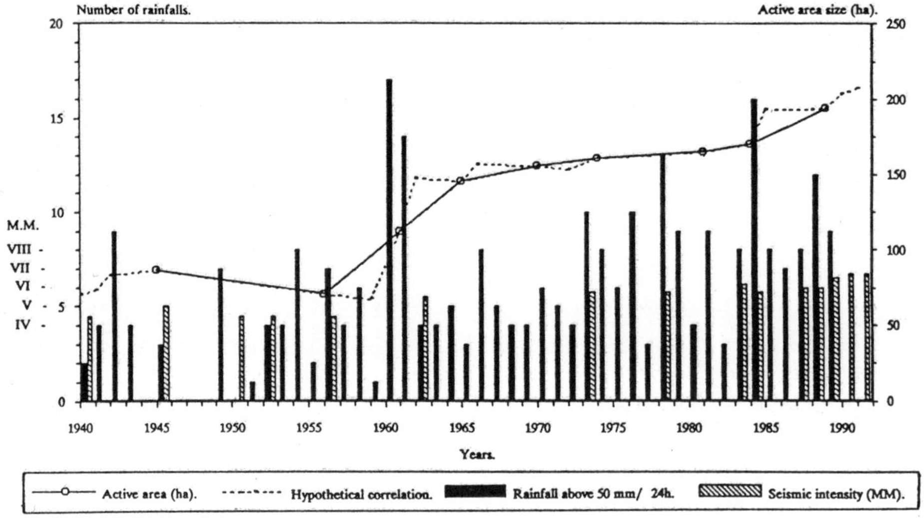

Figure 4 : Correlation between active area size, seismicity and strong rainfalls. (From S. Mora [4]).

It was also noted that the ground has a strong saturation, increasing with depth and closed to 100 %, even well above the phreatic groundwater level. Water content obviously is high too and in places surpasses the liquid limit, especially near the main failure surface.

Ground classification gives the composition of silt, clayey silt and a small amount of clay. This ground is extremely plastic and compressible. Average figures for cohesion and shear-resistance angle are thus logical at c = 105 kPa, and $\phi = 24°$. The same is true for residual values which are c = 0, $\phi_{res} = 13°$. They are confirmed by the results from standard penetration test, based on the equation for clay of : $15.c_u / N = 0.1$ à 0.2

These results were : - between 0 - 20 m : N = 5 ---> $c_u = 50$ kPa

- between 20 - 44 m : N = 19 ---> $c_u = 200$ kPa.

As for the underlying B horizon, cohesion and shear-resistance angle values are : c = 250 kPa, and $\phi = 30°$.

2.2.3 Hydrogeological data

Groundwater level is generally found at a depth of 3 to 7 m in farm wells, but a dozen of springs are known as well (Fig. 2).

The fine grainsize causes the permeability of the ground to be quite low and most groundwater flow is through fractures. Hydraulic conductivity is about 6.10^{-6} m/s, which indicates that it takes about 4 days for a strong rainfall to infiltrate down to the aquifer. This recharge takes mostly place between May and November, with infiltration rates varying from 62 to 307 mm/month.

Geotechnical properties	Minima	Maxima	Means
Water content (w, %)	44	70	59
Liquid limit (w_l, %)	57	89	71
Plastic limit (w_p, %)	20	65	40
Plasticity index (PI)	30	50	41
Degree of saturation (S, %)	81	100	99
Void ratio (e)	1.1	1.9	1.6
Total unit weight (γ, kN/m^3)	15.2	18.5	16.5
Solid unit weight (γ_s, kN/m^3)	26.3	27	26.6
Sand percentage (%)	16	24	20
Passing No. 200 sieve (%)	70	88	79
Cohesion (c, kN/m^2) *	55	175	105
Shear-resistance angle (ϕ, °) *	20	29	24
Residual shear-resistance angle (ϕ_{res}, °) *	9	19	13
Standard penetration resistance 0-20 m (N)	3	18	5
Standard penetration resistance 20-44 m (N)	10	34	19
Classification: Silt; Clayey silt; Clay. High compressibility and plasticity, soft, colour coffee.			
Nota: Phreatic groundwater level between 3 and 7 m. * Data obtained from U.U. tests.			

Table 2 : Geotechnical characteristics for the A horizon.

2.3 EVALUATION OF PURISCAL LANDSLIDE INSTABILITY

2.3.1 Static feedback analysis of geotechnical characteristics

The large dispersion of geotechnical properties is due to spatial variability of the soil materials, as well as to the reduced number of tests. This led us to make a feedback analysis of landslide instability in order to validate the mean values shown in Table 2.

This analysis was done for static conditions (i.e. without earthquake) by using the perturbations method. Hypotheses adopted for the model were as follow :

- The failure plane is fixed in the model and has an irregular shape (Fig. 2). It envelops small secondary moving masses and it is located in the A horizon, a few decimetres above the boundary with the B horizon. In this manner, only the A horizon characteristics will play a role in the safety factor variations.

- Under static conditions, and in the absence of strong rainfall, the ground is supposed to be in a state of near stability ($1.2 < F_S < 1.5$).
- Calculations were carried out under total stress, for various values of cohesion and shear-resistance angle, with a total unit weight of 16.5 kN/m^3.

Results gave c = 25 kN/m^2 and ϕ = 8° for a safety factor F = 1.4 (Fig. 5). They are closed to the residual mean values achieved from tests.

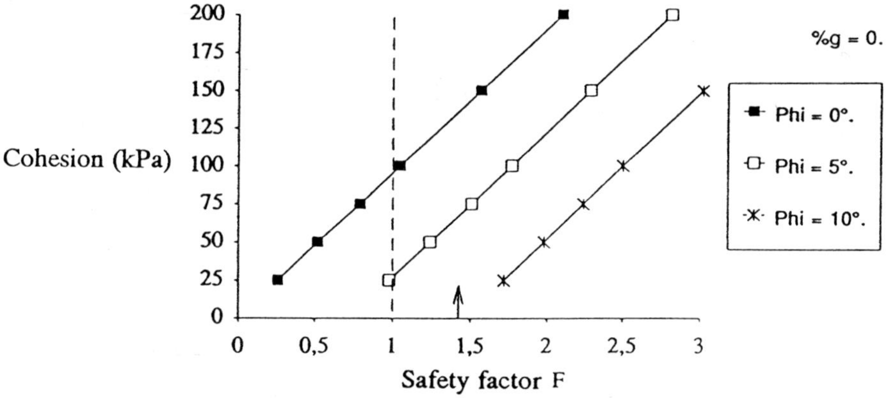

Figure 5 : Safety factor variations versus cohesion for various shear-resistance angles.

2.3.2 Critical acceleration calculation

2.3.2.1 Pseudostatic approach

A supplementary destabilizing force is then introduced into the model as an horizontal seismic acceleration, oriented down the slope. Accelerations of 5%, 10%, and 15% of gravity were tested. These values correspond to those actually felt in Puriscal, during the 30/06/90 earthquake. The aim was to determine the critical acceleration N for which F = 1, which means that the block will be sliding.

Figure 6 plots safety factor variations versus cohesion and seismic acceleration, for shear-resistance angles of 5° and 10°. In the first case, N was 0%, and in the second case, it was 8.5%, for a cohesion of 25 kN/m^2. This means that the critical acceleration corresponding to c = 25 kN/m^2 and ϕ = 8° is N = 5.1%.

Figure 7 plots safety factor variations versus the shear-resistance angle and seismic acceleration, for c = 25 kN/m^2. The critical acceleration, read for ϕ = 8° and F = 1, was N = 5.0%.

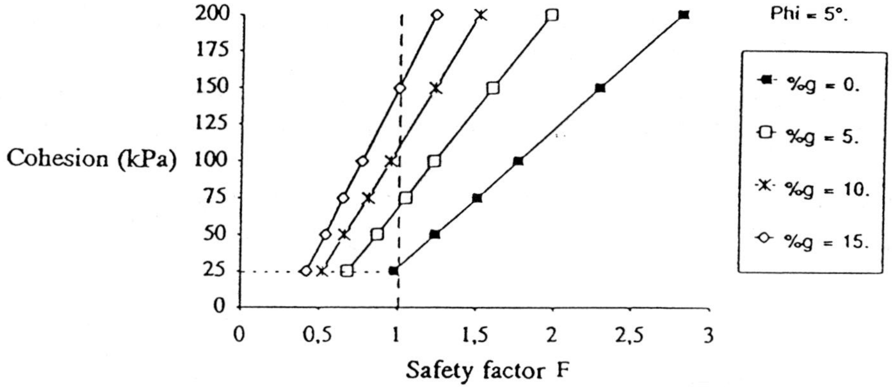

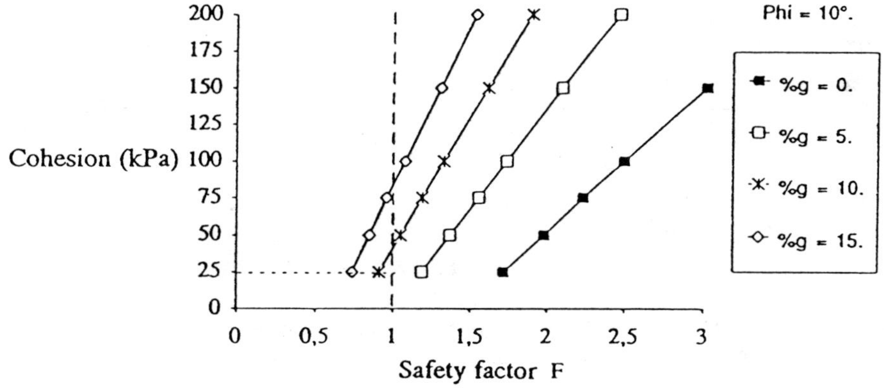

Figure 6 : Safety factor variations versus cohesion for different seismic accelerations.

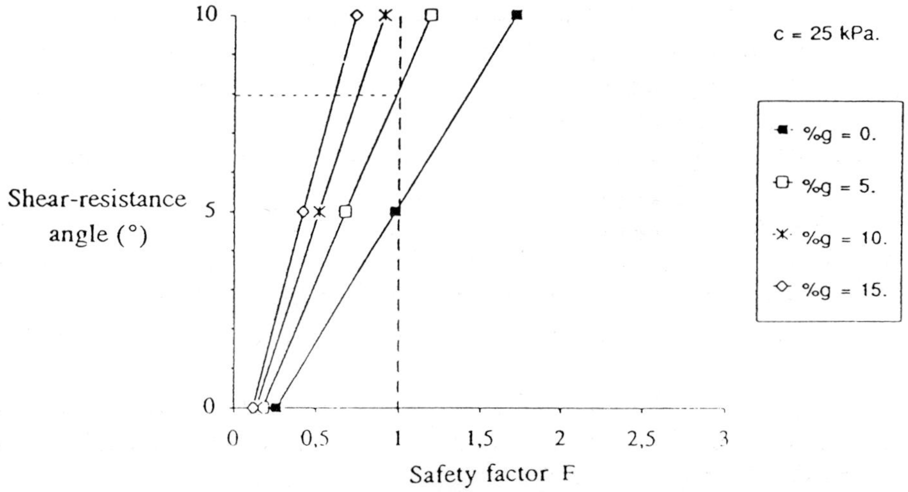

Figure 7 : Safety factor variations versus shear-resistance angle for different seismic accelerations.

2.3.2.2 Newmark relation

Calling β the average slope and F_S the safety factor under static conditions, Newmark showed that : $N = (F_S - 1) . \sin \beta$

In the Puriscal landslide case, we considered that the ground is in a state of near stability ($F_S = 1.4$), and the average slope is $\beta = 8°$. The critical acceleration calculated is thus : N = 5.6%. This value is approximately in accordance with the result of the pseudostatic approach.

2.3.3 Maximum cumulative displacement calculation

The first method is based on Newmark's chart and determine a standardized displacement. We used the modified chart of Franklin and Chang (Fig. 8) and the earthquake of 30/06/90 characteristics. Maximum ground acceleration was 16% and critical acceleration was previously estimated to 5%. So, N/A = 0.31. We obtained a standardized displacement of about 80 cm.

Relation between standardized maximum displacement and actual maximum displacement is given by the scaling relationship : $u_m = u_s . (V^2/V_s^2).(A_s/A)$.

Newmark standard values are $A_s = 0.5$ g and $V_s = 30$ in/s and correlation between maximum acceleration A and maximum velocity V [8] led to a simplified relation : $u_m = u_s . 0.097\ A^{0.7}$ (A is in m/s²).

The result is an average displacement along the main failure plane of Puriscal landslide estimated to 10 cm.

The second method is based on accelerogram. During an earthquake, the safety factor can several times drop below 1 for short moments. Therefore, for a given accelerogram and a critical acceleration N, it is possible to determine the maximum cumulative displacement of a slide mass, by two integrations. Unfortunately, we did not have an accelerogram of the earthquake of 30/06/90, but we generate synthetic accelerograms from the seismic source characteristics: epicentral distance of 7 km, Ml = 5.0 and duration of 3 to 4 s. The Newmark analysis realized confirmed our previous values of the order of 10 cm.

There is agreement of this result with the actual displacement measured on open fractures (cf. 2.1.2). However, 1 m displacements were locally observed in the field, that can be caused by differential amplifications of the movements at the top of slide masses, and by heterogeneities in some places.

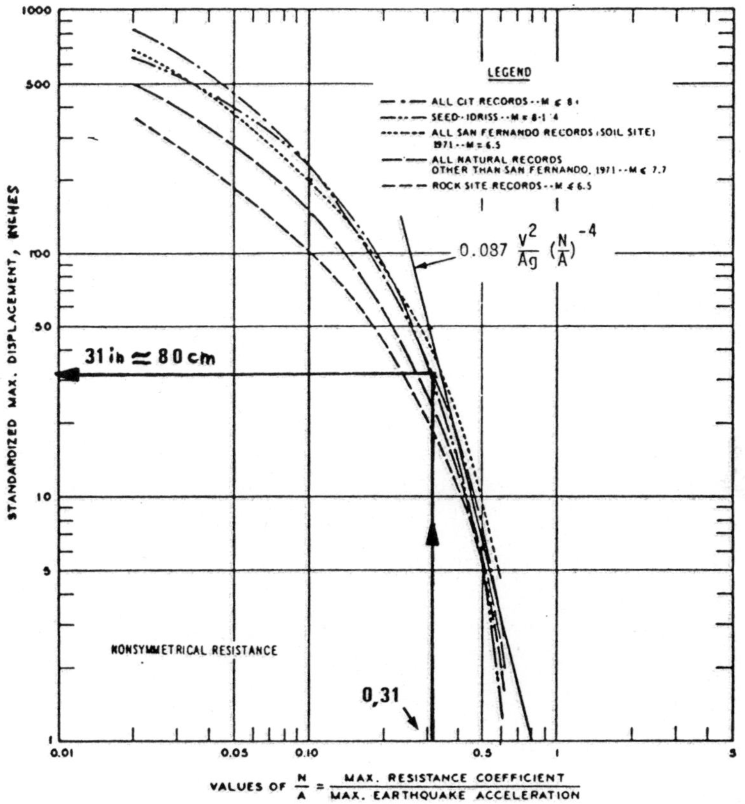

Figure 8 : Chart for maximum cumulative displacement.
(From Franklin and Chang, in Richards and Elms [9]).

3 Conclusions

A first analysis of the behaviour of Puriscal landslide (Costa Rica) is done, using pseudostatic and displacement (Newmark) approaches, and first geotechnical parametres obtained from one borehole and geophysical and geotechnical tests. It presents some results in accordance with observations made in the city of Santiago.

A more detailled study should be carried on concerning :

- a systematic analysis, by aerial photographs of the different successive stages of the landslide.

- a better knowledge of short and long term geomechanical properties of the soil located in the rupture zone.

- a stress/strain analysis using finite element method (for instance, Gefdyn code, [10] and [11]), which would allow a better comprehension of surficial displacements and predictable damages in the city of Santiago de Puriscal.

Bibliography

[1] Seed H.B., (1979) Considerations in the earthquake-resistant design of earth and rockfill dams. *Geotechnique*. vol. 29, n° 3, pp. 215-263.

[2] Newmark N.M., (1965) Effects of earthquakes on dams and embankments. *Geotechnique*. vol. 15, n° 2, pp. 139-160.

[3] Franklin A.G. and Chang F.K., (1977) *Earthquake resistance of earth and rock-fill dams, Report 5 : Permanent displacement of earth embankments by Newmark sliding block analysis*. Miscellaneous paper S-71-17. Soil and Pavement Laboratory, U.S. Army Engineer Waterways Experiment Station.

[4] Mora S., (1990) *Analisis geotecnico preliminar del deslizamiento de Puriscal, Provincia de San José, Costa Rica*. Comision Nacional de Emergencia. Escuela C.A. de Geologia, Universidad de Costa Rica. p. 59.

[5] Mora S., Leandro G., and Leon B., (1991) *La dinamica del deslizamiento de Puriscal, Costa Rica*. V Seminario Nacional de Geotecnia, 1° Encuentro Centroamericano de Geotecnia, 15,16,y 17 de Mayo 1991. Memoria. Ed. Univ. de Costa Rica. p. 20.

[6] Varhson G. et al., (1990) *Aspectos climatologicos relacionados con la morfodinamica externa en la zona de Puriscal, Costa Rica*. Escuela de Ciencias Geograficas, Univ. Nacional, Heredia, Costa Rica. Internal paper.

[7] Rojas-W., (1990) *Parametros del enjambre sismico de Puriscal*. Red Sismologica Nacional. ICE-UCR, Escuela C.A. de Geologia. Internal paper.

[8] Trifunac M.D., and Brady A.G., (1976) Correlation of peak acceleration, velocity and displacement with earthquake magnitude, distance and site conditions. *Earthquake Engineering and Structural Dynamics*. vol. 4, n° 5. pp. 455-472.

[9] Richards R., and Elms D.G., (1979) Seismic behavior of gravity retaining walls. *Journal of the Geotechnical Engineering Division*. ASCE, Vol. 105, n° GT4, Proc. paper 14496, pp. 449-464.

[10] Modaressi H., (1987) *Modélisation numérique de la propagation des ondes dans les milieux poreux anélastiques*. Thèse de docteur de l'Ecole Centrale de Paris.

[11] Aubry D. et al., (1986) *GEFDYN : Logiciel d'analyse du comportement mécanique des sols par éléments finis avec prise en compte du couplage sol-air-eau*. Rapport EDF.

[12] Mora S., (1989) *Extent and social-economic significance of slope instability in Costa Rica*. Proc. of the 28th Int. Geological Congress : Symposium on landslides. Washington D.C.

Thème 2

MICROZONATION AND VULNERABILITY TO LANDSLIDE HAZARD

REAL TIME MONITORING OF A SAMPLE SLOPE IN A SEISMIC AREA.

Raffaello NARDI *, Alberto POCHINI *, Mauro ALLAGOSTA*

Abstract : This report illustrates the instruments, main features and results so far obtained, in connection with the real time monitoring of a landslide event in a seismic area. Such surveillance lies within the scope of research promoted by the "Gruppo Nazionale per la Difesa dai Terremoti del Consiglio Nazionale delle Ricerche" (The National Group for Defense against Earthquakes of the National Research Council); its ultimate goal is the achievement of salient data regarding the slope deformation in ordinary conditions as well as in the case of seismic stresses, with the simultaneous registration of data related to the seismic input.

Riassunto: Vengono presentati dagli autori la strumentazione, le caratteristiche principali ed i risultati, sinora ottenuti, nella sperimentazione di monitoraggio in tempo reale di un fenomeno franoso in area sismica.

Tale monitoraggio, realizzato nell'ambito di ricerche finanziate dal Gruppo Nazionale per la Difesa dai Terremoti del Consiglio Nazionale delle Ricerche, ha come finalità quella di permettere l'acquisizione dei dati salienti delle deformazioni del versante in condizioni normali ed in caso di sollecitazioni sismiche, con la contemporanea registrazione dei dati relativi all'input sismico.

Introduction

The monitoring system for unstable slopes, implemented in the Camporaghena slide (District of Massa), in the Northern Apennine seismic area, known as Lunigiana/Garfagnana, as well as the slide survey and its physical "modelling" are along the lines of research suggested by S.G.L. 1.3.2 of the Gruppo Nazionale per la Difesa dai Terremoti - C.N.R.

Several groups are employed in this research, each of them with specific knowledge into the geological, geotechnical and geophysical branches. Their main goal is to define the behaviour of unstable slopes during the seismic occurrence. The acquisition of data during the earthquake and after it, constitutes the most important feature of such surveillance.

* Dipartimento di Scienze della Terra, Università degli Studi di Pisa
via S. Maria, 53 - 56126 Pisa

Research Procedures

With its 1988-89 executive program, the Gruppo Nazionale per la Difesa dai Terremoti - C.N.R triggered off a project for research into unstable slopes, aimed at defining simplified procedures for general use, which ought to be also geared to describe the behaviour of said slopes, under seismic stresses.

Thus an experiment was started in 1990 on the Camporaghena (Comano, Massa) sample slide, in the Northern Apennine seismic area, also known as Lunigiana-Garfagnana.

Research was articulated in several phases. The first was the geological and geomorphological "framing" of a sufficiently extended area for a better understanding of the basic features of the slope in which the landslide had occurred. A geognostic campaign ensued with the intent of characterizing the materials involved in the slide; then a structural assessment in detail was conducted and eventually the installation and operation phases were established for the surveillance system; the latter was purpose-planned and built. Such system represents the core of the apparatus by which the landslide is monitored, in operation on the slide at the moment.

Geological and geomorphological features

The slope is the seat of a rather complex gravitative phenomenon, which includes several quiescent slides, either potential or operative, scattered over a large area (about 0.35 sq.km.). From the slope landslide map, it is quite clear that such slope is affected by an old quiescent slide of remarkable size, upon which more recent events have supervened: some of them are still active [1].

The typology of the various movements, to which the overall phenomenon may off-hand be traced back, is that of a rotational multiple slip [2, 3].

This area was studied by means of adequate conventional techniques [4, 5], i.e. surface surveys, photogeological and photogrammetric analyses, structural analysis of detail. Research was later on focussed on the main, movements in action between the two towns; a geognostic campaign was conducted, together with refraction seismic surveys, inclinometric checking and so on. The results thus obtained are summarized in fig.1, and in the interpretative section of fig.2.

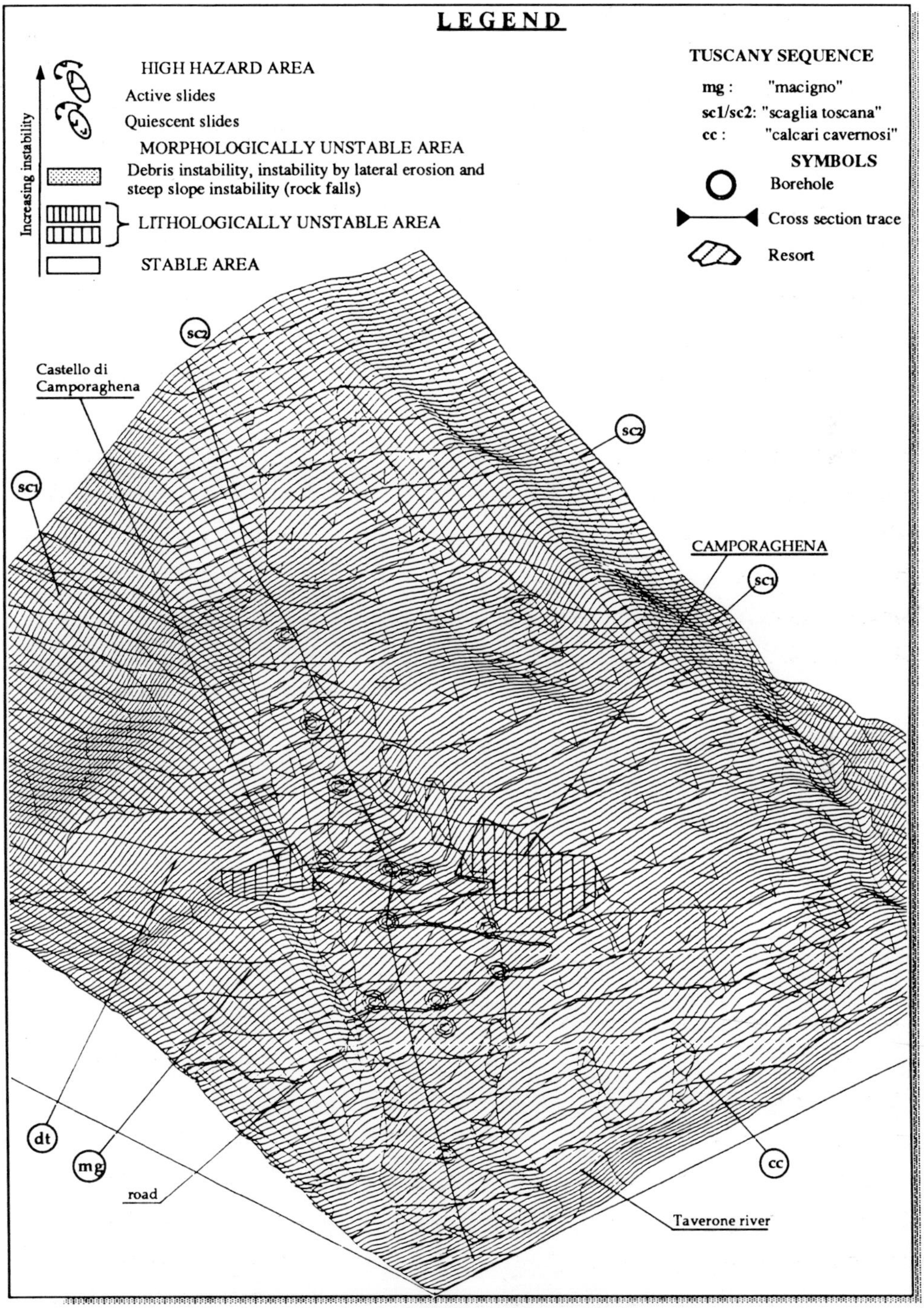

Fig. 1 - 3 D model of the landslides with geomorphological elements.

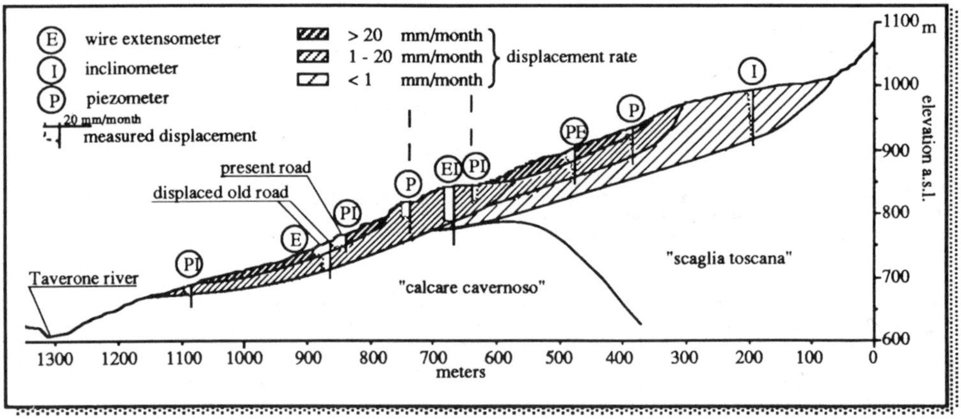

Fig. 2 - Interpreted cross - section of the landslide subdivided according to the displacement rates. The geological formation, istrumentation type and location and corresponding recorded displacements are also shown.

The causes predisposing to such a vast phenomenology are to be found in the marked energy of the hillside relief, in the erosion provoked by surface waters, in the lifting of the area, in its marked rainfall which reaches 2.500 mm/year peaks and in its seismicity, in addition, of course, to the lithologies involved.

In synthesis, it must be noted that the present lithologies belong to the unmetamorphic Tuscan Succession, and more precisely the formations present in the slope are, in geometrical superposition order and from bottom, the "calcari cavernosi" (cc); the "scaglia toscana" (sc) and "macigno" (mg).

The "calcari cavernosi" formation includes cell-limestone and dark grey dolomite of Upper Triassic, with associated sedimentary polygenic breccias of Tertiary. Locally the formation includes important layers of chalk. The "Scaglia toscana" is present in the middle and lower part of the hillside with a member characterized by prevailing shaly facies, with marls, marly limestone and siliceous limestone interposed (sc_1); further up a member is present with prevailing limestone-marl lithologies (sc_2). The formation age can be attributed to Lower p.p. Cretaceous-Oligocene gap.

The slide event, on its whole, concerns almost exclusively this formation; a few samples collected at levels with the worst mechanical features have yielded the following values:

- density $\delta = 18{,}6 \div 21{,}5$ KN/m^3, $\delta_d = 17{,}5 \div 18{,}5$ KN/m^3
- Atterberg limits- $W_l = 32 \div 35\%$; $W_p = 21{,}8 \div 22{,}1$ % $I_p = 10 \div 13$ %
- shear strength - $\phi = 22.5° \div 27.5°$, c' <= 4 KN/m^2.

Trials of rock samples have shown the following features:
- density $\delta = 26{,}1 \div 26{,}7$ KN/m^3
- uniaxial compressive strength 75 ÷ 144 MPa [6].

The interpretation of the down hole trials has supplied the following data [7].

h	Vp	Vs	γ	μ	E	k	σ	vp-γ	vs-γ
m	m/s	m/s	g/cm^3		Gpa	Gpa	Gpa	N/scm^3	N/scm^3
6,2	1210	470	1,9	0,41	1,21	2,28	0,43	2,3	0,89
6,1	770	310	1,9	0,4	0,53	0,9	0,19	1,47	0,59
11,2	1200	590	2,15	0,34	2,07	2,13	0,77	2,58	1,28
21,4	1660	480	2,15	0,45	1,47	5,4	0,5	3,58	1,03

The "macigno" formation which is present on the slope summit and in the vicinity of Castello di Camporaghena, is constituted by quartzose and feldspar sandstone in alternation with shale and siltstone. The period is Middle/Upper Oligocene/Upper Oligocene.

The area surveyed also presents Quaternary soils constituted by: paleovalley deposits (pall), terraced alluvions (at), recent alluvions (all), current alluvions (all), debris and top soil (dt).

The overall structure of the slope pertains to the South-East section of an extended unsymmetrical Apennine-oriented anticline, whose southern side is bounded by fault systems, which are also Apennine-oriented. The structural survey has identified the following major discontinuity groups: 180/63 272/65 294/50 55/80 240/70 (direction/immersion).

The stratigraphic sequence outlines the lack of some elements of the Tuscan Succession which is elsewhere attributed to stratigraphic shortcomings and/or laminations. It might here be linked with diapiric phenomena, as was shown for similar situations in surrounding areas [8, 9, 10].

This area is prone to seismic stresses, that are anything but negligible. The last earthquake, in which most severe stresses were registered, occurred on 7/9/1920 and covered the totality of Lunigiana-Garfagnana territory (IX-X degree MCS and with 6,1-6,2 Mk). Peak values were reached in the Camporaghena area (VIII-IX degree MCS) [11].

On that occasion the landslide movement brought about a quantity of new cracks at different heights, together with earthflows and the over turning of large-sized rocky elements, on both the distance between the two towns, where movements are at present more accentuated, as well in the territory immediately on the East of Camporaghena.

Damage to buildings resulted heavier on the Camporaghena south-west boundary, where the collapse of buildings and even one death casualty were registered.

As a consequence of the 1920 earthquake, important variations in the water circulation were brought about; as a matter of fact, the flow of a few water springs was reduced until it disappeared; this is specially true for those springs situated at about 1000 m a.s.l., at the head of the ancient phenomenon.

For this site, calculations are T=50 years for return-time, predictable intensity I >=7.5 MCS, whereas for T=200 years we have I>8.5 MCS [12,13].

Monitoring

The monitoring implemented over this hillside, is aimed at acquiring the fundamental data which are needed in order that the evolution of the main active phenomenon may be characterized, even in the case of a seismic event, by controlling the deformation processes; on the basis of the analyses performed, they are expected to be of different extent at various depths inside the slope.
The controlling and monitoring instruments that have been chosen on the basis of the previous description and conventional techniques [14, 15, 16, 17, 18] are as follows: surface extensometers, bore inserted multiple extensometers, electrical piezometers, geophones and accelerometers.

The location of each apparatus is shown in fig.3, which schematically also presents the structure of connections between the single instruments and the data acquiring and storing unit. Thus the slope deformations are continually monitored by measuring the surficial displacements with standard invar wire extensometers and the internal deformations with bore-inserted multiple wire extensometers.

The latter are schematically constituted by weights positioned at various depths and then bound to the surrounding matter.

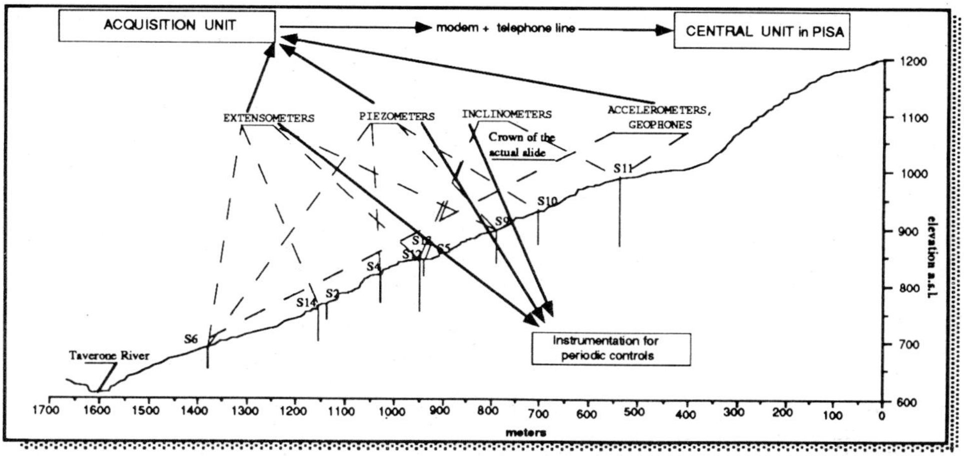

Fig. 3 - Sketch of cross - section, location and type of the istrumentation, and structure of the monitoring system.

Further, they are connected to the surface by cables that lead to a measuring head, in such a way as to check the extent of progressive displacements that occur at various depths, in correspondence with the position of such weights.

The inclinometric distortion initially recorded in the inclinometric pipe of the probe pipe are schematically shown in fig.4 and compared with the recording obtained by the multiple wire extensometers, later fitted into the same pipe, and, in turn, themselves correlated with an operational scheme of instrumentation.

All the multiple wire extensometers have been equipped with a mechanical instrument for periodical readings. For some of these extensometers, the instruments have been replaced by potentiometric transducers capable of converting the displacement into an electric signal and of being elaborated and recorded on a magnetic support. Those positioned immediately above and below the deformation surfaces have been chosen for acquisition. Such surfaces have been intensified by means of preliminary measurements performed with an inclinometric, servoaccelerometric probe and periodical readings carried out at each measuring head.

As shown in fig.3 the signal transmitted by the potentiometric transducers arrives at the data acquisition and storage unit, which is also reached by signals concerned with the triaxial accelerometers and triaxial geophones. The latter are positioned on both the landslide surface and deep into its interior. In addition, other sensors are connected with this unit with the intent of gathering further significant data; they consist in a rain gauge, a thermometer and sensors for checking the correct operation of the system.

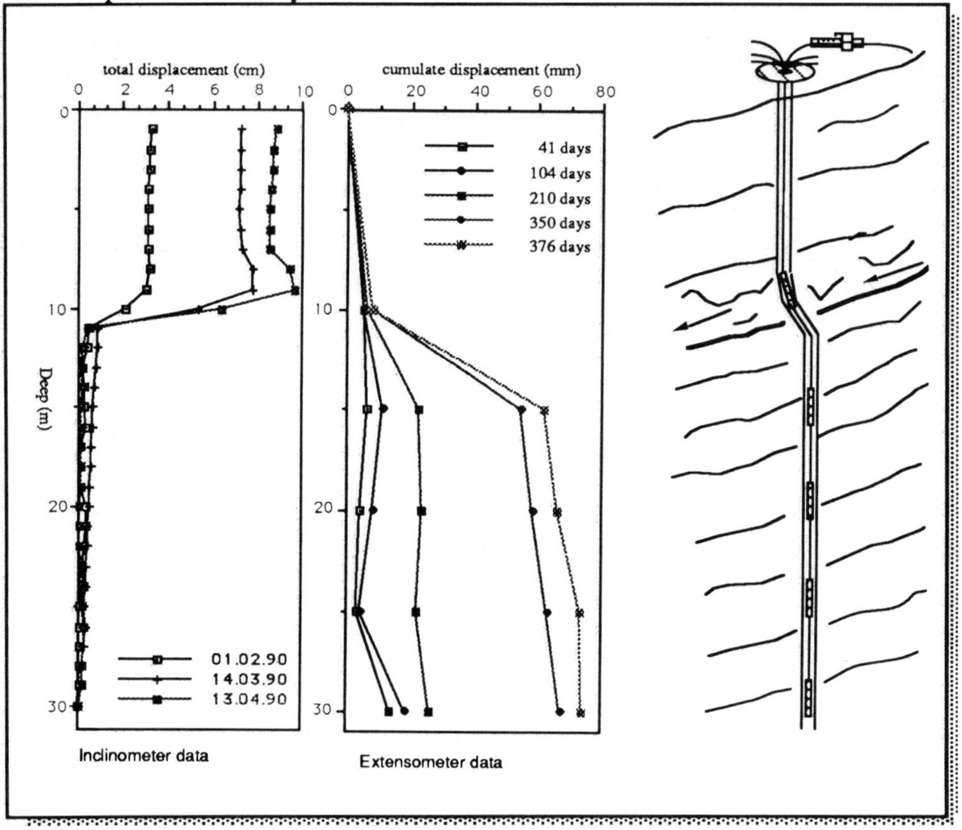

Fig. 4 - Correlation between inclinometer and extensometer displacements and scheme of a multiple wire extensometer.

The fundamental features of the sensors under surveillance are summarized in the table 2

The data acquisition unit has been realized on the basis of industrial technologies applied to numeric control, in such a way as to maintain the greatest flexibility and cost efficiency. It pilots the sensors, performs a preliminary check of incoming data and stores them into a data base, for transmission to a remote unit, such as the central calculation unit at the Department of Earth Sciences of the University of Pisa. A program is employed which also acts as "remote control" over the acquisition unit (in our case, the program "Carbon Copy tm" has been utilized).

N°	Instrumentation	N° channels	type sensor	range	resolution
4	hole wire extensiometer	4	potentiometer	0-20 cm	0,048 mm.
3	hole wire extensiometer	3	potentiometer	0-20 cm	0,048 mm.
2	hole wire extensiometer	2	potentiometer	0-20 cm	0,048 mm.
2	hole wire extensiometer	2	potentiometer	0-20 cm	0,048 mm.
3	superficial wire extensiometer	3	potentiometer	0-20 cm	0,048 mm.
4	piezometer	4	strain gage	0-10 atm	0,0024 atm
1	pluviometer	1	potentiometer	10 mm	0,22 mm.
1	termometer	1	thermocouple	-30/+100°C	0,1 °C
1	voltmeter	1	-	0-20 V	0,005 V
1	triaxial geofhone at top hole	3	acoustic	+- 500 mV	0,5mV.
1	triaxial geophone at bottom hole	6	acoustic	+- 500 mV	0,5mV.
1	triaxial accelerometer	3	piezoelectric	+- 0,5 g	0,002 g
1	triaxial accelerometer	3	piezoelectric	+- 0,5 g	0,002 g

Table 2 - Fundamental features of the installed sensors.

The acquisition unit is constituted schematically, by a personal AT class computer, fitted with a 48 channel conversion A/D card and relative interfaces for the various sensors; the dynamic range is 12 bit, capable of discriminating 4096 levels.

For the acquisition unit handling, a program has been realized, written in "C language" (Landslide Seismic Real-Time Monitoring) that takes case of all the operations of the unit itself.

The scanning on all channels and conversion are carried out by this unit in the actual time of 200 microseconds, each channel is therefore read every 200*48=about 9600 microseconds. As a consequence, this system allows a scanning on each sensor at 1/100 of a second.

This potential has been deemed suitable for characterizing such a complex case like the one in question. Acquisition times and number of sensors are such as to provide data on the variation of a large number of physical magnitudes at intervals that are thoroughly sufficient for a full definition of a phenomenon brought about by a seismic event. To this purpose, 32 channels fitted with a buffer that allows the registration of the pre-triggered conditions are utilized for a maximum-speed acquisition whereas in the other instances of "low" acquisition, the data of all sensors in action are registered.

The data acquisition and sensor piloting operations are carried out by applying a strategy of pre-determined threshold. This means that the acquisition unit performs a first automatic elaboration of incoming data in real time, in other words, the readings carried out sequentially every 200 microseconds are compared with the established conditions:

- for each sensor type (intensity threshold of signal for geophones and accelerometers, displacement speed threshold or deformation in extensometers, etc.);

- for groups of sensors analyzed according to a logic and (simultaneity of events) for instance, the threshold of a surface geophone and threshold of a deep geophone, in such a way as to distinguish abnormal situations and thus to adopt the adequate strategies for memorising data.

The program even allows the identification of five alarm strategies, i.e. the aforesaid conditions recorded by various groups of sensors, can yield different recording procedures on groups of sensors; at the moment, three recording strategies have been sorted out:

- State of Alarm 0 or "default", concerning the displacement conditions, identified as "normal" on the basis of conditions registered to date, with acquisition on all the active sensors, with 5 min. recording time;

- State of Alarm 1, further to getting over the displacement speed threshold, established for each extensometric sensor with recording from each sensor or group of sensors with, for instance, speed of one second;

- State of Alarm 2, set off by a seismic or acoustic signal which is above the established threshold (for geophones, is at the moment been established at 100 mv, whereas the value of 0,05 g. has been chosen for accelerometers, with data recorded on 32 channels, at the expected maximum speed of 1/100 se.

All the threshold values, the groups of channels linked with the various interrogation logics and those groups on which recording is in various cases activated, can be defined and changed in function of the conditions surveyed.

Checking and monitoring results

During the starting operations, the deformation of the several landslide parts were checked by means of inclinometric surveys and periodical readings at the wire extensometers.

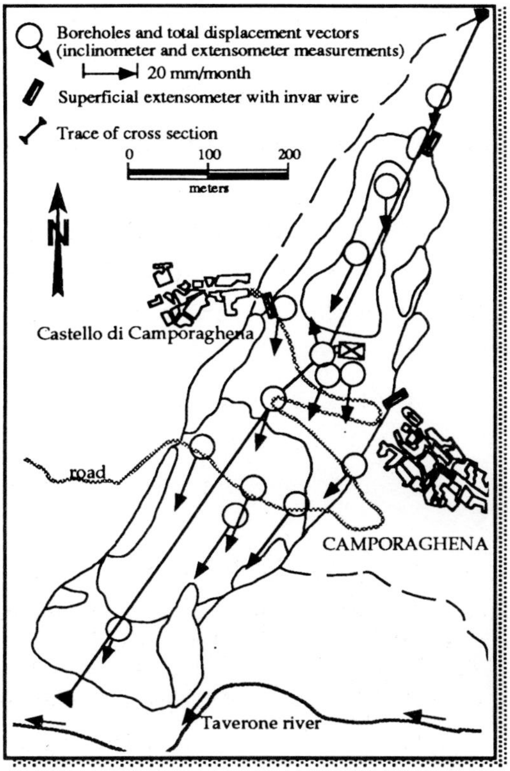

Fig. 5 - Displacement sketch with displacement vectors from inclinometers and estensometers data.

This enabled us to improve our knowledge of the relationship between the several slip surfaces. The main data obtained from this assessment are shown in fig.5

Continuous monitoring was at the same time started, with the acquisition of one datum every 5 minutes from all sensors; that speed was deemed more than sufficient for analyzing the landslide progress in static conditions. This proved to be most useful in comparing the consistency of results gathered at the experimental sensors with that of the results collected at the periodical control instruments; in checking the correct functioning of software; in identifying alarm thresholds most suitable in the case at issue, and, eventually, in improving the system-handling program.

Acquisition is now yielding significant data on displacements provoked by stresses that act in environmental "normal" conditions. More specifically, it has allowed us to highlight an increase of the deformation speed, due to the heavy rains in November 1991 (fig.6).

As concerns the threshold for activating monitoring at the maximum speed, should a seismic event take place, they have been established on the basis of the site seismicity. Values more suitable for the recording of even lesser events, but still as useful, are looked for, by recording from the accelerometers and geophones, the maximum values concerning the present acquisition intervals, in such a way as to identify either the environmental noise level, or potential micro-earthquakes that may possibly be compared with recordings of the National Seismic Network stations. These values will make it possible to identify the alarm threshold more

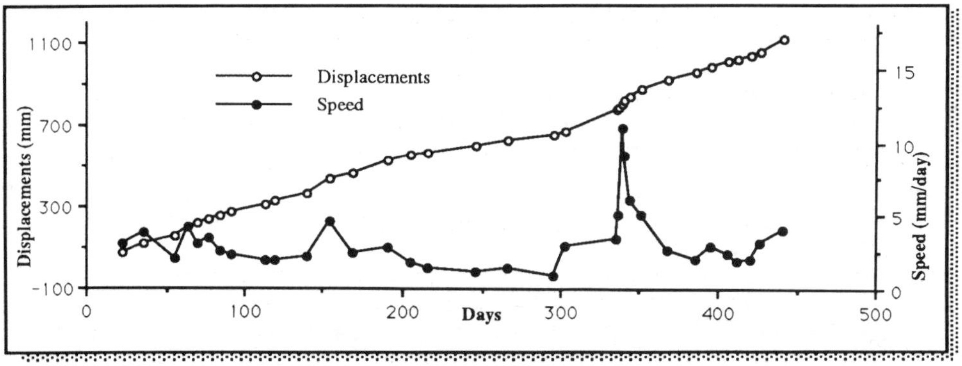

Fig. 6 - Example of the elaboration of the data from a hole wire extensometer

suitable for the complete recording of those earthquakes that cause significant stresses in the slope. Such data might supply the necessary elements for the formulation of simplified behaviour models that would also take into account the local amplification of effects, possibly recorded by the instruments.

Acknowledgements

Si ringraziano i Sigg. Marcello Gini e Franco Colarieti per la fattiva collaborazione alla realizzazione delle attrezzature.

Ricerca finanziata dal Gruppo Nazionale per la Difesa dai Terremoti del C.N.R. e con il contributo della Provincia di Massa

References

[1] D'Amato Avanzi G., Nardi R., Pochini A., Puccinelli A., Trivellini M. (1988) - La frana di Camporaghena (MS). Quaderni di Geologia Applicata, 2, M.P.F. Ed. Lucca

[2] Varnes D. J. (1978) - Slope movement types and processes . In Transportation research board, Special Rept. N. 176, Nat. Acad. of Sciences

[3] Carrara A., D'Elia B., Semenza E., (1985) - Classificazione e nomenclatura dei fenomeni franosi Geol. Appl. e Idrogeol., 20, (2), Bari, 1987

[4] Pellegrino A., (1986) - L'analisi dei movimenti franosi per la progettazione degli interventi di stabilizzazione. (Relazione Generale). Atti XVI Conv. Naz. Geot. A. G. I. , Vol.3, pp. 121 - 150, Bologna.

[5] Schuster R. L., Krizek R. J. Ed. (1978) - Landslides analysis and control. Transportation research board, Special Rept. N. 176, Nat. Acad. of Sciences

[6]Cancelli A., Crosta G., Nardi R., Pochini A. (1992) - An Example of combinated geological and geothecnical investigations for a landslide in seismic area Sixth International Symposioum on Landslides I.A.E.G. 10-14 feb. 1992 Christchurch, New Zealand.(in stampa)

[7] Carrara E., Patella D., Roberti N. - Misure di sismica down - hole sulla frana di Camporaghena in Lunigiana. Atti del convegno annuale del G.N.D.T. del C.N.R., 24/26 Giugno 1990 Pisa (In stampa).

[8] Azzaroli A. (1950) - Tettonica della catena principale dell'Appennino lunigianese tra il passo di Pradarena e il passo del Lagastrello. Boll. Soc. Geol. It., V. 69, p.367

[9] Krampe K.D. (1964) - Zur geologie des Hochappennins zwischen Secchia und Enza. Diss. Freie Univ. Berlin.

[10] Andreozzi M., Casanova S., Chicchi S., Ferrari S., Patterlini P., Pesci M., Zanzucchi G. (1987) - Riflessione sulle evaporiti triassiche dell'alta Val Secchia (RE). Mem. Soc. Geol. It., V. 39, p.69

[11] Ferrari G., Patacca E., Petrini V., Scandone P. (1985) - The Garfagnana earthquake of September 7 1920. In Atlas of isoseismal maps of italian earthquakes. C.N.R., Bologna.

[12] Cattaneo M., Eva C., Giglia G., Merlanti M., (1986) Pericolosità sismica della Garfagnana. in Progetto Terremoto in Garfagnana e Lunigiana. C.N.R.-G.N.D.T., Regione Toscana.

[13] Cattaneo M., Eva C., Giglia G., Merlanti M., (1986) Evoluzione strutturale e modello sismotettonico dell'appennino Nordoccidentale in Terremoto in Garfagnana e Lunigiana. C.N.R.-G.N.D.T., Regione Toscana

[14] Bruzzi D., Pezzetti G., Piccoli S., (1986) - La strumentazione geotecnica nel monitoraggio dei movimenti franosi profondi Atti XVI Conv. Naz. Geot. A. G. I. , Vol.1, pp. 153 - 164, Bologna.

[15] Hanna T. H. (1985) - Field instrumentation in geotechnical engineering. Trans Tech. Publications, FRG.

[16] Pilot G. (1984) - Instrumentation and warning system for research and complex slope stability problems.Atti "IVth Int. Symp. on Landslides", V.1, pp. 275-306, Toronto.

[17] Ribacchi R. (1986) - Il controllo della stabilità dei pendii (Relazione Generale). Atti XVI Conv. Naz. Geot. A. G. I. , Vol.3, pp. 403 - 452, Bologna.

[18] Zaruba Q., Mencl V. (1982) - Landslides and their control. Developements in geothecnical engineering, vol 31. Elseviers Scient. Publ. Comp., II edition.

PRESENTATION OF THE RESEARCH PROJECT : "ANALYSIS AND MITIGATION OF EARTHQUAKE TRIGGERED LANDSLIDE HAZARD AFFECTING DAMS, ROUTES AND LIFELINES".

Denis AUBRY*, Jean-Marc CREPEL** and Ezio FACCIOLI***

Abstract : A global research project involving 11 participants from 5 European countries has been initiated in 1989 within the framework of EPOCH Research Program of the EC. The basic objective is to propose a global methodology for slope stability analysis under seismic loading conditions, thanks to the sharing of various practical and theoretical experiences of all participants.

Résumé : Un projet de recherche integré, regroupant 11 participants dans 5 états européens, a été présenté en 1989 à la CCE dans le cadre du programme de recherche EPOCH. Le but essentiel est de proposer une méthodologie globale pour l'étude de la stabilité des pentes sous sollicitaions sismiques, grâce au regroupement des compétences pratiques et théoriques de tous les participants.

1. General overview

Large areas in earthquake-prone countries are exposed to widespread slope instability hazards. Significant earthquakes in such areas can trigger a large number of earth- and/or rock-slides, some of them capable of causing major damages to major structures, important roadways, railway lines, dam reservoirs, aqueducts, etc as experienced to varying extents within Europe in Italy, Yugoslavia or Greece during the destructive earthquakes of Friuli (1976), Montenegro (1979), Irpinia (1980), or Alkyonidis (Corinth, 1981). The social and economic consequences justify a serious, multidisciplinary effort both to develop the appropriate tools of analysis and to devise relevant mitigation strategies. This provides the basic motivation for the proposed project.

This project has been accepted for financial support by the European Communities within the framework of European Programme On Climatology and natural Hazards (EPOCH). The duration is planned to be 24 months, starting from December 1991. For that reason, no important result is yet available, but the objectives and the detailed work program are presented here.

---

* Professor, Ecole Centrale de Paris, France

** Manager, Software Engineering Department, Coyne et Bellier, Paris, France

*** Professor, Technical University (Politecnico), Milan, Italy

This project rallies 11 participants from 5 different countries :

- Coyne et Bellier Consulting engineers - Paris, FRANCE (acting as coordinator)
- Studio Geotecnico Italiano S.r.l. - Milano, ITALIA
- Greco Géomatériaux (CNRS) - FRANCE
- Aristotle University of Thessaloniki - GREECE
- Geognosis S.A. - Thessaloniki, GREECE
- Computational Dynamics Research Ltd - Swansea, UNITED KINGDOM
- Centro de Estudios y Experimentacion de obras publicas - Madrid, SPAIN
- Politecnico di Milano - ITALIA
- The Victoria University of Manchester - UNITED KINGDOM
- SPEA Ingegneria Europea SPA - Milano, ITALIA
- ISMES SPA - Bergamo, ITALIA

The objective of the project is to develop the methodology and the quantitative tools for :

- the assessment of landslide hazards affecting major structures, transportation routes and lifelines in the event of a strong earthquake;
- the detection of the key factors influencing slope stability ;
- the prediction of permanent deformations occurring as a result of land and rock sliding and their probable effects on the infrastructure considered;
- a mitigation strategy against landslide risk, encompassing both prevention measures and post-event actions;
- a new methodology for limiting seismic risk on motorway bridges.

This will be achieved by means of :

- a general method for earthquake hazard evaluation, allowing for flexible use of historical and instrumental earthquake data as well as deterministic approach;
- an automated procedure for landslide vulnerability zoning based on relative hazard assessment, and capable of identifying representative landslide scenarios in different environmental conditions;
- laboratory testing of geomaterials under earthquake loading to produce a set of consistent tests;
- constitutive modelling of geomaterial behaviour under complex loading;
- development of new algorithms to improve the speed-up of computations and their accuracy;
- use of GEFDYN and SWANDYNE programmes, general computer codes for finite element analyses in geomechanics;
- improvement of simplified methods in order to incorporate key factor effects;
- analytical evaluation and experimental research on seismic characteristics of motorway bridges.

2. The work program

The development of the project is envisaged to cover seven major tasks :

- Methodology for the evaluation of the earthquake hazard, i.e. seismic macrozoning, in the region occupied by the lifeline,

- Landslide hazard zoning and methods of identification of the most vulnerable zones; this will result from a combination of non-seismic and seismic criteria, implemented through an appropriate information system,
- Development of advanced numerical and constitutive models for stability analysis of slopes under seismic loading,
- Dynamic modelling of slope instability scenarios, identified as representative scenarios, aimed at improving knowledge of phenomena and to exhibit main mechanisms and important parameters,
- Implementation of previous results to improve simplified dynamic methods,
- Analysis and experimentation of seismic vulnerability of motorway bridges,
- Guidelines for remedial measures and mitigation strategies.

2.1. EARTHQUAKE HAZARD EVALUATION (TASK 1).

Effective peak ground acceleration (PGA) and velocity (PGV) will be at first used to represent the severity of earthquake ground motions. Additional parameters that may be needed in subsequent phases of the project include significant duration and Arias intensity. The basic function proposed for the seismic macrozoning of a linear region, such as those considered herein, consists of one or more "constant risk" curves, expressing the values of PGA and PGV for given values of return period T. Periods between 50 and 200 years are appropriate for expressways and pipelines, while larger values should be used for railway lines and dam reservoirs.

As an alternative to the well-known probabilistic method for calculating the earthquake hazard, which originated from a classical paper by Cornell (1968), a simpler "historic-statistical" approach will be developed, relying as much as possible on historical seismicity, and especially on recent macroseismic maps of important historical earthquakes in a region. This will allow for maximum exploitation of the rich historical records existing in several EC countries (as opposed to the scanty strong-motion data, if at all available), while minimizing the usage of attenuation relationships which are known to introduce significant uncertainties in the results. The region of analysis is reduced to a set of closely spaced, discrete points and for each of them a frequency-intensity, or frequency-PGA curve will be constructed.

A third approach (of deterministic nature) will also be included for seismic hazard evaluation, best suited for regions where the geotectonic information is more reliable than seismic catalogues and historical seismicity data. This approach will be similar to that used in the nuclear industry for the determination of design earthquakes.

The site intensity values will come both from attenuated epicentral intensities and from direct readings on intensity maps of single earthquakes. The frequency-intensity curves will be obtained with due consideration for completeness intervals of the input catalogue of epicenters, elimination of fore-and aftershocks. Consideration will also be given to exclusion criteria which rule out the occurrence of landslides for combinations of magnitude and distance exceeding some critical values.

From the set of frequency-PGA, or PGV, curves of all sites, the "constant risk" curve is simply obtained by joining all PGA values corresponding to a selected frequency of occurrence (approx. inverse of return period). The PGA values will be representative of ground motions on average or "standard" ground conditions, i.e. typically devoid of local site effects, to be considered in a later phase (if necessary). The constant risk curve for a given T, to be corrected on the basis of seismotectonic criteria for possible gaps caused by the limited time span of the earthquake catalogue, will give a clear, synthetic picture of the seismic hazard exposure of the investigated region.

A more advanced approach will also be considered, based on the estimate of permanent displacements through simplified Newmark type models. The "constant risk curve" in terms of PGA or PGV will then be replaced by a set of curves ideally associated to different types of slopes through different critical accelerations.

The output of this task will be a report describing the methodology in detail, including at least one example of application, and the software for the interactive treatment of historical and instrumental seismicity data both by the historic-statistical and the probabilistic method, the elaboration of constant risk curves and the graphic display of significant input data and results. **The findings** of this task will be especially helpful in task 4.

2.2. CRITERIA FOR LANDSLIDE HAZARD ZONING (TASK 2).

The procedure for executing this task requires the realization of a computer based data bank for storing all the information which can be associated with landslide occurence in a given region, and which will allow for cross-referencing. The basic subjects to be entered in the databank are : geology, morphology, hydrogeology, hydrology, land use. Earthquake related information will be additionally entered both in the form of PGA (or other pertinent parameter) and in the form of relevant site intensity data directly available from single historical earthquakes, moreover distance of sites from seismic source zones could be taken into account.

The following steps will be subsequently performed in an ideal application :

- selection of stretches of a region which are homogeneous from the viewpoint of stratigraphical and general geological characteristics;
- identification of active and quiescent landslides and of areas of potential instability within the region, to be included in a geomorphological map, which is then digitized;
- identification and surveying of factors which can (even indirectly) influence the sliding events in the area : e.g. slope steepness, surface geology, seismic hazard, rainfall, etc.;
- transfer of the previous information onto raster digital maps, (whith dimension of the cells related to the scale and the accuracy of the original maps: e.g 25x25m cells have been considered suitable for an analysis with 1:5000 maps) i.e. a map of slope values, a map of land use, of specific rainfall, etc;
- application of an automated information system to explore the relationships between the parameters illustrated in the previous maps and landslide hazard.

This process is expected to provide information about factors or the combinations of factors which have major influence on slope stability for the investigated zones. In essence, the objective is to elaborate landslide zoning criteria that can be inferred from the features of the individual areas by means of statical analyses. This should make possible to classify as potentially unstable some areas which are not affected by present instability, but nevertheless exhibit the same critical combinations of factors which are found in stretches where numerous landslides have occurred.

The output of this task will consist of identifying some criteria as far as dynamic stability is concerned, based on specific case histories. One example of application will be provided. **The findings** of this task will be especially helpful in task 4.

2.3. DEVELOPMENT OF ADVANCED NUMERICAL AND CONSTITUTIVE MODELS (TASK 3).

Quantitative prediction of effects of earthquakes on unstable slopes requires a full modelling of geomaterial behaviour in the non-linear range, coupling of solid skeleton with pore water and development of zones where strains localize and concentrate, causing failure surfaces to develop.

For an accurate study of this phenomenon, it is necessary to:

(i) Introduce a mathematical model for fluid-skeleton interaction

(ii) Define quantitatively the behaviour of soil and rock masses under complex loading paths

(iii) Develop efficient numerical tools, especially inside finite element programmes.

Therefore, research efforts will be directed along these lines.

First of all, laboratory data concerning the main features of soil behaviour under earthquake loading on reference materials will be gathered, and complementary tests will be conducted. The aim is to provide enough data to identify basic phenomena for the constitutive modelling sub-task. Tests will include cyclic triaxial and hollow cylinder both under drained and undrained paths. Interface behaviour and response of soil in zones where very high gradients of strain are present will also be analysed in ring torsion and plane strain apparatuses.

The results of these tests will be used to develop and validate more accurate constitutive equations describing soil, rock and interface behaviour when subjected to earthquake loading. Particular attention will be paid to model the effects of rotation of principal stress axes, anisotropy and the role of the initial shear stresses which deserve special attention in the case of slope analyses. Efforts will also be devoted to integrate features of developed models into a global constitutive model.

Accurate numerical schemes will be used to reproduce the behaviour of slopes under earthquake loading, taking into account interaction between skeleton and pore fluid, both for saturated and unsaturated materials. The essential phenomenon of dynamic strain localization seems to be difficult to model, although it is a key point in determining failure surfaces, the safety factor, and also **the amount of permanent displacements** and **the volume of soil** involved in the potential sliding.

The output of this task will consist of a databank of selected tests on reference materials to constitute a set of calibration tests for constitutive modelling, a choice of constitutive models in the form of "constitutive matrix" subroutines which could be switched directly on the computer codes, and efficient numerical techniques able to capture failure surfaces to accurately reproduce slope response to earthquake. **The findings** of this task will be especially helpful in task 4.

2.4. DYNAMIC MODELLING OF REPRESENTATIVE SLOPE INSTABILITY SCENARIOS (TASK 4).

Dynamic modelling will be carried out on representative landslide scenarios, selected by the methodology of tasks 1 and 2. The finite element computer codes GEFDYN and SWANDYNE, which already have many capabilities to deal with dynamic pore pressure effects, will be supplemented with the numerical techniques of task 3, and will be used as the main computational tools of this task.

Generally, 2-dimensional modelling will be used to make parametric studies easier, but some 3-dimensional models will be analysed with GEFDYN code if necessary (as in the case of pipelines). Rigid blocks with sliding friction or continuous media will be used, depending upon the particular phenomenon to be investigated (rock mechanics or soil mechanics).

In each case, **the purpose** will be :

- to better assess the applicability of nonlinear finite element technique to the seismic stability analysis of natural slopes, which up to now is not firmily established;
- to improve knowledge of rupture mechanism (necessary for improving simplified methods);
- to identify important parameters required by such calculation (necessary for recommending or selecting in-situ measurements);
- to quantify the role of specific configurations, including site effects (subsoil profile features or topography), or mitigation measures such as concrete tunnels, anchoring, drainage well, etc. (also connected with task 7).

An important role is played by earthquake excitation. This will be typically described by acceleration time-histories, chosen from the library of real weak and strong motion records and constrained by :

- the PGA value read on the appropriate constant-risk curve for the site of interest;
- similarly obtained values of additional parameters apt to measure the overall severity of ground shaking identified in task 1.

The findings of this task will be especially helpful in task 5 and task 7.

2.5. IMPLEMENTATION OF SIMPLIFIED DYNAMIC METHODS (TASK 5).

Geomechanics models fall into the class of data-limited problems which require a very different modelling approach from that currently used in other areas. It follows that in everyday use one cannot apply sophisticated models in analysing dynamic slope instability and that there is a need to define **rational simplifications.** More particulary, complete calculations (as those described under task 4) cannot be performed in all cases, because they are generally too heavy, or not enough data is available. Simpler methods should be used to obtain fundamental "design" quantities rapidly, like permanent downslope displacements to be compared with serviceability or collapse limits of the investigated facilities.

After carefully analysing the kinematics of the failure modes and discussing the weighted roles of the main parameters, such as pore pressure, simplified methods will be developed either by improving existing ones (like the Newmark method) or by creating new ones. Extensive use of advanced postprocessing software will be necessary for this purpose, using essentially graphic facilities. In any case, all the proposed methods will stem from dynamic analysis and will not be extensions of pseudo-static analysis (as used currently).

The output of this task will be a set of tools (mainly small computer programmes) which could be used easily, i.e. with as little input data as possible, for a first design. The **validity domain** of such tools will be specified, to indicate when the hypotheses cannot be applied and more sophisticated models are necessary. **The findings** of this task will be especially helpful in task 7.

2.6. SEISMIC VULNERABILITY OF MOTORWAY BRIDGES (TASK 6).

The overall objective of the present task is to gain a more thorough understanding of the seismic behaviour of highway bridges, and consequently trigger some progress of the engineering knowledge in the reliable evaluation of the seismic vulnerability of existing structures. Some feedback regarding the general earthquake-resistant design criteria and specifications would also be expected. Within this context, the realization of a sophisticated monitoring system - specially developed for highway overpasses - should be stressed, because it would as certain the future availability of data on the actual seismic response of the structures of interest.

Increased understanding of the seismic behaviour and the potential vulnerability of highway bridges would also offer the possibility of cost-effective retrofitting existing structures (enhancement of their seismic resistance). Considering the serious damage a single earthquake may cause to the transportation system structures of seismic regions, economic return on funding invested in this area is though to be high.

This task follows four main lines :

- Data-base on highway bridges earthquake-resistant design, construction and seismic vulnerability : **the output** being special technical reports on description of highway bridges structures in seismic regions data-base, codification of seismic vulnerability levels and description of possible antiseismic devices, togother with an examination of European design rules and a comparison with national standard;
- Highway bridge seismic monitoring system : **the output** consisting in reports describing the transducer and monitoring system installation project, the monitoring system hardware and software, the monitoring system capabilities, and a laboratory demonstration;
- In situ dynamic excitation experiments : **the output** will be the description of experiment set-up, testing technique, raw data presentation, analyses procedures, final results with in situ demonstration;

- Mathematical model and numerical analyses : **the output** will describe the analysis to calibrate the structural model and the geometrical and mechanical parameters defining the mathematical model.

2.7. REMEDIAL MEASURES AND MITIGATION STRATEGIES (TASK 7).

The remedial measures against landslide hazards depends strongly on the type of facility investigated, and a hierarchy of actions will be considered, depending also on the results of the previously considered tasks. For example, the main practical implication methodology of task 2 will consist of carrying out more specific geological and geotechnical investigations in certain stretches, or else placing appropriate monitoring instrumentation (such as clinometers) in others. On the other hand, the prediction of specific permanent displacement values exceeding, for example, the serviceability limits of an expressway pavement at sites where symptoms of instability are already obvious, will call for substantial intervention, e.g. with anchored retaining walls, etc. At last, the output of task 4 will permit the effect of some remedial measures to be predicted according to the kind of problem encountered and the cheapest and the most efficient one to be recommended.

The output of this task will be the general strategy of the application of the results of the project and of the associated remedial measures, organized on a convenient format, such as matrix format, with columns identifying the "problems" and the rows the most appropriate "actions". The actions will be specified in full detail, including instrumentation, analysis, design, cost and time factors affecting them.

3. General organisation and role of participants

As shown in Table 1, each participant will contribute to several of the tasks described above, according to his main area of competence and to the most frequent applications. Each task is coordinated by a "task leader", who is in charge of orienting the work, promoting exchanges between participants during inter-meeting periods, and regrouping the results of the task.

Each meeting is also under the responsibility of a "local committee" which will be successively in charge of checking the general progress of all the work.

	TASKS						
Participants	1	2	3	4	5	6	7
01 COB	P	P	P	P	L		L
02 SGI	L	L		P	P		P
03 GRECO		P	P	L	P		P
04 AUTH	P	P		P	P		P
05 GEOGNOSIS		P		P			P
06 CDR			P	P			P
07 CEDEX			L	P			P
08 PM			P				P
09 UM			P	P	P		P
10 SPEA						L	P
11 ISMES						L	P

Table 1. Distribution of the participants amoung the different tasks
(L) : Task Leader **(P)** : Participant

4. Expected achievements

The whole project is aimed at introducing a new and comprehensive approach to slope instability analysis under earthquake loading both from a theoretical and engineering standpoint. Improvement in methods of analysis and design are expected, especially from the implementation of simplified dynamic models, which are intended for use by standard structural and geotechnical engineers. The use of complex modelling to develop and validate such simplified methods is essential in an engineering field with limited data available.

In addition, a global and multidisciplinary approach such as that previously described will stimulate the exchange of pratical experience and knowledge on slope stability and associated remedial measures.

THE ROLE OF SEISMICITY IN LANDSLIDE MAPPING

Eric LEROI[*], Monique TERRIER[*], Pierre MOUROUX[*]
Olivier ROUZEAU[**] and Jean-Pierre ASTE[*]

Abstract : The mapping with a GIS of unstable areas based on a stability index, enables to quantify the passage from static (no earthquake) to dynamic conditions. The results obtained at local and regional scales are promising, both for the methodology used and the expected planning results.

Résumé : La cartographie sous SIG des zones instables, à partir d'un indice de stabilité permet de quantifier le passage des conditions statiques (hors séisme) aux conditions dynamiques. Les résultats obtenus à deux échelles (régionale et locale) sont prometteurs quant à la méthodologie utilisée et aux objectifs attendus (planification préventive).

1 Introduction

Landslides occur because of the conjunction in time and space of several types of factors, such as :

- permanently unfavourable factors, like geology, geomorphology and drainage patterns ;
- temporary triggering factors, like rainfall, seismic shocks and certain types of human activity such as earth movements.

Concerning earthquakes as being at the origin of landslides, two questions present themselves : what was the slope stability before the shock, and what will be the effect of this shock on the slope in terms of increased instability, which can be mapped ? We have tried to answer these questions by using the results of our ongoing work on the RIVET project (French acronym for "Integrated Research on Slope degradation in Mountainous Terrain), which is part of the European EPOCH research programme.

---

* BRGM Marseille.

** BRGM Orléans.

The objectives of this project, a cooperative effort between several italian (ERSAL, IATA and CSIM) and french (BRGM, CETE, ADRGT and IRIGM) organizations, were manifold. They aimed at developing our knowledge on slope degradations phenomena in mountain areas, and the rules for determining the risks generated by such phenomena and the preventive action to be taken. In addition, an operational methodology was to be defined, for using such knowledge as part of a Geographic Information System (GIS), and for proposing recommendations to prevent such risks.

2 Method for determining the hazard

2.1 THE VARIOUS APPROACHES

Determination of which areas are exposed to phenomena of slope instability can be done under static conditions, in which case the inertial forces unleashed during an earthquake are not considered, or for dynamic conditions. Three different approaches are possible :

a - A "naturalistic" approach based on expert knowledge, for which the "rules" of assumed instability are defined from the start, based on criteria of variable subjectiveness. This is an integrating approach that, however, can be not very explicit as it uses many subjective empirical factors.

b - A "naturalistic" approach based on a feedback analysis of actual events, using a statistical analysis when it is possible. In this case, the rules of assumed instability are defined afterwards, and they are applied according to the principle of event similarity (same causes - same effects).

c - A "mechanical" approach based on the forces applied to a given mass. Under dynamic conditions, three types of methods can be used for evaluation of the stability : rupture methods (pseudo-static), displacement-to-rupture methods, and general stress-deformation methods ; all require knowledge of several parameters such as the geotechnical properties of the materials, and lead to a calculated local or general stability coefficient.

As far as the naturalistic approach is concerned, it is difficult to quantify the passage from static to dynamic conditions, particularly for modifications that take place of the level of zones. For this reason, it is necessary to be able to work on a quantifiable model, for calibration of the "models" used in the naturalistic approach.

2.2 THE INFLUENCE OF SCALE ON THE STUDY

All workers agree that the understanding of slope-instability phenomena requires a knowledge of unfavourable factors on a large scale (problems of water circulation, determination of geotechnical properties, etc). However, it is equally obvious that regional preventive planning studies integrating landslide risks, for economic reasons cannot be done at a large scale. To overcome this problem, the habit has grown of carrying out small-scale regional studies, underlining the location of vulnerable areas that would require particular attention, where specific large-scale studies are necessary.

However, the question to be answered is whether this change in scale is a simple zoom effect, with a loss or gain in information, or whether the information become distorted.

3 The means deployed

The methodology adopted is based on a **multi-criteria analysis** of the phenomena, using a Geographic Information System, or GIS. The main interest of a GIS lies in the fact that it is possible to carry out automatically, and in rigorous and "objective" fashion certain fastidious operations, like:

- data acquisition by means of digitizing, scanning and data transfer from other computer files ;
- data management, localized and thematic, as vector or raster maps, the latter being a meshed image whose pixels have specific values ; vector maps consist of a cover of subjects with geometric and thematic attributes, and whose basic features are point, line, polygon or character chain ;
- data processing or the transformation of initial data to extract or create new ones that answer well-defined constraints ;
- data restitution on screen or on paper and at a given scale.

The ideal tool, unfortunately, does not exist as yet, and the mapping exercise reported here required the use of three GIS packages combined in a single processing unit.

In view of its didactic design and its integrated coupling with a digitizing table, the ILWIS system [1] enabled rapid data entry from maps ; however, although this software manages both vector and raster data, the fact that it is designed for use on desktop computer ('PCs') limits the number and size of the files, and increase the time needed for calculation. Because of these two restrictions, we imported and processed the data on SynerGIS [2], which functions on a workstation. The ARC/INFO [3] system enables the rapid processing of vector data ; it was used in parallel with ILWIS and SynerGIS, to obtain multiple covers (maps of geotechnical data) from a cover of initial lithology polygons.

4 Application of the method

4.1 THE TEST AREA

The study area lies in the French West Alps, in the commune of Sinard that forms part of the Trièves region (Fig. 1). Geologically, this region lies in the southern Sub-alpine chains of the external Alps. It is dominated to the West by the limestone ledges of the Vercors, to the south by the Vocontian ridges of the Diois, Barronies and Dévoluy, and to the east by the external crystalline massif of the Mure. The region lies in a great depression that is part of the Drac drainage basin, and was repeatedly invaded by Alpine glaciers. The well-developed Recent, Würmian and Rissian deposits have a typical structure and texture, of the glacio-lacustrine and moraine types, the latter being the site of major landslides.

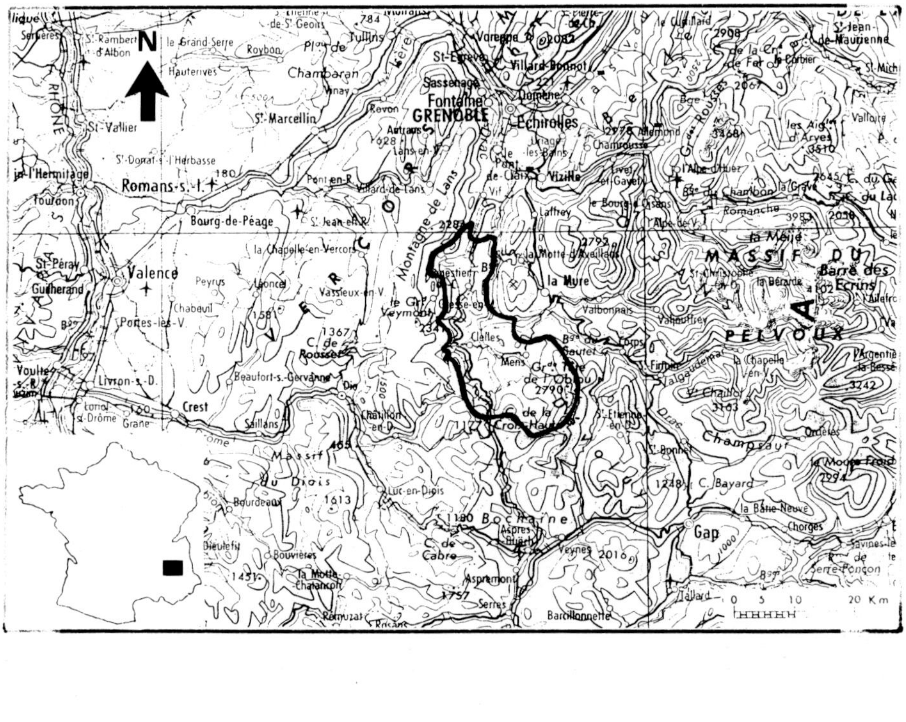

Figure 1 : Location of the Trièves region.

4.2 THE BASIC DATA

Basic data comprise information from thematic maps, satellite images and aerial photographs. On the regional scale, this are the 1:50,000- scale geological maps by BRGM and the digital elevation model (DEM) that was created from stereoscopic Spot images. At the local scale, such basic data are an (unpublished) geological map at scale 1:20,000 and a DEM based on standard overlapping aerial photographs by IGN.

From these basic data, geotechnical- and slope-data maps were made. The geotechnical data were defined from descriptions of the various units shown on the geological maps; they are cohesion (C) and the friction angle (ϕ) under well-drained conditions, the wet volumetric weight (γ), the thickness of the units (H)and the elevation of the water table above the rupture surface (H_W) (Fig. 2). For each of the two DEMs a slope map was calculated, after which smoothing for each pixel was carried out, with functions in $1/d^n$, where 'd' is the smoothing radius around the pixel and 'n' is a whole power value. In our case, 'n' was taken to be 1, and 'd' 10 metres for the local scale and 100 metres for the regional scale.

LITHOLOGY	CODE	THICKNESS (H) in m	WATER LEVEL (Hw) in m	COHESION (C) in KN/m2	FRICTION ANGLE (ϕ) in degrees	DENSITE (γ) in KN/m3
Unsorted scree	e1	3	0	0	40	18.5
Bedded scree	e2	2	1	0.5	35	19
Scree whit large blocks	e3	2	0	0	40	18
Displaced scree	e4	3	2	0.5	35	19
Alluvium and recent dejection cones	Fz-jz	3	2	0.5	35	19
Alluvium and old dejection cone	Fyx-Jyx	3	0	1	35	20
Würm-age moraine	g1	5	3	1.5	30	19
Riss-age moraine	g2	5	3	1.5	35	20
Landslide	Gs	3	2	1	20	19.5
Quaternary tuff	u	5	3	5	30	20
Late Cretaceous limestone	c6-7	5	0	5	40	21
Early Creaceous marl / limestone	n6-7	5	0	5	32	21
Early Cretaceous massive limestone	n4-5	5	0	20	45	23
Early Creaceous marl / limestone	n1-3	5	0	5	35	22
Malm-age massive limestone	j6-9	3	0	20	45	23
Dogger-age marly limestone	j5	5	0	5	30	19
Dogger-age black marl	j1-4	5	0	3	25	19
Lias-age marly limestone	l6	5	0	5	35	21
Lias-age limestone	l2-5	5	0	15	45	23
Triassic dolomite and gypsum	t	3	0	5	35	20

Figure 2: The lithological units and their geotechnical properties.

Each thematic information item was coded as an image. At the regional scale an image consists of 2 million pixels, each representing a surface area of 20 x 20 m. This information takes 8 million bytes of memory. At the local scale an image contains 4 million pixels, each covering 2 x 2 m and stored as 16 million bytes.

4.3 THE COMPUTATION MODEL ADOPTED FOR F_d

As a first step, the local safety coefficient 'Fd' was considered, which should rather be seen as a stability index. The following assumptions were made (Fig. 3):

- computation is of the pseudo-static type and considers a rupture plane that is parallel to the slope, at a depth H from the surface;
- increases in pore pressure and decreases in shear strength remain low, in order to exclude liquefaction phenomena;
- a groundwater level 'H_w' is defined for each lithological unit, and for a given unit the height of this level above the rupture plane is seen as constant and independent of morphology;
- initial runoff is parallel to the slope;
- a seismic-acceleration coefficient 'K' was introduced that makes an angle δ with the horizontal.

This led to the following results:

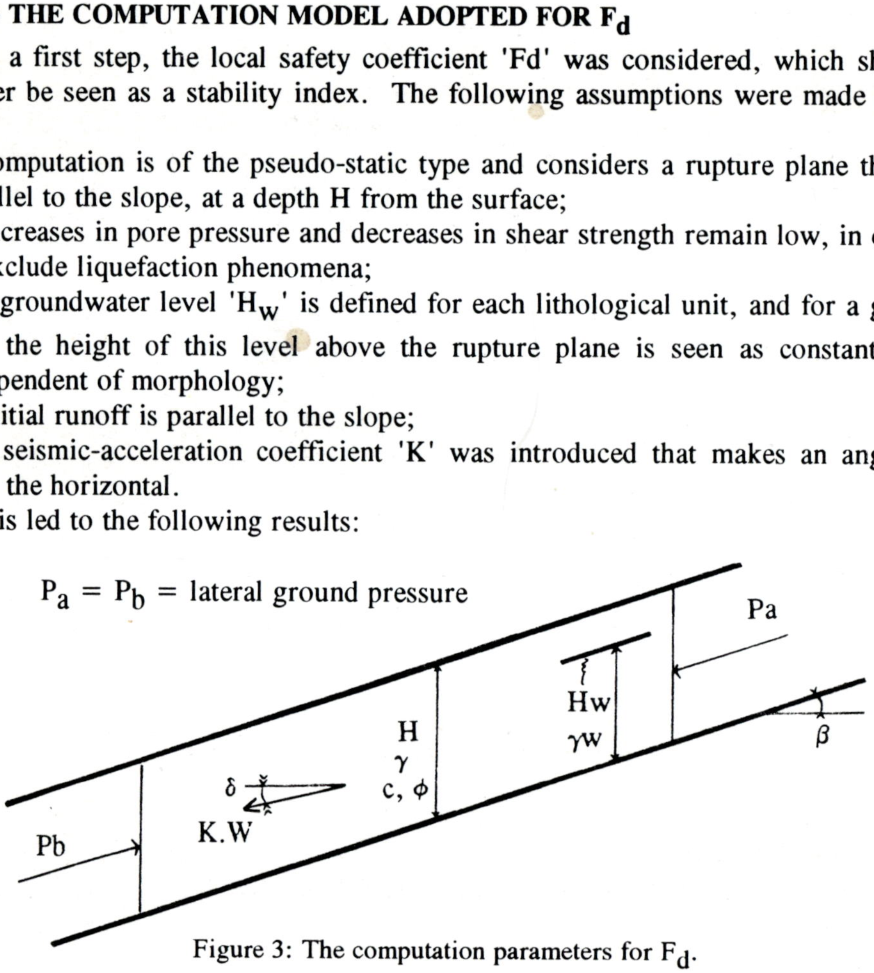

Figure 3: The computation parameters for F_d.

which led to:

$$F_d = \frac{C/(\gamma.H) + [(1-(\gamma_w.H_w)/(\gamma.H)).\cos\beta - K.\sin(\beta-\delta)].\cos\beta.\mathrm{tg}\phi}{\cos\beta.(\sin\beta + K.\cos(\beta-\delta)]}$$

4.4 METHODOLOGY

The methodology adopted for mapping landslides is summarized in the flow chart of Figure 4.

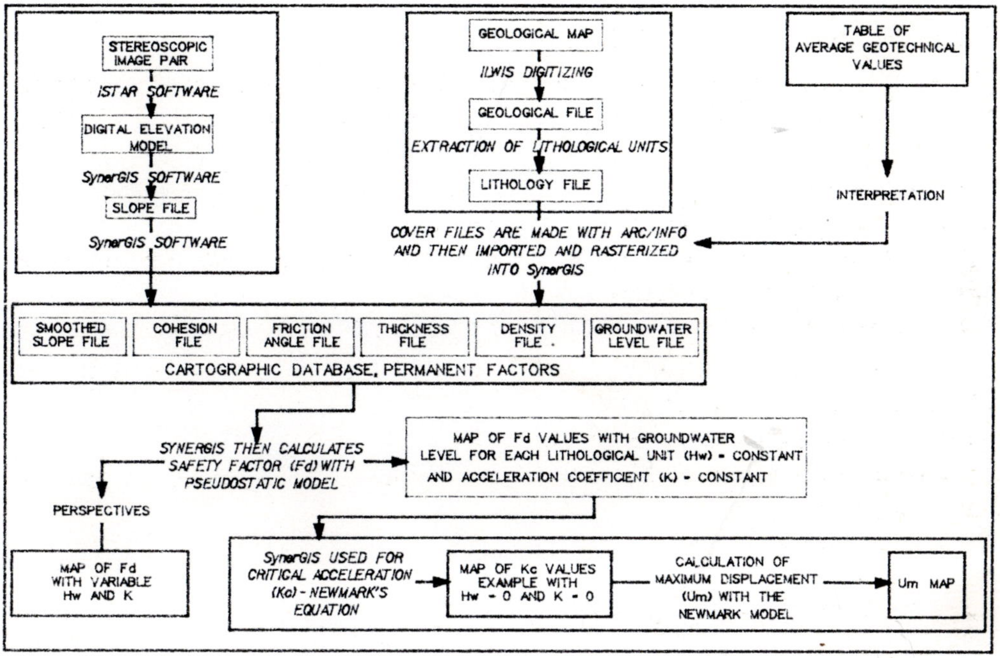

Figure 4: methodology developed for stability computation

4.5 EXAMPLES

The case illustrated in Figure 5 shows the methodology adopted for computation of F_d, taking an acceleration coefficient $K = 0.1$ and using the H_w values shown in Figure 2.

5 Results

5.1 COMPUTATION OF F_d

The safety coefficient F_d was calculated for different K and H_w values, and using the same hypotheses for both the regional and local scales (Figures 6 to 11).

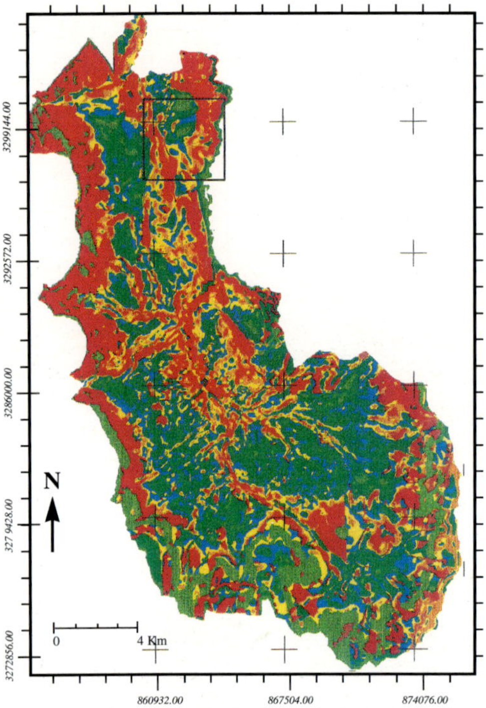

Fig. 6: Trièves region: computing F_d with $H_w = 0$ and $K = 0$.
Fig. 7: Sinard commune: computing F_d with $H_w = 0$ and $K = 0$.

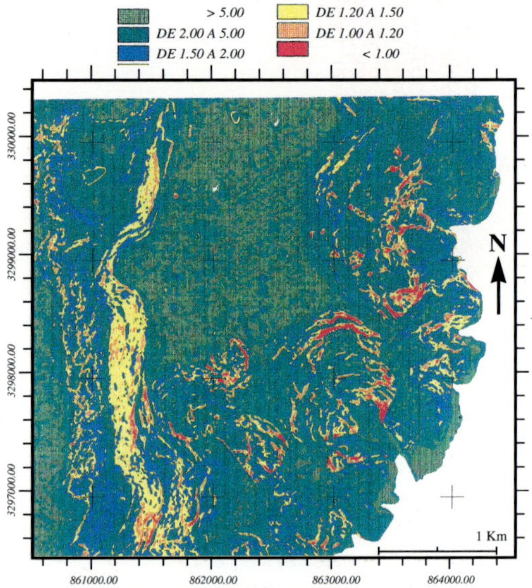

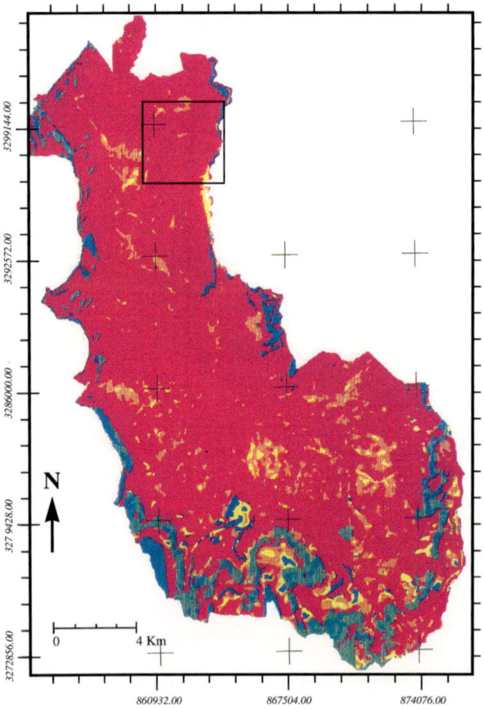

Fig. 8: Trièves region: computing F_d with $H_w = 0$ and $K = 0.2$.
Fig. 9: Sinard commune: computing F_d with $H_w = 0$ and $K = 0.2$.

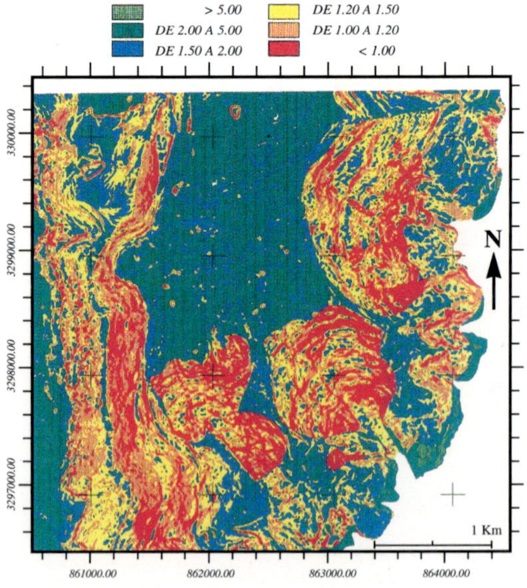

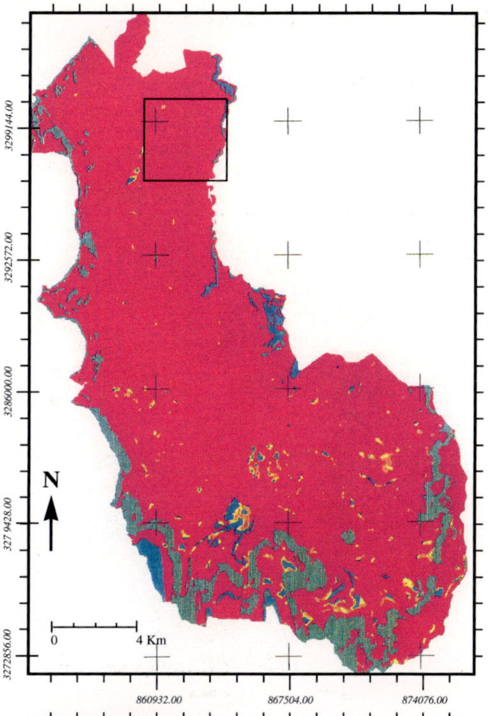

Fig. 10: Trièves region: computing F_d with $H_w = 0$ and $K = 0$.

Fig. 11: Trièves region: computing F_d with $H_w = H$ and $K = 0$.

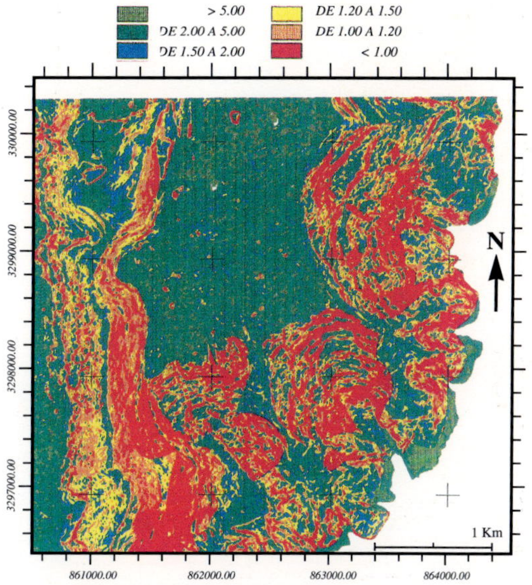

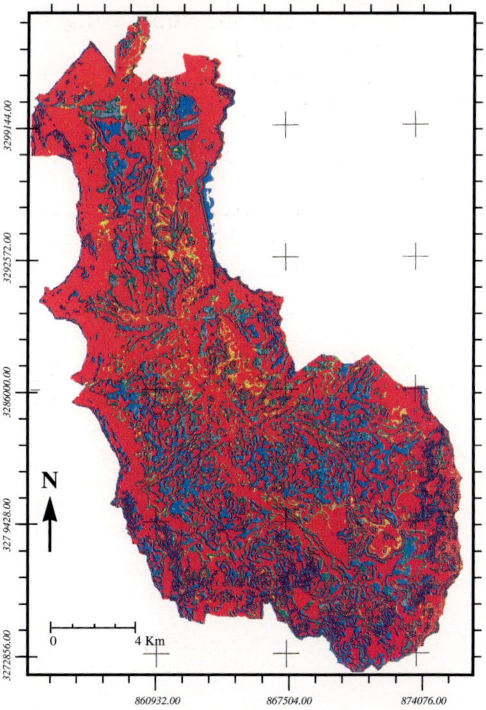

Fig. 12: Trièves region: computing K_c with $H_w = 0$ and $K = 0$.

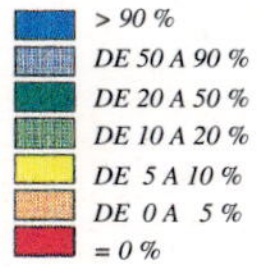

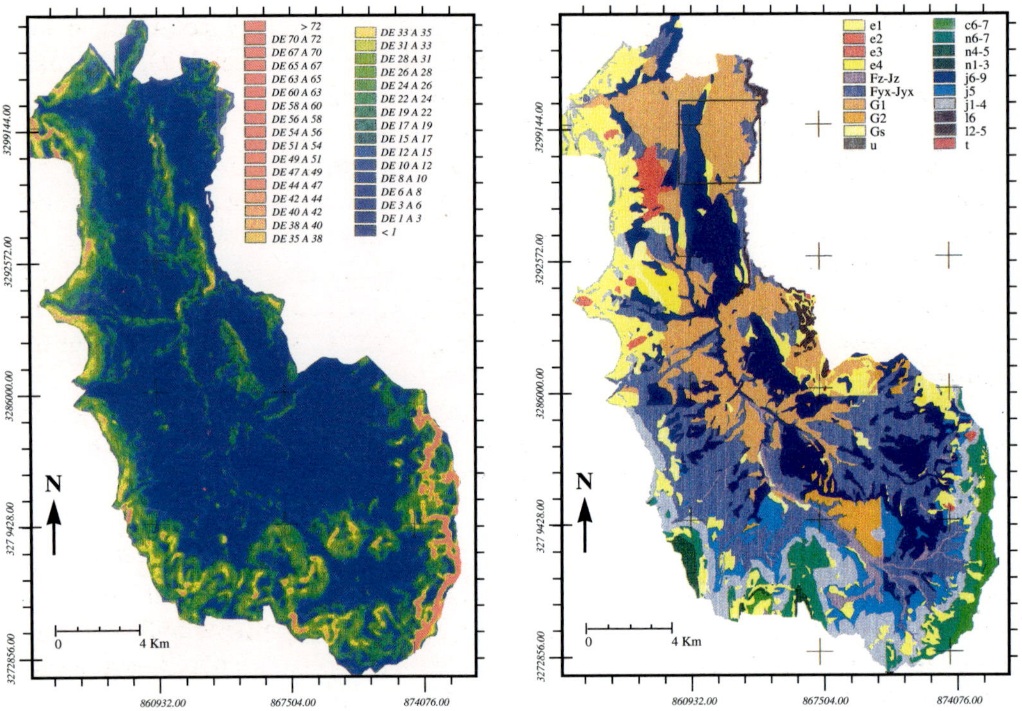

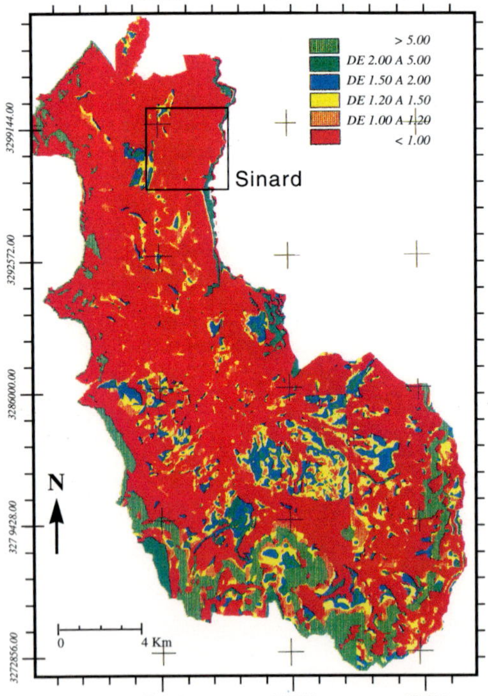

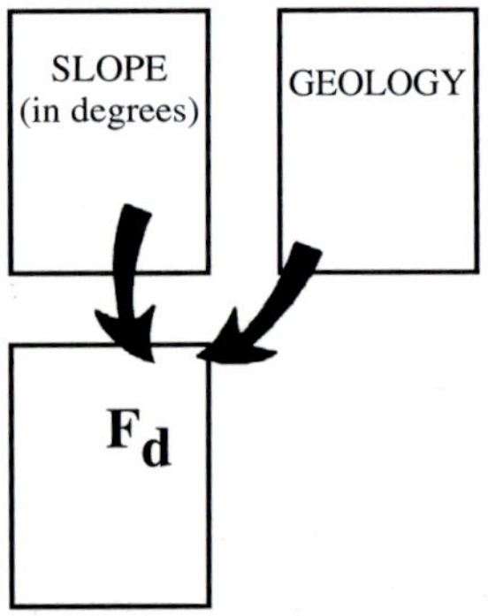

Fig. 5: Trièves region: combining geological and slope maps for computing F_d. Example shows a variable H_w and $K = 0.1$.

In addition to the fact that the maps obtained are generally comform to our understanding of the instability of the study area (Figs. 6 and 7), the main interest of this exercice is the analysis of changes in scale.

It is clear that that the local map is much less penalizing than the regional one, although their information never is in contradiction. Unstable areas are more clearly shown, but no obvious reversal was noted when compared whith the tendancies apparent on the regional scale.

Unsing the working hypothesis of K = 0.2 for the area as a whole (Figs. 8 and 9), the unstable areas become much greater in number, especially at the regional scale, but this hypothesis is very pessimistic as will be explained in the following chapter.

On the local scale, the unstable areas are clearly defined and are in perfect agreement with observed features. However, it should not be forgotten that the determination of F_d is strongly influence by the geological map used, which contains implicit information on instability.

The last hypothesis (Figs. 10 and 11) considers the groundwater to be at ground surface for all rock types. the result is a strong increase in unstable areas, particularly in the western part of the local map where slope scree is found. This hypothesis, too, is highly penalizing and not very realistic, but it enables to compare the two types of landslide trigger, which are hydraulic and seismic.

5.2 CALCULATION OF K_c

The critical acceleration 'K_c' corresponds to the value of the acceleration coefficient K for which Fd = 1. It can be calculate with Newmark's empirical equation [4], which is:

$$K_c = (F_s - 1)\sin\beta$$

where F_s corresponds to K = 0.

The case studied assumes K = 0 and Hw = 0 (Fig. 12). It can be seen that areas unstable under static conditions are represented by null values for K_c; the contrast between stable and unstable areas is well marked, as the critical acceleration values are distributed mostly fall between null values and those that are higher than 75% of gravity.

6 Conclusions and perspectives

The results from the first work are promising, not only for the methodology used but also for the means deployed. However, the results now must be validated.

First of all, this should involve tests with other values for the basic data, particularly by modulating the K and H_W values. More realistic hypotheses, enabling to vary these two parameters with morphology (for site effects) and discontinuities (structural or textural), will lead to obtaining less penalizing results.

Another point requiring analysis is the interdependence between basic data, so as to avoid any bias in the interpretation; this is presently the case with the geological data used in the system, as such maps contain implicit data for the location of landslides.

Furthermore, the trend should be confirmed that seems to indicate that unstable areas are optimized and reduced when going from the regional to the local scale. If this is not confirmed, the interest of small-scale zonation must be reconsidered, because of the fact that its conclusions might be systematically contradicted on the local (large) scale.

The perspectives for future development are two-fold at present: A stability analysis with the Newmark method would lead to obtaining values for the maximum displacement envisaged for a certain slope. Second, a more quantified naturalistic approach, based on analysis by means of shape recognition and on the use of fuzzy sets [5], would lead to an analysis based on the knowledge of determinant factors.

Bibliography

[1] ILWIS, GIS system,International Institute for Aerospace Survey and Earth Sciences (ITC), 7500 AA Enschede, the Netherlands.

[2] SYNERGIS, GIS system, Bureau de Recherches Géologiques et Minières, 45000 Orléans, France.

[3] ARC/INFO, GIS system, Environmental System Research Institute, California 92373 USA.

[4] Newmark N.M., (1965) Effects of earthquakes on dams and embankments. *Géotechnique*, vol 15, p.139-160.

[5] Juang C.H., Lee D.M. and Shen C., (1992) Mapping slope failure potential using fuzzy sets. *Journal of geotechnical engineering*, vol.118, p.475-495.

THE PRACTICE OF SLOPE STABILITY ENGINEERING IN SEISMIC AREAS IN THE SOUTH-EAST OF FRANCE

Jean-Pierre MENEROUD* et Jean-Louis DURVILLE**

Abstract : The current practice of slope engineering in seismic areas in the S.-E. of France is reviewed. Although numerous landslides may be triggered by an earthquake, strikingly very few dynamic calculations of slopes are being carried out. However some seismic microzonation maps, including slope stability, have been done in the last years.

Résumé : On examine la pratique française récente et actuelle dans la prise en compte de l'action sismique en stabilité des pentes. Le bilan est plutôt décevant : bien que le Sud-Est soit à fort relief et de sismicité moyenne, le dimensionnement sismique des talus est encore très peu répandu ; en revanche plusieurs microzonages incluant la stabilité des pentes ont été réalisés.

1. Introduction

During most of large earthquakes in moutain areas there are a lot of seismic induced slope failures like rockfalls, rockslides, lateral spreadings, etc. For example, due to Loma Prieta earthquake in California, more than 1000 landslides and rockfalls were documented in the Santa Cruz Mountains, and during the 1990 Menjil (Iran) and 1991 Costa Rica earthquakes, numerous landslides were triggered.

Many slope movements take place every year, without any seismic triggering, in the south-east of France, due to the recent tectonic movements and the important erosion activity. For instance the Roquebillière slide which caused the death of 19 people in 1926, the 1952 Menton debris slide where 11 people died, or the La Clapière landslide (50 hm^3), now threatening the village of Saint-Etienne-de-Tinée.

A large part of this region is classified as seismicity degree II (the maximum degree in the continental France), and experienced actually some large earthquakes in the past : in 1564, an earthquake of maximal intensity X (MSK) struck the Vésubie valley, 40 km at the north of Nice, and it seems that some large rockfalls occurred and destroyed hamlets like Valdeblore [1] : figure 1 ; in 1887 the ligurian earthquake caused also many slope instabilities. Besides we can observe in the Alpes-Maritimes some very large rockslides which probably occurred during earthquakes in prehistoric times, like for instance in the cretaceous marls and limestones near Breil village in the Roya valley, which is following an active fault, or in the limestone cliffs of the Vésubie gorge. These large phenomena are between rockslide and collapse tectonic structures.

---

* CETE- Méditerranée, Laboratoire des Ponts et Chaussées de Nice.

** Laboratoire Central des Ponts et Chaussées, Paris.

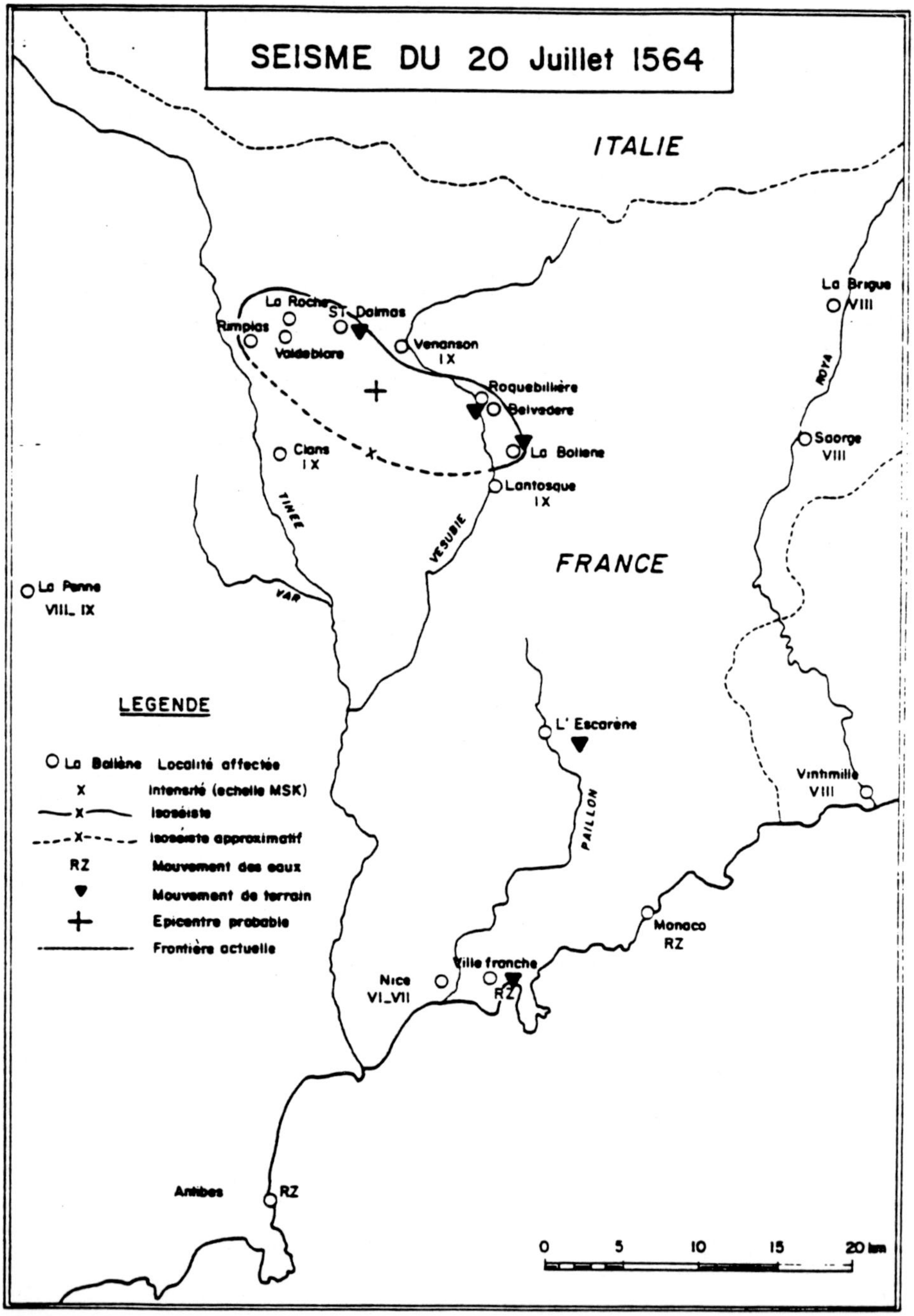

Figure 1. Isoseismal map of the 1564 earthquake, with location of the main probable induced landslides (picture taken from [1]).

Thus both earthquake and landsliding hazards are threatening the south-east of France and we try below to investigate how in the practice of slope engineering the seismic aspect has been taken in account.

2. Types of seismic induced slope failures

The most commonly observed slope instabilities during earthquakes are, according to Keefer [2] : rockfalls, debris slides and superficial soil slides, rock slides along discontinuities, lateral spreading in soils, etc. These failures have two main causes : the inertia forces during the earthquake and the weakening of soil properties (particularly due to increase of pore pressure that may cause sand liquefaction).

Observations of slope instability during major earthquakes have shown that certain failures have special particularities due to the dynamic action. They may occur in morpho-geological conditions that are hardly, or not at all, suspect in the absence of earthquakes [3]. This phenomenon may be caused particularly by site amplification effects around convex topographical forms, or by partial or complete liquefaction of sandy layers in river banks (fig. 2). At present no tried and proven diagnostic method exists that completely accounts for such specific movements, and the methodologies generally used treat them only very imperfectly. On the other hand the pseudo-static method, very simple to use, may be applied to rockslides and to soils that do not undergo pore pressure increase.

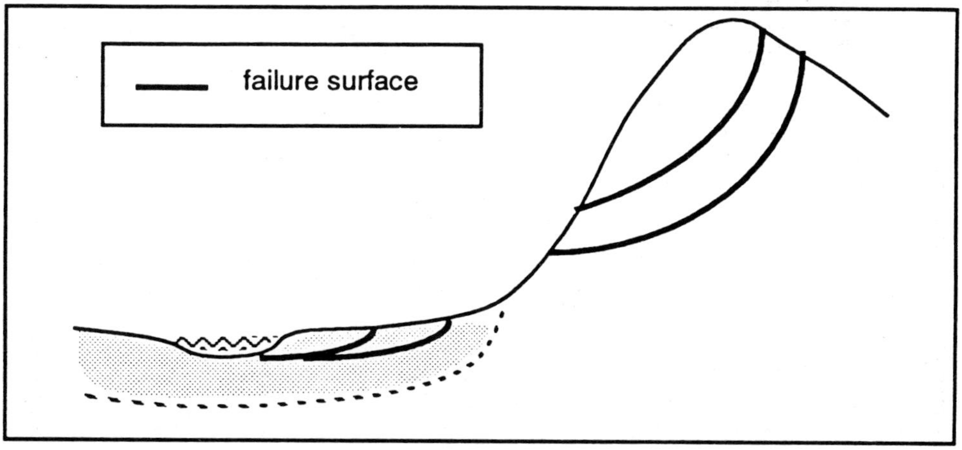

Figure 2. Sketch showing some typical failures induced by earthquakes.

3. Current practice in slope stability surveys

3.1. SOIL SLOPES

In soil slopes the seismic condition has been rarely used : for example no seismic coefficient was used in the works of the recently completed Aix-en-Provence - Sisteron motorway. In the same manner, the stabilization of some large lanslides that endangered the A8 motorway between Nice and Menton in tne seventies didn't take account of any seismic effect.

For the Fontvieille terreplein (a large hydraulic fill set up in the sea in the Principauté de Monaco) built in the late sixties behind a large dyke, some calculus were made in the late seventies, after the construction, to verify the seismic stability. A pseudo-static process was used in this survey using the Fellenius method with a horizontal acceleration equal to 0.125 g (it is probable that today a value of 0.15 g would be used) and a seismic excess pore pressure on the slide surface. Liquefaction is here the main problem for the stability because there is a thin sand layer underneath the fill, and sand into the fill too.

During a research program in 1983 on microzonation, a pseudo-static calculation was applied to the Roquebillière slope, in the Alpes-Maritimes ; a seismic coefficient of 0.2 was used (it should be 0.125 following the 1990 AFPS[1] recommendations, but an amplification effect due to surficial layers is probable) conditions and the same drained shear strength than in the static case : the results are given in table I and show the drastic reduction of the safety factor when adding the seismic and the hydraulic effect.

Circles of failure	F	0,0 g	0,2 g	Water table
half-deep	1,25	x		no
	0,78		x	no
	0,98	x		yes
	0,60		x	yes
deep	1,18	x		no
	0,75		x	no
	0,94	x		yes
	0,59		x	yes

Table I. Factor of safety calculated with the Bishop method (Roquebillière slide)

It is only in the seismic zone III (high seismic hazard, cf. [4]), i.e.in the french Antillas, that seismic calculations for slope stability are currently used ; for instance, a recent calculus of anchors and stabilization in soil cuts and fills in the Martinique island used a seismic coefficient of 0.17, following the AFPS recommendations [5] i. e. a pseudo-static method with two cases of seismic loads : $\sigma_H + \sigma_V$ and $\sigma_H - \sigma_V$, where σ_H, the horizontal load, is taken as one half of the nominal acceleration , and with a factor of safety related to shear strength of soil equal to 1.

---

[1] Association française de Génie parasismique.

The situation is different if we consider earth dams : for instance the design of the Serre-Ponçon dam (near Gap) was done with a pseudo-static coefficient of 0.1g around the year 1960.

3.2. ROCK SLOPES

It is a matter of fact that in the french practice the seismic conditions are rarely used for rock stability problems. For instance in the design of alpine motorways with high cuts and often with stabilization works, no seismic coefficient was used. It is interesting to consider as an example the first french motorway in montainous country, built in the late sixties, in the Menton basin ; the Careï viaduct (fig. 3) was designed with a 0.1g horizontal acceleration, but the piles were founded on a downslope dipping flysch, and no seismic load was used for the slope stabilization.

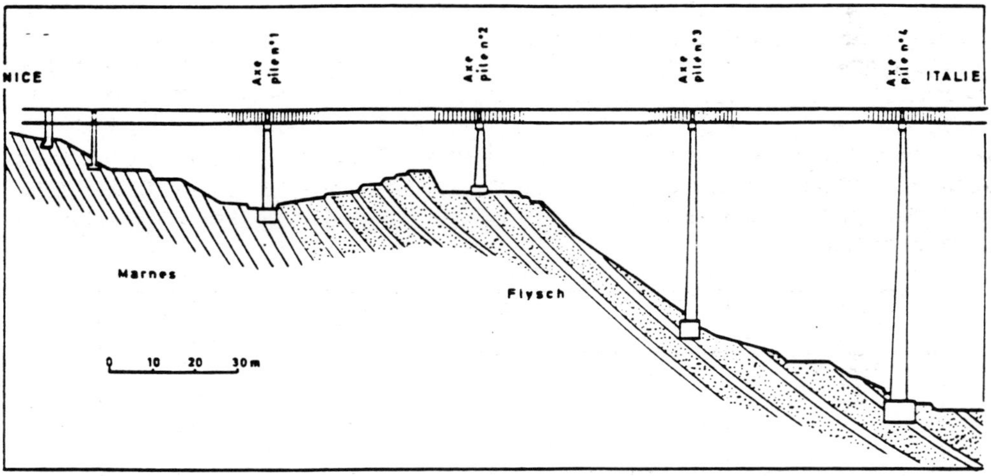

Figure 3. Careï viaduct on the A8 motorway near Menton (from [6]).

Nevertheless this situation may be now changing : for instance in a recent work in the Roya valley, located in one of the most seismical area of the country, near an active fault, calculations of anchors used to stabilize a rock slope were made in accordance to the french "Recommandations AFPS " [5].

It is clear that these recommendations are more concerned with soils than with rocks. It is interesting however to consider an application to rock stability using the rules given by the recommendations. For a typical case of a wedge failure from a cliff 19 m high and dipping 70° (figure 4), we find the safety factors given in the table II. In this case the static safety factor Fs is 1.50 with peak values and 1.02 with residual values. In rock stability it is often advised to use residual values, so we compare the static value 1.02 to those given in the table. One may observe that, for normal buildings in the french seismicity zone Ib and probably in the zone II also, the seismic condition will not modify significantly the situation if Fs >1.5 is wanted in static conditions and Fd >1.0 in the seismic case : the lowering of the safety factor does not excess 20 %.

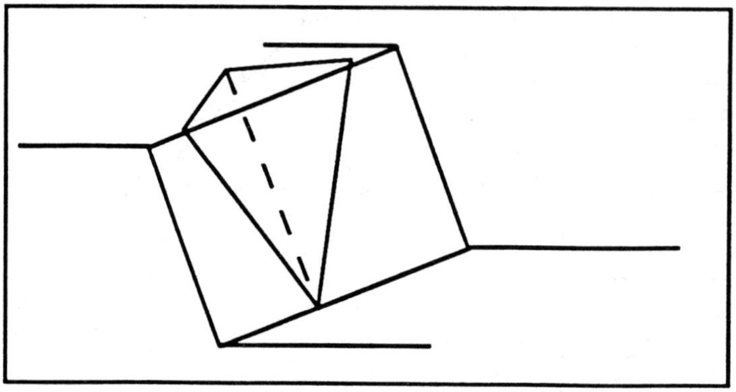

Figure 4. Wedge failure used for the table II .

a_N / g	0	0.15	0.25	0.30	0.35	0.40
$\sigma_H + \sigma_V$	1.02	0.90	0.82	0.78	0.76	0.72
$\sigma_H - \sigma_V$	1.02	0.88	0.80	0.74	0.70	0.66

Table II. Factor of safety in a wedge failure problem, calculated from five nominal accelerations ; two kinds of loads are considered.

A second example is given below, with the calculation of anchors in a 2D-plane failure (figure 5). A 6 m high cut is designed in schists dipping at 30° ; the friction angle is 20° and the discontinuity planes are supposed to possess a high dilatancy.

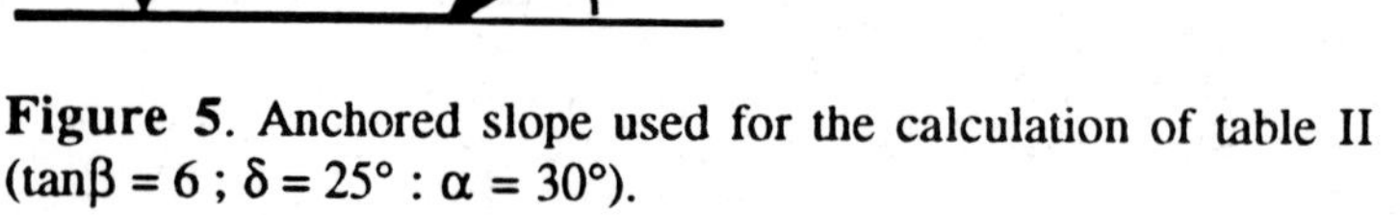

Figure 5. Anchored slope used for the calculation of table II ($\tan\beta = 6$; $\delta = 25°$: $\alpha = 30°$).

Without any seismic load, the calculation shows that the anchors have to bring a tensile force of 245 kN / ml to get the factor of safety 1.3 generally adopted. Table

III gives the traction force (in kN / ml) in the anchors at the limit equilibrium in order to obtain a factor of safety of 1 with the seismic loads given by the AFPS Recommendations. We can conclude again that in the zones of low or moderate seismicity the seismic condition is easier to fulfil than the static one.

a_N / g	0 (F = 1.3)	0.15	0.25	0.30	0.35	0.40
$\sigma_H + \sigma_V$	245	198	240	261	282	303
$\sigma_H - \sigma_V$	245	188	223	241	258	276

Table III. Forces (kN/ml) needed in the anchors to ensure a factor of safety of 1 with five nominal accelerations and two kinds of loads.

4. Slope stability in microzoning

In the recently developed program of the PER (Plans d'exposition aux risques), the seismic microzonation of some towns (like Menton or Digne) incorporated slope stability. But we have to notice that in most cases the technical hazard mapping did not come to effective regulations. Recently the microzonation of the Principauté de Monaco was carried out with the same method.

A simplified pseudo-static method was defined [7], the result of which is to move the limits between hazard zones. The first step is to draw up a "classical" hazard map, which is to say, without taking seismic conditions into account, with four hazard degrees. Accounting for the seismicity can be done as :

- adding specific phenomena such as lateral spreadings along banks,
- an increase of the hazard level for a given zone and phenomenon,
- an extension of areas that might be affected by rockfalls.

For landslides in unconsolidated ground for example, we apply a simplified pseudo-static method to determine the increase of the hazard in a given zone. The main hypotheses are :

- soil strength (C,ϕ) and pore pressure are assumed to be invariable during the earthquake,
- the different hazard zones are supposed to correspond to different intervals for the factor of safety,
- a ratio of 1.15 between the factor of safety in static and in dynamic conditions,
- the pseudo-static load is equivalent to a rotation of gravity, and changes the limit angle of slopes or the safety factor (figure 6).

Figure 6. Microzonation of Menton : landslide induced hazard map (partly).
Tha map A (above) shows a zonation realised in static conditions and the B one shows zonation in dynamic conditions. Each zone is identified by a safety factor. It can be observed that in zone I the safety factor equals 1.5 on the A map and 1.0 on the dynamic one.

5. Conclusions

Roughly speaking, from the author's experience, seismic loads are not yet used for slope stability problems in the french practice today, even with simplified methods, although a region like the Alpes-Maritimes is typically prone to earthquake-induced landslides. We must notice however that microzonation maps included induced landslide hazard, but it is a little too soon to draw conclusions about the practical efficiency of these surveys.

The dynamic effect of the seismic vibration on slopes is rather complex, and, although several methods exist, it is not quite satisfying for the engineers : some methods are often considered as oversimplified (pseudo-static and Newmark methods), other are very sophisticated and need data about soil properties which are generally not available (non linear constitutive laws for example). This may explain the reluctance of the designer to take account of earthquake conditions, especially when specific phenomena may occur.

Another reason of these inadequacies could be that there is at this time in France no regulation dealing with this field of engineering. Nevertheless for bridges also there are no regulations, and there is some empirical agreement to carry out dynamic calculations about large viaducts in seismic areas.

A last explanation of the situation could be that in the areas of low and moderate seismicity (zones Ia and Ib, even II, from the french seismic zonation) the dynamic condition in slope stabilization works is not more constraining than the static one, as the examples presented seem to show. But this fact should be verified in many other cases and with other methods (Newmark,...).

As far the pseudo-static method is concerned, the following hypotheses should be also thoroughly examined :

- which value of the seismic coefficient ? generally a fraction of the maximal acceleration is used, which may depend on the magnitude, the epicentral distance, etc. [8].
- which factor of safety ? Some people recommend to use a ratio of 1.15 between static and seismic factor of safety ; others propose to use the same value in both cases ; the AFPS suggests to adopt a seismic factor of safety 1.
- which shear strength should be used (peak, residual value ; undrained or drained,...) ?
- which hydraulic conditions ? In a probabilistic approach, the worse hydraulic conditions, used in a static calculation, should note be used simultaneously with the earthquake load.

We draw attention upon the fact that these questions have to be considered simultaneously ; otherwise you may get positive or negative answer to your problem, depending on the hypotheses...

According to the authors' point of view, research programs should be developed including *in situ* observations and testing different methods of calculation, but the available simple method should be recommended or required in any contract related to slope works. In particular, the case of rock slopes should be mentioned in the french future regulation.

Bibliography

[1] Vogt J. (1979). Les tremblements de terre en France. *Mémoire du B.R.G.M.* n° 96, 220 p.

[2] Keefer D.K. (1984). Landslides caused by earthquakes. *Bulletin of the Geological Society of America*, vol. 95, p. 406-421.

[3] Durville J.-L., Méneroud J.-P. (1982). Phénomènes géomorphologiques induits par le séisme d'El Asnam, Algérie ; comparaison avec le séisme de Campanie, Italie. *Bull. liaison P. et Ch.*, n°120, p. 13-23.

[4] D.R.M. (1986). Nouveau zonage sismique de la France. *La Documentation française*, 146 p.

[5] A.F.P.S. (1990). Recommandations pour la rédaction de règles relatives aux ouvrages et installations à réaliser dans les régions sismiques. *Presses de l'ENPC*, 182 p.

[6] Amar S., Méneroud J.P., Pilot G. (1971). Fondations des viaducs. In : Autoroute de Menton, *Bull. Labo. Ponts et Chaussées*, Spécial R, p.117-132..

[7] Bard P.Y., Méneroud J.P., Durville J.L., Mouroux P. (1987). Microzonage sismique. Application aux plans d'exposition aux risques (PER). *Bull. Labo. P. et Ch.*, n° 150-151, p. 130-139.

[8] Durville J.L., Méneroud J.P., Mouroux P., Simon J.M. (1985). Evaluation de l'aléa sismique local - Microzonage. In : *Génie parasismique*, chap. III-7. Presses de l'Ecole nationale des Ponts et Chaussées, p. 239-264.

INSTABILITY OF SUPERFICIAL SEDIMENTS IN "BAIE DES ANGES" (NICE, FRANCE); INFLUENCE OF SEISMICITY.

Thierry MULDER *, Jean-Paul TISOT *, Pierre COCHONAT ** and Jean-François BOURILLET **

Abstract : The *Baie des Anges* provides many areas of sedimentary instabilities firstly, by the high values of the slope gradient, secondly, by the local existence of under-consolidated subsurface sediments. The offshore hazard probability is increased by taking into account of seismical horizontal acceleration in a pseudo-static model. These different hazards should have to be taken into account in the future development of the bay.

Résumé : La baie des Anges présente de nombreuses zones d'instabilités sédimentaires en raison à la fois du fort gradient de pente et de la présence locale de sédiments sous-consolidés en surface. La prise en compte de la sismicité par l'intermédiaire d'accélérations horizontales (modèle pseudo-statique) augmente fortement le risque d'instabilité dans la baie. Ces risques devront, à différents degrés, être pris en compte dans l'aménagement futur du site.

1 Geological and geotechnical setting

The *baie des Anges* is located in the center of the Ligurian continental margin. The continental slope morphology is irregular. The continental shelf is reduced and a depth of 1000m is reached at only 5 km from the coastline. It is cut by two canyons dug by the Var river and the Paillon river (fig. 1).
The bottom is constituted by plio-quaternary sediments above the messinian erosional surface. They essentially are carbonated silty clay and, after a quick deposition phase, are actually submitted to many instabilities ([1], [2], [3], [4], [5]). The 1979 Nice airport slump occurs in this context ([6]). The slope gradient values are strong and some values above 40° can be sometimes observed (fig. 2).
The bottom of the canyons shows a coarser sedimentation. The average saturated specific weight of the silty-clays is 19.3 kN/m^3.

---

* Laboratoire de Géomécanique, ENSG, Nancy, France.

** IFREMER, Centre de Brest, Géosciences Marines

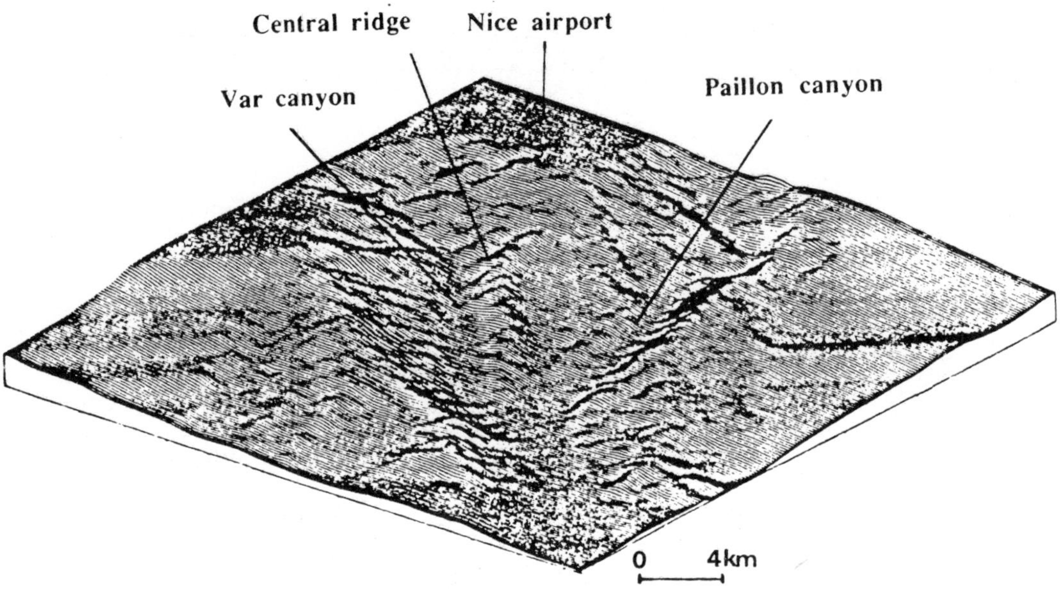

Figure 1 : 3D diagram of the submarine bottom in the *baie des Anges* (resulting from IFREMER data).

The silty-clays are homogeneous both in their sedimentological and geotechnical features and in their mechanical behavior. Nevertheless, a precise characterization of the consolidation state of the sediments provided by the Küllenberg cores, allowed to establish four sediment families (fig. 3). Each of them is characterized by a linear relation between the undrained cohesion c_u and the depth (z) ([7], [8]) .

Family A : Under consolidated sediments; $c_u = 1.35z$;
Family B : Normally consolidated sediments; $c_u = 1.2z + 5$;
Family C : Slightly overconsolidated sediments; $c_u = 2.5z +5$;
Family D : Overconsolidated sediments; $c_u = 2.5z + 15$;
This classification allows the geotechnical mapping used in the undrained condition study (fig. 4).

Under drained conditions, we applied the c' and ϕ' values (c' : drained cohesion and ϕ' : friction angle) defined by the Mohr-Coulomb criterion : τ (kPa) = 5 + σ'.tan 35°.

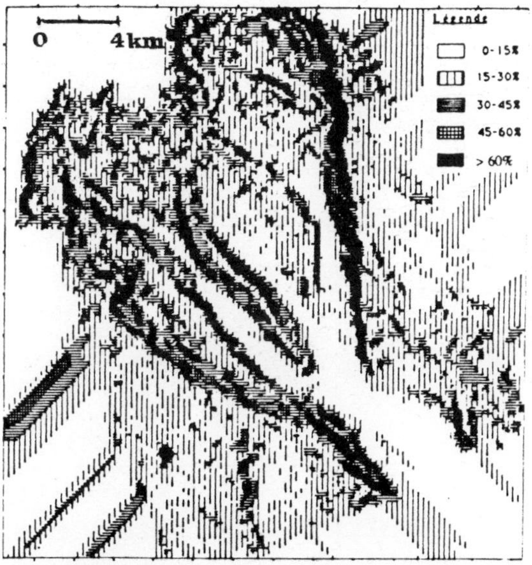

Figure 2 : Map of the slope gradient in the *Baie des Anges* (resulting from IFREMER data).

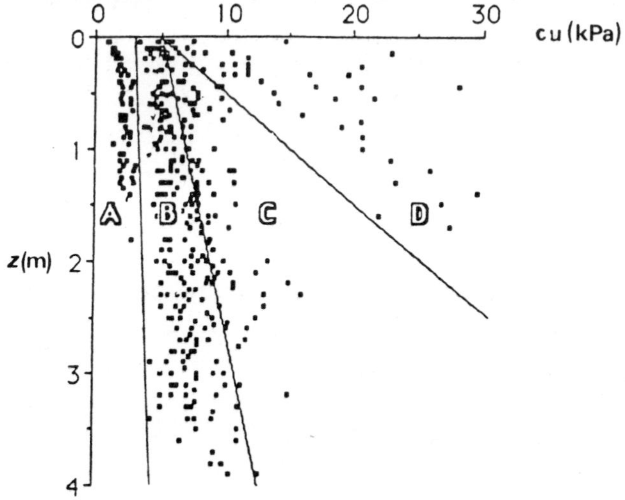

Figure 3 : Geotechnical classification of the sediment, according to its consolidation state.

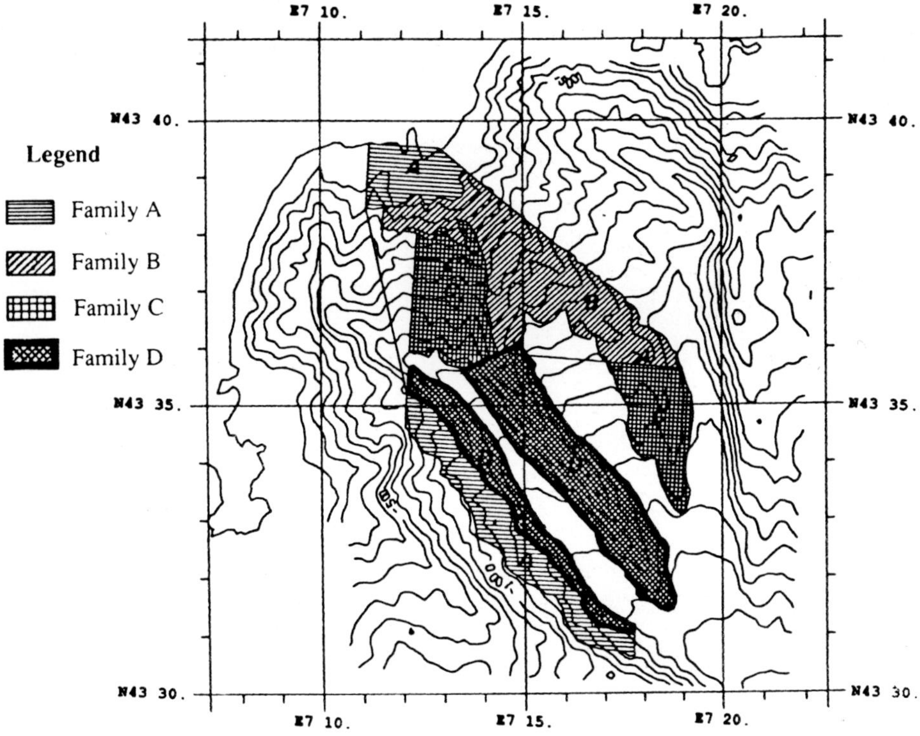

Figure 4 : Map of the sediment families according to the geotechnical classification.

3 Seismicity of the bay and evaluation of the seismical parameters

The active tectonic of the bay is marked by frequent earthquakes. Some of them show a value of magnitude greater than 6. The magnitude 4 is frequently reached (October 4th, 1985, M = 4.3, may 1st, 1984, M = 4, December 26th , 1989, M = 4,3 ([9]). The bay is submitted to the more intense seismical activity of the neighbouring Ligurian sea ([10]).

Many empirical relations give the horizontal acceleration as a function of the value of the Mercalli Modified (MM) intensity (fig. 5). For our calculations we used the global Murphy and O'Brien relation .

Assuming this law, figure 6 shows a 40 % decrease of the security factor value for a 10° slope value in the bay submitted to a 0.05.g value of the horizontal acceleration (F = 3,5 → F = 2) . The addition of the vertical acceleration can be neglected. According to these results, it is indispensable to make an evaluation of the seismical influence on offshore hazards in the *baie des Anges*.

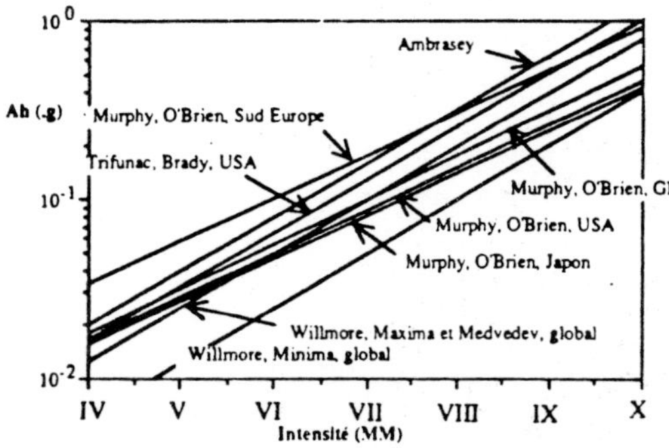

Figure 5 : Correlatives laws between the ground horizontal acceleration and the seismical intensity (MM) (according to [11], [12], [13], [14] and [15]).

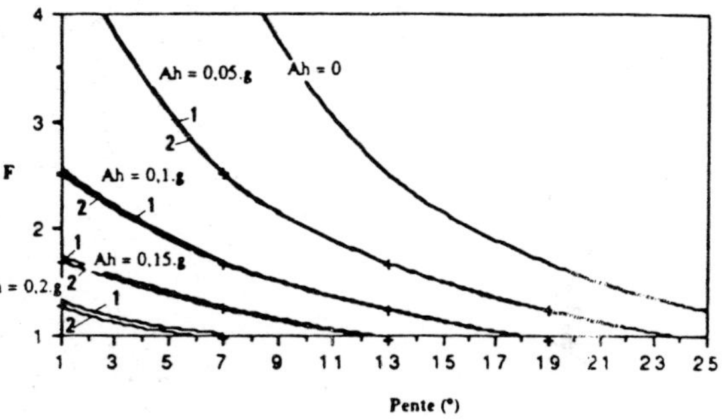

Figure 6 : Influence of the horizontal and vertical acceleration on the security factor F (1 : influence of Ah; 2 : influence both Ah.and Av).

4 Calculation model

An approach of the stability can be done assuming the infinite plane sliding surface model. The observation of scars of old failures on the submarine bottom shows that this model is a good approach of the reality. Circular slumps can appear but they are generally submitted to the existence of a scarp caused by an anterior plane slump.

Two kinds of analysis can be chosen : a long-term analysis (drained conditions) using the drained cohesion, c' and the friction angle, ϕ'. A short-term analysis using the undrained cohesion, c_u. In this last case, it is possible to introduce dynamic parameters as the excess pore pressure, Δu, and the seismical accelerations.

Assuming Δu, Ah and Av as static parameters, it is possible to introduce them in a drained model (pseudo-static model) (fig. 7). This model represents

the possibility of occurrence of punctual events in a long-term model. ([16]). The theoretical equation [1] of the strength equilibrium can be separated into three components (eq. [2]), A, B and C where A represents the static factor (eq. [3]), B (eq. [4]) and C (eq. [5]), the influences respectively of the excess pore pressure and the seismical accelerations ([17], [18], [19], [20]).
In an offshore site, the knowledge of the soil parameters and of the mechanical behaviors is insuficient to applied any dynamic model, but pseudo-static models are appropriated to make a regional offshore mapping.

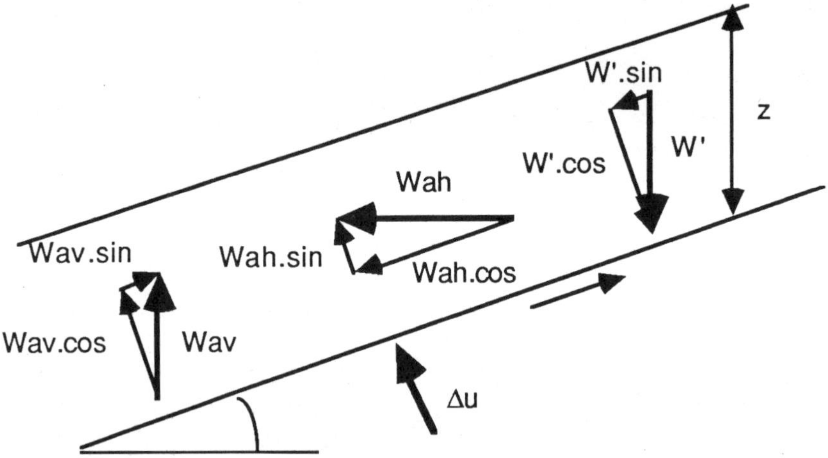

Figure 7 : strength equilibrium in the pseudo-static infinite plane sliding surface model.

$$F=\frac{c'+(\gamma'.z.\cos^2\alpha-\Delta u-\gamma.z.Ah.\sin\alpha.\cos\alpha-\gamma.z.Av.\cos^2\alpha)\tan\phi'}{\gamma'.z.\sin\alpha.\cos\alpha+\gamma.z.Ah.\cos^2\alpha-\gamma.z.Av.\sin\alpha.\cos\alpha} \quad [Eq.1]$$

where γ = total specific weight, γ' = buoyant specific weight, c' = drained cohesion, ϕ' = friction angle,
α = slope gradient, Δu = interstitial overpressure, Ah = horizontal acceleration, Av = vertical acceleration
z = depth of the failure surface, F = security factor.

F = A - B - C [eq.2] where A = static coefficient, B = excess pore pressure coefficient, C = seismical coefficient

$$A=\frac{c' + \gamma'.z.\cos^2\alpha\ .\ \tan\phi'}{\gamma'\ .\ z.\ \sin\alpha.\ \cos\alpha} \quad \text{[eq.3]}$$

$$B=\frac{\Delta u\ .\ \tan\phi'}{\gamma'.z.\ \sin\alpha.\ \cos\alpha} \quad \text{[eq.4]}$$

$$C=\frac{\gamma.\tan\phi'\ (Ah+(Av.\tan\alpha-Ah).(((\Delta u-c')/\tan\phi')/\gamma'.z\ ))}{\gamma.\sin^2\alpha\ (\gamma'/\gamma + Ah.\ \cot\alpha\ -\ Av)} \quad \text{[eq.5]}$$

We will restrict our study to the quantification of the horizontal seismical acceleration influence.

5 Stability study in a static state.

Many calculations have been made for different depths of the failure surface, from 5 to 15m according to the height of the scars observed on the submarine bottom or to the values calculated from the sedimentary hiatus corresponding to an old slump ([21]). The stability calculation under drained conditions (fig..8a), with an average Mohr-Coulomb criterion for the whole bay, shows that the superficial sedimentary deposits are stable for a depth of the failure surface varying from 5 to 15 m. Under drained conditions, the F isovalues map is equal to the slope gradient values map within about a constant.

The study under undrained conditions (fig. 8b) shows that the F values are directly linked to the geotechnical zonation. The critical state is firstly reached for the underconsolidated sediments situated in front of the Nice airport and on west side of the occidental crest. The stable state is reached for a depth of the failure surface value smaller than 2m.

5.1 THE *A POSTERIORI* ANALYSIS

For a given event, the 1989/26/12/ earthquake (fig..9), the estimated acceleration values deduced from the isoseimal lines allow to calculate the C values and then, the resulting F values in the bay. The epicenter of this event is situated 20 km in the south of the bay. So, this earthquake seems to have not triggered any sedimentary disorder in the bay. In the opposite case, a posterior sonar study of the bottom in the theorically-perturbed areas might allow the validation or the modification of the numerical data.

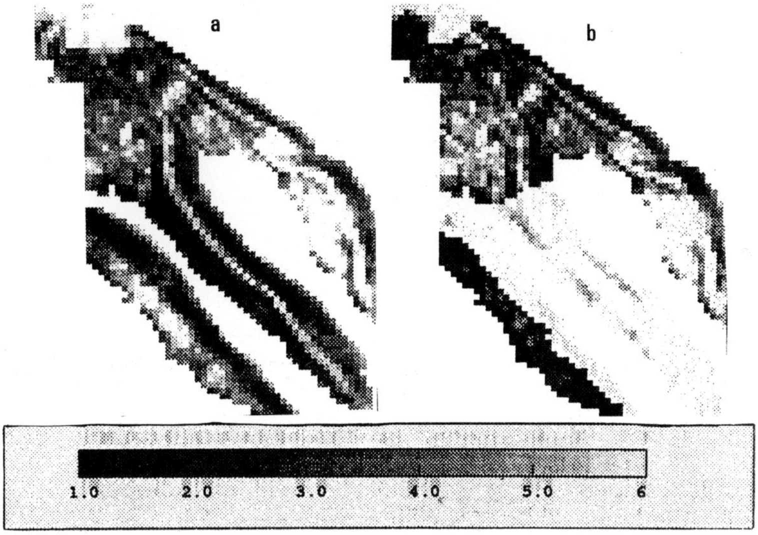

Figure 8 : F values in a static state. a : drained conditions (z = 10m) ; b : undrained conditions (z = 1m).

5.2 THE *A PRIORI* ANALYSIS

It is possible to make an evaluation of the possibility of slump for a seismical event with a given return period. We have chosen three return periods : 100, 200 and 1000 years. The centennial, bicentennial and millennial accelerations and C values have been deduced from the seismical event map. Then, the new F values have been calculated by the subtraction of C from the statical values of F (A values) (fig. 10).

6 Interpretation

The results show that the slumps in the bay can be separated in two groups (fig. 11):

Firstly, the instabilities in relation with the internal characteristics of the site (geomorphology, sedimentological features and geotechnical behavior). They can be classified into two different groups :

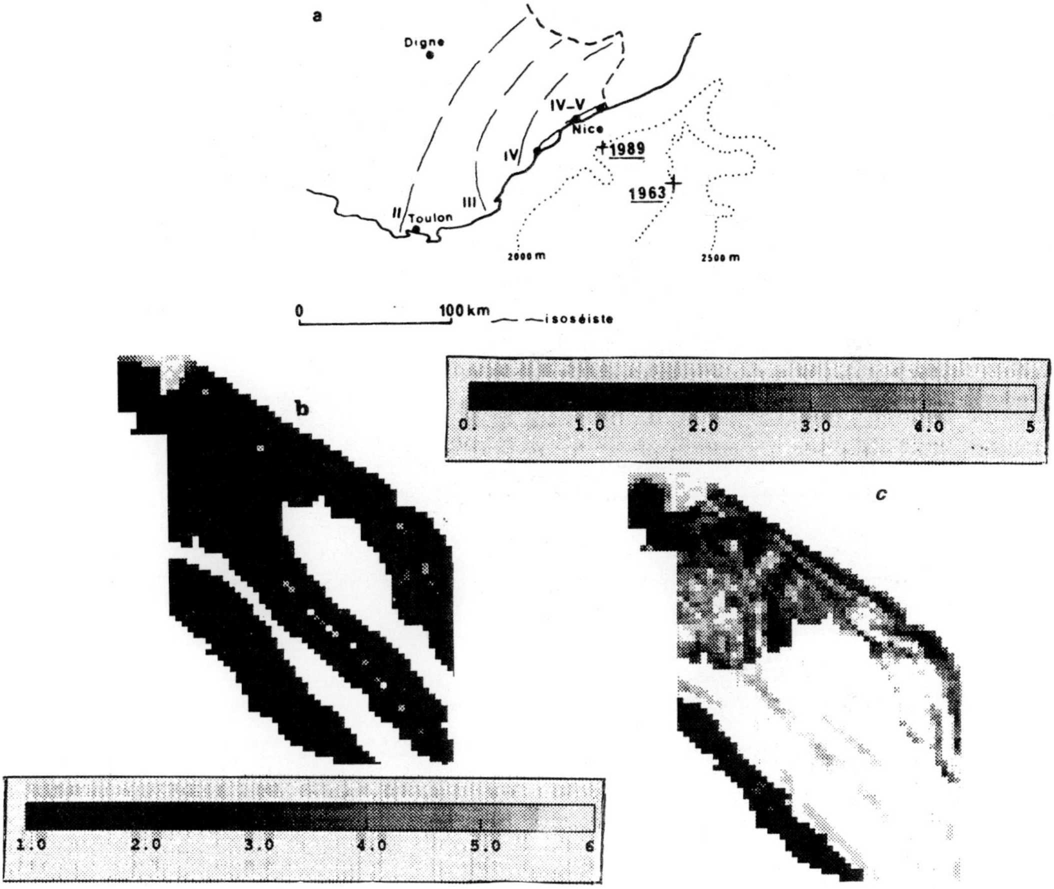

Figure 9 : 1989/26/12 earthquake. a : isoseimal lines (after [9]) ; b : C isovalues map ; c : F isovalues map (b and c : undrained conditions, z = 1m).

- the top-continental shelf instabilities, linked with the under-consolidated state of the 1 or 2 first meters of sediments.Their slumped volume is small but they are in the area where the human offshore infrastructures might be built. Because of their location where the sedimentation rate is high, they have probably a cyclic occurrence with a short return period.

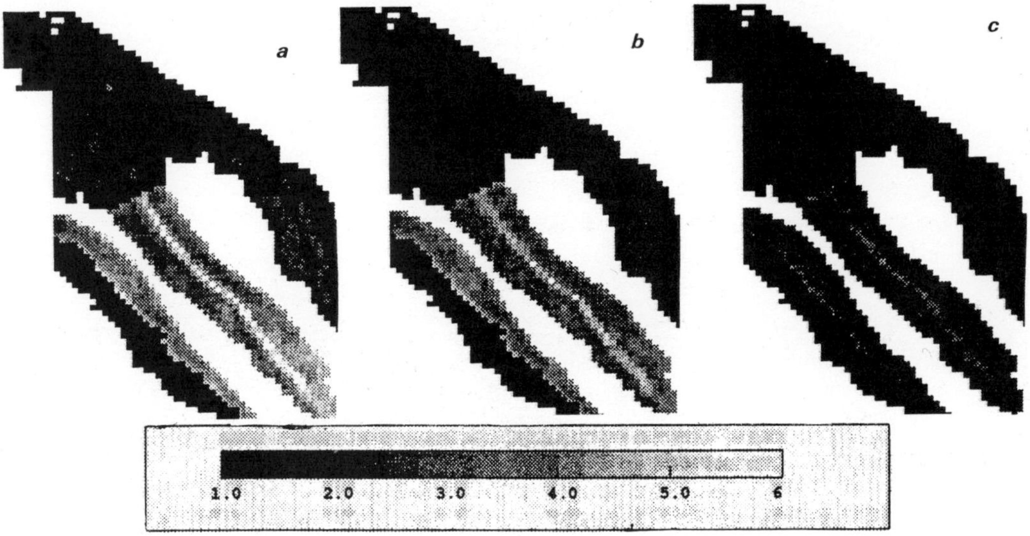

Figure 10 : values of F calculated with a centennial acceleration (a), a bicentennial acceleration (b) and a millennial acceleration (c). (undrained conditions, z = 1m).

- the ridge-side instabilities, affecting the normally-consolidated or the over-consolidated sediments. The depth of the failure surface is greater than 5m and the slumped volume is important. Located in an area with high slope gradient values, this kind of slump will preferentially degenerate into turbidity currents. They can run along an important distance (probably more than 100 km) before their deposit in areas where the slope gradient reaches some very low values (1°).

Secondly, the instabilities resulting of the seismical activity. A 0.12.g acceleration is enough to generate sedimentary disorders in the bay, first, on the sides of the central ridge. The created instabilities will preferentially be ridge-side instabilities. A millennial acceleration generates disorders in the whole bay.

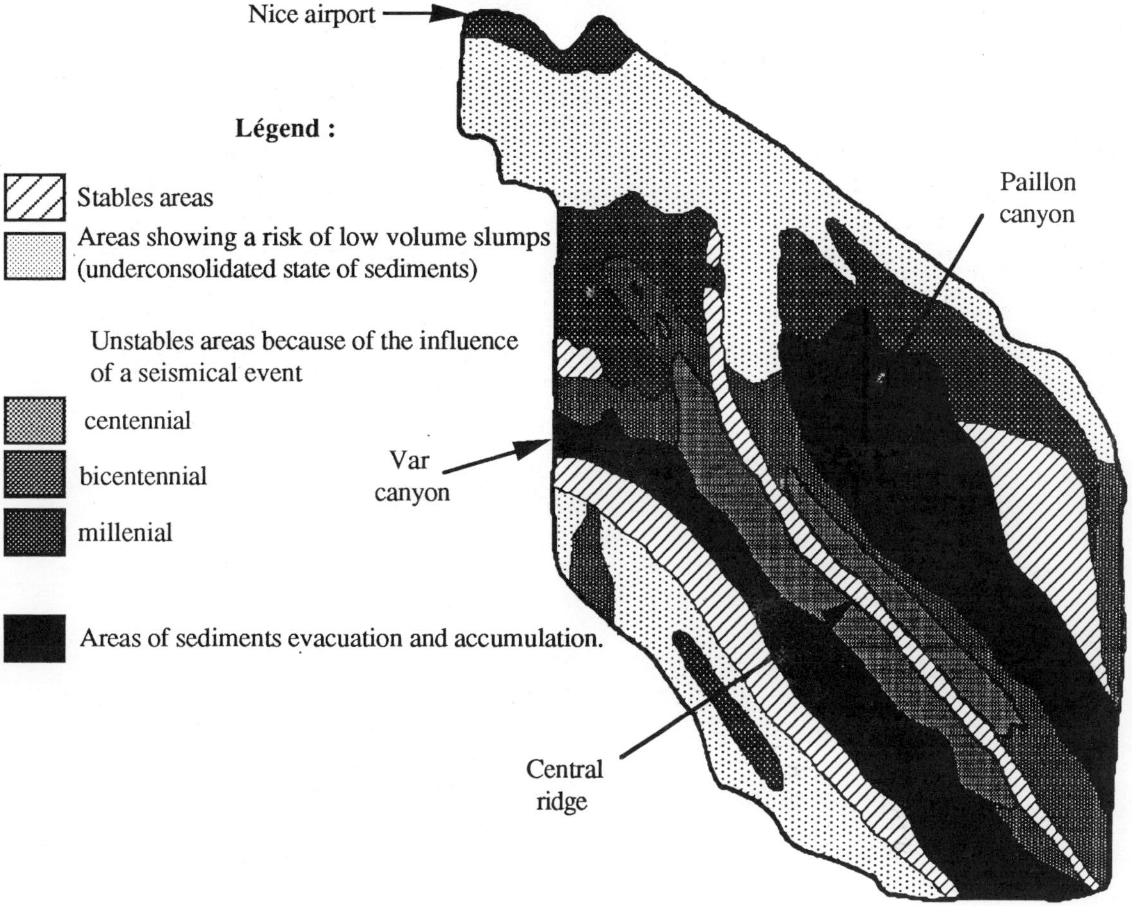

Figure 11 : synthetic map of offshore hazards in the *baie des Anges.*

7 Conclusions

The *Baie des Anges*, as the Mississipi delta ([22] and [23]) can be regarded as an example for the offshore instability studies. Here, the hazards have essentially a mechanical source. The use of the infinite plane sliding surface model provides a simplified but good description of the observed slumps on the submarine bottom. The interpretation under drained and undrained conditions are complementary. The pessimistic interpretation under undrained conditions shows the influence of the soil properties. An interpretation under drained conditions shows the influence of the slope gradient value.

A global instabilities study should have to take into account the creep phenomenon and the liquefaction risk. Further investigations have to be done in the bay to provide a better knowledge of the sediment and *in-situ* investigations will allows to be sure of the non-disturbance of the materials.

References

[1] Auffret G.A., Auzende J.M., Gennessaux M., Monti S., Pastouret L., Pautot G. and Vanney J.R., (1982). Recent mass wasting processes on the provencal margin (Western Mediterranean). Marine Slides and other mass movements, Proc. of a NATO Workshop on marine slides and other mass movements, Algarve, Portugal, 15-21 dec. 1980, Ser. 4, Vol. 6, S. Saxov et J.K. Nieuwenhuis ed., p. 53-58.

[2] Cochonat P, Schieb. Th., Guillaume J., Kerbrat R., Tisot J.P., Auffret G.A. and Muller C., (1988). Geotechnical and sedimentological properties of Nice Slope and submarine Var Canyon deposits. AAPG Convention.

[3] Savoye B., Cochonat P., Ollier G., Auffret G.A. and Bourillet J.F. (1989). The submarine Var Canyon (French Riviera) : Detailed study of the activity of a modern canyon. 10th regional meeting of sedimentology, Budapest, april 24-26, 1989.

[4] Schieb Th, Tisot J.P., Cochonat P. (1990). Propriétés géotechniques et étude de la stabilité des sédiments marins de la baie des Anges (Nice, Alpes Maritimes). A.S.F., 2ème congrès Fr. de Sédimentologie.

[5] Savoye B. and Piper J.W. (1990). The Messinian event on the margin of the Mediterranean sea in the Nice area, southern France. Marine Geol., 97, p. 279-304.

[6] Malinverno A., Ryan W.B.F, Auffret G., Pautot G. (1988). Sonar images of recent failure events on the continental margin of Nice, France. Geol. Soc. of America, Special paper, p. 59-74.

[7] Cochonat P., Dodd L., Bourillet J.F. and Savoye B. (1992). Geotechnical characteristics and instability of submarine slope sediments, the

Nice slope (N-W Mediterranean sea), a case study, (submitted to Marine Geotech..
[8] Schieb Th. (1992). Faciès géotechniques et état de consolidation des sédiments de la pente continentale niçoise (Baie des Anges),Thèse doct. I.N.P L. Nancy. 190 pages.
[9] Ritz J.F., Hoang-trong P., Rebai S., Philip H., Herquel G. (1990). Le séisme du 26 décembre 1989 en Méditerranée, au large de la côte d'Azur: tectonique active en compression, perturbation des contraintes et inversion tectonique au niveau d'une marge continentale. C.R.Acad.Sci. Paris, t.310, Série II, p. 1505-1512.
[10] Rehault J.-P., Bethoux N. (1984). Earthquake relocation in the Ligurian sea (Western Mediterranean): geological interpretation. Marine Geol., 55, p. 429-445.
[11] Willmore P.L. et Karnik K V. (1971). Manuel pratique des observatoires sismologiques. Centre Sismologique International, Edimbourg.
[12] Ambrasey N.N. (1978). Middle East - A Reappraisal of the Seismicity. J. Eng. Geol., Vol.11, p.19-32.
[13] Mac Guire R.K. (1978). Seismic Ground Motion Parameter Relations. J. of the Geotech.Eng.div, GT4, p. 481-490.
[14] Despeyroux J.(1990). Construction en zone sismique. Niveaux d'agression à considérer pour l'application des recommandations de l'AFPS relatives aux constructions à édifier dans les régions sujettes à séismes. Cahier technique de l'AFPS, n° spécial, "accélérations nominales", 48p.
[15] Godefroy P. et Levret A. (1985). Nature et définition de l'aléa sismique. Génie parasismique. Presses de l'Ecole Nationale des Ponts et Chaussées, Paris, pp. 159-177.
[16] Karlsrud K., Edgers L. (1982). Some aspects of submarine slope stability. Marine slides and other mass movents, Saxov et Nieuwenhuis edt., Plenum Press., N.Y., p. 63 - 82.
[17] Tisot J.P., Mulder Th. (1990). Méthodes d'études de la stabilité des talus sous-marins, Studia Geotechnica et Mechanica, Vol. X, N° 3-4, 3ème Colloque Polono-Français, 15-17 mai 1990, Wroclaw, p. 53-66.
[18] Mulder Th. et Tisot J.P. (1991). Méthode générale d'étude de l'instabilité des sédiments superficiels en milieu marin, 3ème congrès Français de Sédimentologie, Brest, 17-20 Novembre 1991.
[19] Mulder Th. et Tisot J.P. (1992). Méthode d'analyse des instabilités en mer. Exemple de la baie des Anges, Nice, France, 3ème R.S.T, Toulouse, 15-17 avril 1992.

[20] Mulder Th. (1992). Aspects géotechniques de la stabilité des marges continentales. Application à la baie des Anges, Nice, France. Thèse INPL Nancy.

[21] Mulder Th., Cochonat P., Schieb T. et Tisot J.P. (1992). Estimation de l'épaisseur des glissements sous-marins à partir de l'état de consolidation des sédiments. Application à la baie des Anges (S.E. de la france). C.R.A.Sc (submitted).

[22] Prior B. and Coleman J.M. (1978). Disintegrating retrogressive landslides on very low angle subaqueous slopes, Mississipi delta. Marine geotech., Vol 3, N°1, p. 37 - 60.

[23] Prior B. and Sunhayda J. N. (1979). Application of infinite slope analysis to subaqueous sediments instability, Mississipi delta. Eng. Geol., 14, p. 1-10 .

Thème 3

MODELIZATION AND COMPUTATIONAL ASPECTS

MODELLING OF THE SEISMIC RESPONSE OF SLOPES BY RIGID BLOCKS CONNECTED WITH ELASTIC JOINTS

Roberto ANDRIGHETTO* and Ezio FACCIOLI**

Abstract: A procedure for calculating earthquake-induced displacements in unstable natural slopes is presented, based on a generalization of the well known model which replaces the sliding mass by a single rigid block lying on an inclined plane. In the proposed method, the sliding mass is modelled by an ensemble of rigid blocks separated by joints with elastoplastic behaviour. This description allows to take into account both topographic irregularities and a limited deformability of the mass, entirely concentrated at the joints. Applications of the method to the analysis of two real slopes located in areas of high seismicity are also illustrated.

Riassunto: Viene illustrato un procedimento per il calcolo degli spostamenti permanenti causati da terremoti in pendii naturali instabili, basato su un'estensione del noto modello che assimila la massa franante ad un blocco rigido posto su un piano inclinato. Nell'estensione proposta la massa viene modellata per mezzo di un insieme di blocchi rigidi separati da giunti a comportamento elastoplastico. Questa schematizzazione consente di tenere conto da un lato delle irregolarità topografiche, dall'altro di una limitata deformabilità della massa in movimento, che viene interamente concentrata nei giunti. Viene presentata l'applicazione del metodo alla previsione del comportamento di due pendii reali posti in aree di elevata sismicità.

1. Introduction

Failures of unstable slopes due to earthquake ground motion can result either from an increase in the forces acting on the soil mass, or from a reduction of resistance due to pore pressure buildup, permanent displacements, loss of strength, etc.; they often occur immediately after the seismic shaking, but sometimes, e.g. for slopes in overconsolidated clays, weeks or months later [1,2,3].

Two parameters are alternatively used to quantify the degree of stability of a natural slope in seismic conditions: the factor of safety obtained from a pseudo-static approach, and the permanent displacement induced by the earthquake, estimated by a dynamic analysis. These quantities can be evaluated by different methods: the simplified ones [4] are typically based on strong assumptions

* Junior Engineer, Studio Geotecnico Italiano, Milan, Italy.

** Professor, Technical University of Milan, Department of Structural Engineering.

regarding the slope of the sliding surface, the forces acting on soil and the mechanism of motion, while the sophisticated ones make typically use of finite elements [5,6]. Although the simplified models are the most used in practice for computational convenience and the reduced number of parameters, predicting the seismic response of actual slopes often poses severe difficulties by both the previous approaches, because of complicated or unknown geometry, geotechnical and hydrogeological characteristics, seismic site effects.

The work presented in this paper is an attempt at extending a widely used simplified method for evaluating earthquake induced permanent displacements in slopes. It has been developed in order to overcome the difficulties related to the study of moderately disrupted soil masses, which behave as an ensemble of blocks. Landslides with a high degree of internal disruption will not be considered.

In the following we will first recall the single-block approach, which is at the basis of our model; this is illustrated next and, finally, some examples of application to the study of real landslides in seismic areas are presented.

2. Single block model for calculating seismic displacements

In the approach proposed by Newmark [7] for evaluating seismically induced permanent deformations in earth dams and embankments, also applicable to natural soil slopes, a soil mass moving under earthquake excitation on a sliding surface is represented by a rigid block on an inclined plane, subjected to the inertia force (Fig. 1). Extensions of this model allow to treat in a simplified way some of the complexities of real landslides, by assuming for the slip surface a circular or logspiral shape, or by introducing topographic irregularities [3,8].

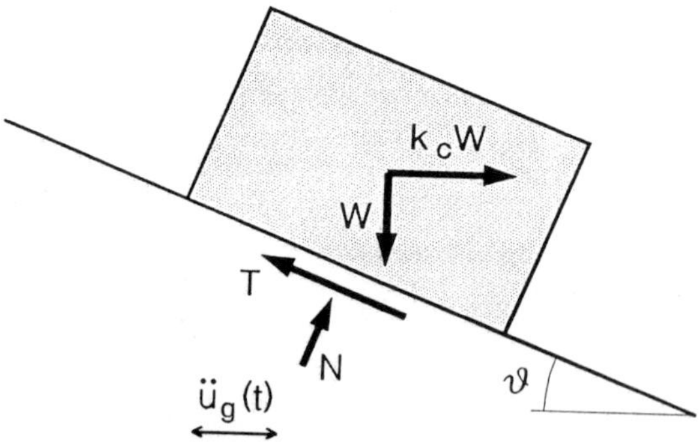

Fig. 1 : Geometry of sliding mass for the single-block model, and forces acting in limit equilibrium.

The two main elements which substantially influence the result of the analysis are the geometrical model for the slope, and the value of the critical acceleration coefficient k_C; both depend on the geometry of the failure mechanism. The critical acceleration coefficient is defined as the ground acceleration value for which the soil mass is in a condition of incipient movement, divided by *g*. This coefficient condensates in a single number all the information on geometrical, geotechnical and loading conditions. Formulas for k_C for inhomogeneous slopes with irregular sliding surface are given in [3]. A different approach, recalled in the following paragraph, has also been proposed for determining k_C, which considers the possibility of shear failures inside the sliding mass, and therefore of the mass itself being subdivided into several blocks [9].

The main assumptions of the single-block model require the absence of significant loss in shear strength during seismic shaking, and the soil mass behaving as one coherent block.

3. Multi-block model for calculating seismic displacements

For landslides not easily amenable to a single coherent block, it is natural to resort to a multi-block representation of the unstable mass to match more closely the actual sliding mechanism and the topographic surface. Herein, in particular, we describe a triple-block model, that has proved sufficiently adaptable to the description of real landslides; extension to a greater number of blocks is, however, easy. The model simulates the motion of a soil mass subjected to seismic loading by studying the motion of three interacting rigid blocks lying on plane surfaces, and separated by elastic joint interfaces (Fig. 2).

The main assumption for applications is that the failure mechanism must be translational. We suppose that the geometry of the slip surface is known, and that the material of the slope is not subject to loss of shear strength, nor to pore pressure increase during shaking.

3.1 DESCRIPTION OF MODEL

The first step consists of approximating the geometry of the failure surface by three straight segments (Fig. 2). The soil mass between the slip surface and the ground surface will be then subdivided into three slices (or blocks), according to the internal failure mechanism. Since this is generally unknown, the inclination of the surfaces separating the blocks is also unknown, and it is therefore necessary to find the critical values of the angles β_1 and β_2 in Fig. 2. These correspond to the minimum degree of stability and, following [9], we assume that they are associated to the smallest critical acceleration coefficient k_C. The value of k_C is taken to be the same for all the blocks, and is determined by limit equilibrium conditions imposing Mohr-Coulomb failure criterion at the interfaces separating the blocks from the support and from each other. This implies that the blocks start moving simultaneously, i.e. that the strength is mobilized at the same time along the whole sliding surface; this does not occur in reality, as the stresses tend to reach first the yield limit near the foot of the slope. The resulting expression for k_C is:

$$k_C = \frac{\sum_{i=1}^{3} \delta_i [W_i \sin(\varphi_i' - \vartheta_i) + (c_i' l_i - U_i \tan\varphi_i') \cos\varphi_i'] + \sum_{j=1}^{2} (\bar{c}_j' d_j - \bar{U}_j \tan\bar{\varphi}_j') \xi_j}{\sum_{i=1}^{3} W_i \delta_i \cos(\varphi_i' - \vartheta_i)} \qquad (1)$$

where:

$$\xi_i = \sum_{i=j}^{j+1} (-1)^{i+j} \delta_j \sin(\beta_j + \varphi_i' - \vartheta_i) \qquad (j = 1,2)$$

$$\delta_1 = \frac{\cos(\beta_1 + \varphi_2' - \vartheta_2 - \bar{\varphi}_1')}{\cos(\beta_1 + \varphi_1' - \vartheta_1 - \bar{\varphi}_1')} \qquad \delta_2 = 1 \qquad \delta_3 = \frac{\cos(\beta_2 + \varphi_2' - \vartheta_2 - \bar{\varphi}_2')}{\cos(\beta_2 + \varphi_3' - \vartheta_3 - \bar{\varphi}_2')}$$

in which β_j and θ_i are the angles illustrated in Fig. 2, ϕ'_i, c'_i, U_i, l_i denote the friction angle, the cohesion, the water pressure at the base of block *i*, and the length of the base, respectively; $\bar{\phi}'_j$, $\bar{c}_j'$, $\bar{U}_j$, d_j are the same quantities along the surfaces separating the blocks.

The forces acting in limit equilibrium conditions are shown in Fig. 3. Beside the weights of the blocks and the inertia forces (proportional to the acceleration k_C), there are the reactions between the base and the blocks, T'_i and N'_i, and the forces exchanged between the blocks themselves, X'_i and E'_i, respectively tangential and normal to the separation surface. The calculation of k_C must be performed iteratively, by varying the values of β_1 and β_2 until a minimum is found; the corresponding geometry of the failure mechanism will be used in the subsequent steps of the analysis.

Beyond the limit equilibrium condition, each of the blocks is assumed to slide without rotation along its support, under the forces shown in Fig. 4. In addition to the forces acting in limit conditions, there are now also the inertia forces due to relative motion of the blocks, which are considered to act parallel to the support of each block. This approximation violates kinematic compatibility but can be considered acceptable for translational sliding mechanisms with inclination of consecutive supports differing less than 15°- 20°. Moreover, displacements must be small with respect to the block dimensions (≅ 1/100), in order to have

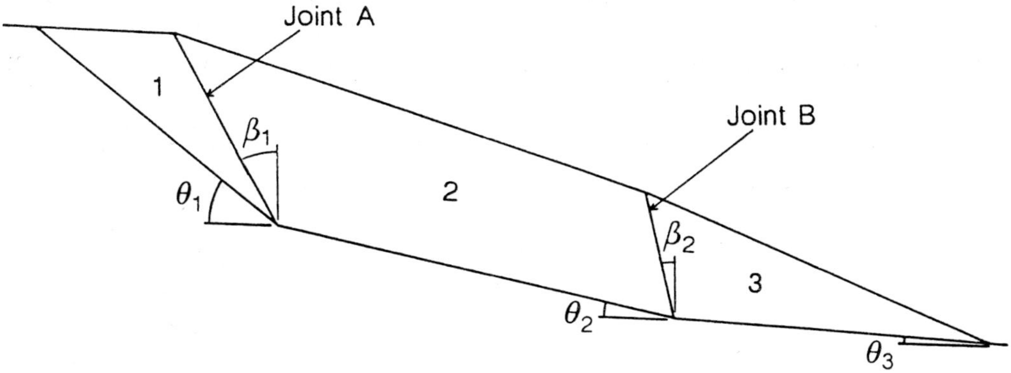

Fig. 2: Triple-block geometrical model of sliding mass.

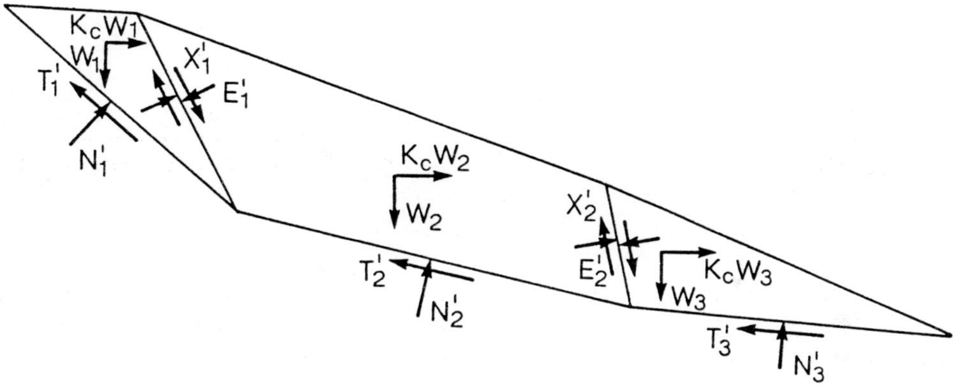

Fig. 3: Forces acting in the triple-block model in limit equilibrium.

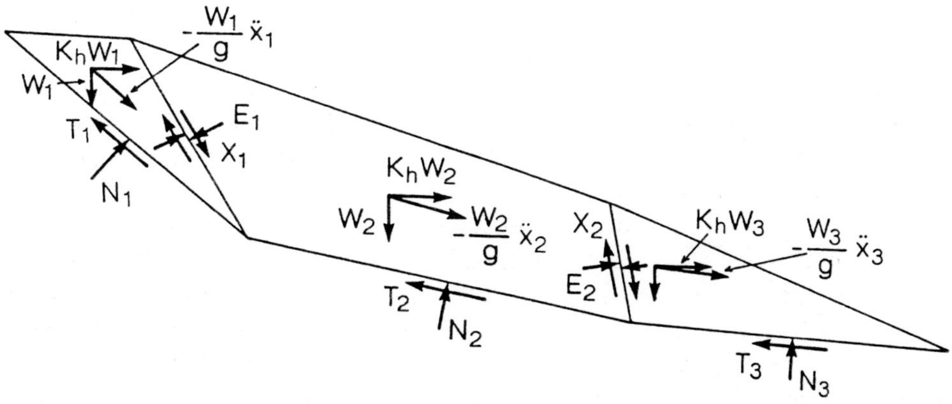

Fig. 4: Forces acting on the blocks during motion caused by the ground acceleration $\ddot{u}_g(t)=k_h(t)\,g$; the relative accelerations $\ddot{x}_1$, $\ddot{x}_2$, $\ddot{x}_3$ are assumed parallel to the base of the blocks.

negligible penetration of the blocks into the base (not taken into account). Another important assumption, usually introduced also in the single block approach [3,8,11], is that the blocks are only allowed to move downslope since their resistance to upslope sliding is sufficiently large to make its occurrence unlikely. This hypothesis evidently leads to maximizing the calculated permanent displacement.

The equations of motion of the three blocks are obtained by subtracting the limit equilibrium equations from those of dynamic equilibrium [8]. During all phases of motion the forces acting at the base of each block in normal and tangential direction are supposed to satisfy the Mohr-Coulomb failure criterion.

It follows from this approach that relative motion depends only on the fraction of ground acceleration and of inter-block forces in excess with respect to the limit conditions, as shown by the dynamic equations of the three blocks which have the form

$$\ddot{x}_i = \frac{g}{\cos\varphi_i'}(k_h - k_C)\cos(\varphi_i' - \vartheta_i) - \frac{g}{W_i\cos\varphi_i'}[(E_{i-1} - E_{i-1}')\cos(\vartheta_i - \varphi_i' - \beta_{i-1}) - (X_{i-1} - X_{i-1}')\sin(\vartheta_i - \varphi_i' - \beta_{i-1}) - (E_i - E_i')\cos(\vartheta_i - \varphi_i' - \beta_i) + (X_i - X_i')\sin(\vartheta_i - \varphi_i' - \beta_i)] \quad (i = 1,3) \qquad (2)$$

E'_i, X'_i, T'_i, N'_i, E_i, X_i, T_i, N_i are the forces illustrated in Figs. 3 and 4, and $k_h = k_h(t)$ is the base acceleration time history (expressed as a fraction of g).

The forces E_i and X_i exchanged between adjacent blocks depend on the deformability of the separation joints. These can be represented by the physical model illustrated in Fig. 5, formed by an elastic spring (a) (prestrained by an amount Δ_0 at onset of relative motion) and a friction element (b). The prestraining of the spring expresses the condition that, at the onset of relative motion, the inter-block force must equal the thrust F_o (Fig. 6) exerted when the adjacent blocks move as a single body. F_o is calculated by the dynamic equilibrium equations.

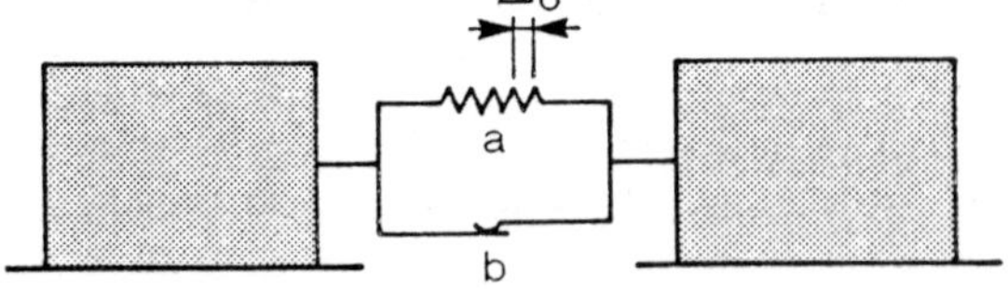

Fig. 5: Physical analogue for the elastoplastic separation joints: (a) is an elastic spring, prestrained by Δ_o at onset of relative motion, (b) is a friction element.

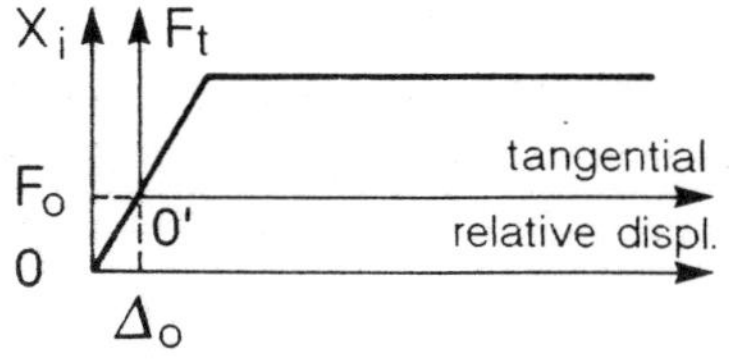

Fig. 6: Effect of prestraining, Δ_o, of the spring on the shear force X_i at the joint.

Dilatance phenomena are disregarded for the separation joints, as well as for the interfaces at the base of the blocks.

To minimize interpenetration between two adjacent blocks, high values are assigned to the stiffness of the joints; this leads to upper bound values for the calculated permanent displacements.

Residual values can be used for the soil strength parameters, for the sake of conservatism. More sophisticated relationships, taking into account the variation of shear strength with displacement and with normal stress, such as proposed in [12], are more appealing for newly formed landslides, while residual values are generally suited for quiescent slides.

The system of equations of motion (2), reduced to the matrix form

$$\boldsymbol{A}\ddot{\boldsymbol{x}} + \boldsymbol{B}\boldsymbol{x} = \boldsymbol{Q} \tag{3}$$

is integrated step-by-step by Wilson's θ method for a given acceleration input, which typically consists of a real strong-motion accelerogram applied to the support. The results are represented in terms of the horizontal projection of the cumulative displacement calculated for each block versus time. In many applications the critical quantity is the final displacement at the foot of the slope, which should be compared with an allowable value. This can be related to the serviceability requirements of structures erected on the slope, but also to the post-earthquake stability of the slope itself; opening of cracks and subsequent water percolation could in fact trigger failure long after the seismic event. There are very few studies on this subject, and the existing ones refer to regional conditions. An example of criteria applicable to natural slopes, proposed by the State of Alaska Geotechnical Evaluation Criteria Committee, is reported in [13].

4. Applications

Two applications of the illustrated triple-block approach are presented, concerning the study of an existing landslide and of a potentially unstable slope, both located in Southern Italy.

The first case history refers to an existing landslide (now stabilized) which affected a motorway in the Southern Apennines; this is one of the Italian regions of high seismicity and the site is about 15 km from the epicentral area of the M_S 6.9, 1980 Irpinia earthquake. A cross-section of the landslide is illustrated in Fig. 7. The sliding surface has been determined from geotechnical borings, and inclinometer measurements, to lie between two formations of high plasticity clays; the base formation consists of structurally complex ("Variegated") clays. The water table level is very close to the ground surface for the whole year, so that it is possible to assume saturated conditions for the analyses. The factor of safety in static conditions is slightly lower than one. The mobilized strength along the sliding surface can be assumed to be very close to the residual one, so that residual values of the resistance parameters are used.

The adopted triple-block approximation to the actual landslide geometry is illustrated in Fig. 8. Based on the results of geotechnical tests, a friction angle $\phi_r = 7°$ and zero cohesion were assumed on the sliding surface, and a slightly larger value ($\phi_r = 8.5°$) along the joints between the blocks.

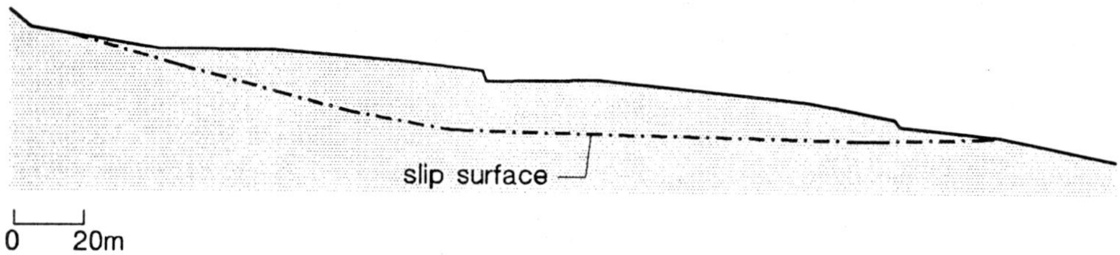

Fig. 7: Cross-section of the landslide in structurally complex clays.

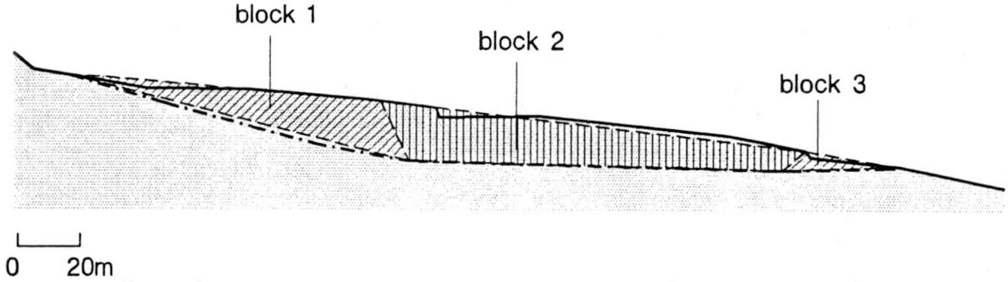

Fig. 8: Representation of the landslide in Fig. 7 by three rigid blocks.

A minimum value of the critical acceleration coefficient k_C= - 0.046 is obtained (corresponding to the inclination of the joints in Fig. 8), which implies that a Newmark type analysis is not feasible in this case. An advantage of the proposed method with respect to the single-block model is that the slope response to a given earthquake can be evaluated even in the presence of a negative k_C, because the motion of the blocks depends mainly on the forces mutually exerted between them.

A seismic hazard analysis was conducted for the landslide site under consideration, since the permanent deformation following a significant earthquake could cause local disruption of an important lifeline. Considering the seismicity of the region, with destructive earthquakes occurring roughly every 100 to 200 years, and the nominal economic lifespan of the lifeline (at most 50 years), the peak ground acceleration (PGA) corresponding to 100 years return period, namely 0.18 g, was considered appropriate for the seismic stability analyses. Some accelerograms with peak value on hard ground close to 0.18 g, recorded during the 1980 earthquake, were used as base excitations. The results of one calculation are displayed in Fig. 9, which shows that the downslope motion of the blocks stops well in advance of the ground shaking, in spite of the negative value of k_C; note that a slight penetration between the blocks is predicted by the analysis.

The calculated permanent displacement indicates that no catastrophic failure of the slope is to be expected. However, substantial motion could take place after the earthquake, because of the redistribution of pore pressure, contributing strongly to the final permanent displacement. It is also noted that a displacement of about 20 cm, as indicated in Fig. 9 for the foot of the slope (block 3), would certainly cause an important loss of serviceability in the affected motorway.

A different, instructive application of the proposed model was made to the analysis of an offshore slope on the Ionian coast of Calabria, in Southern Italy. Extensive offshore investigations (boreholes, SPTs and CPTs) were carried out for the geotechnical characterisation of the slope, within a repowering project for a conventional energy plant located on the sea shore. A conventional static analysis was performed first, indicating the possibility of formation of a rupture surface within the slope, at the boundary between two layers of sandy gravel of different density (Fig. 10).

The potentially unstable soil mass has been modelled by the three blocks illustrated in Fig. 11; the critical mechanism of failure corresponds to internal rupture surfaces dividing the mass in such a way that one of the blocks has almost negligible dimensions with respect to the others. The k_C value, calculated by (1), is equal to 0.049.

Geotechnical tests provide a friction angle of 29° for the material shown in Fig. 10 as very loose sandy gravel, and, conservatively, a vanishing cohesion is assumed.

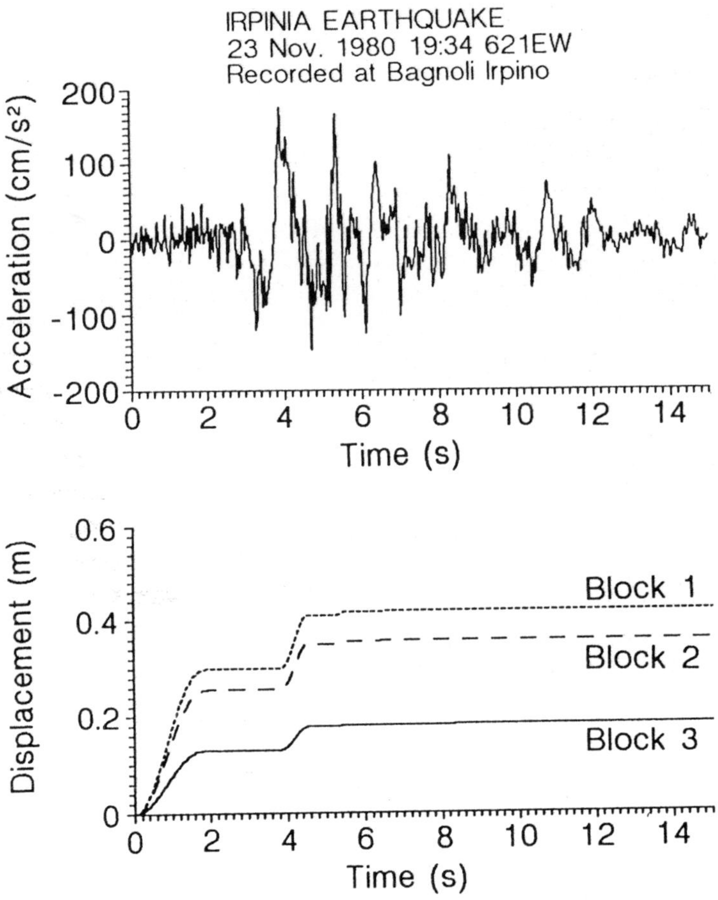

Fig. 9: Input accelerogram and computed permanent displacements for the triple-block model in Fig. 8.

Considering an economic lifespan of 30 years for the adjacent power plant, the earthquake with unit probability of exceedance in 100 years was indicated by the owners as appropriate for the slope stability analysis in operating basis conditions. A PGA=0.10 *g* was estimated through a seismic hazard analysis for this earthquake, and virtually zero permanent slope displacements are obtained for this condition.

A second slope response anlysis was carried out for base excitations with a peak value of 0.27 *g*, estimated to correspond to 500 years return period.

The results of one such analysis, performed using as input a real accelerogram recorded during the 1980 Irpinia earthquake, are shown in Fig. 12. The

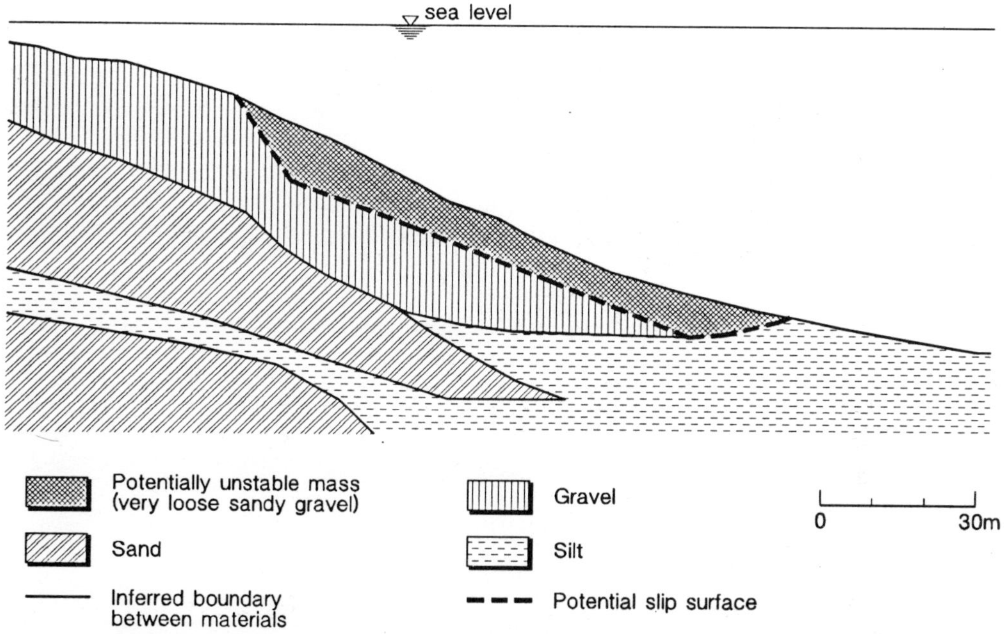

Fig. 10: Simplified cross-section of the offshore slope.

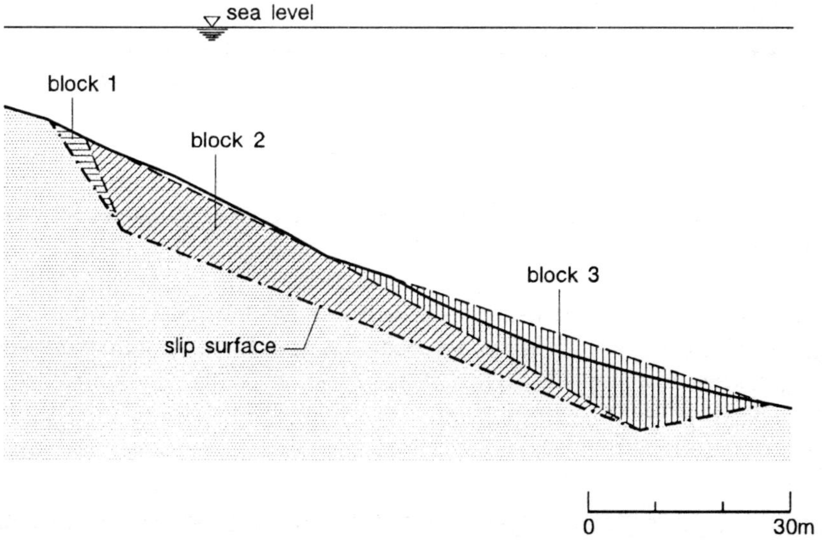

Fig. 11: Approximate representation of the unstable soil mass in Fig. 10 by 3 rigid blocks.

displacement time histories indicate that the motion of the system is governed by the central block; the downslope block acts as a stabilizing element, and moves only due to the thrusting of the central block, while the motion of the upslope block has little influence on the sliding, due to its negligible mass. At the end of shaking, the permanent displacement of the foot of the slope is less than 10 cm.

A single-block analysis has been also carried out for comparison because motion in the presence of three blocks is such that the downslope block moves only if pushed by the central one and, therefore, the two can be approximatively treated as a single mass.

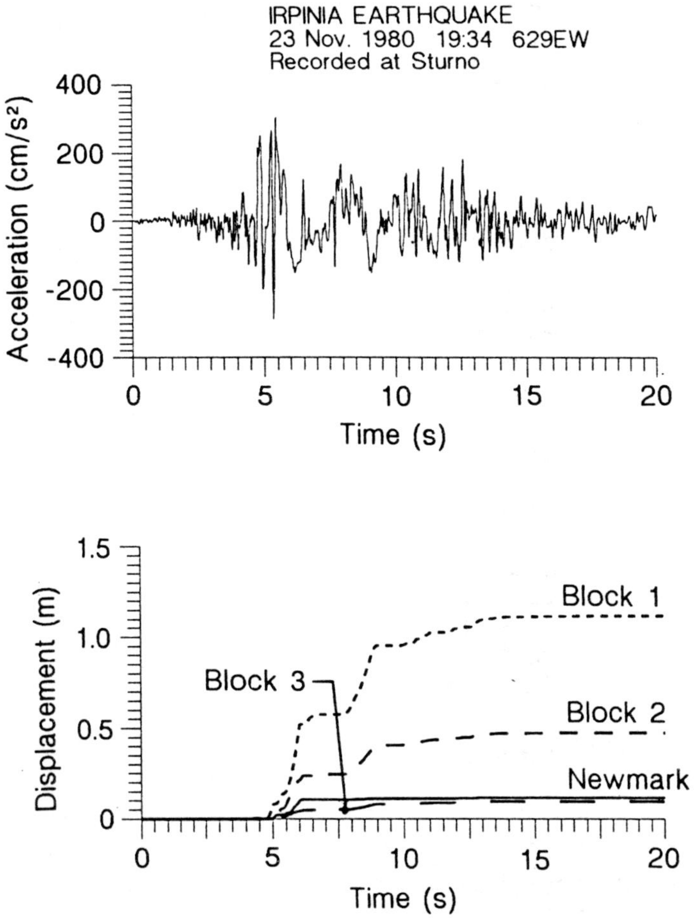

Fig. 12: Input accelerogram and computed permanent dispacements for the triple-block model in fig. 11 and for the single-block Newmark approach.

The previous value of k_C is quite low, due to instability of the upper block, and leads to overestimating the final displacement. Assuming that the whole mass acts as a single block, the value of k_C can be estimated by the corresponding exact solution (for horizontal base acceleration), valid also in the case of an infinite slope:

$$k_C = (Fs - 1)\,\frac{\tan\vartheta}{1 + \tan\varphi'\tan\vartheta} \tag{4}$$

where Fs is the static factor of safety, θ is the inclination of the support (see Fig. 1) and ϕ' is the friction angle [14]. The resulting value of k_C is 0.113 (Fs being equal to 1.38). The permanent displacements obtained are not very different from those of the foot of the slope, evaluated using three blocks, as shown in Fig. 12.

5. Conclusions

From the foregoing analysis it appears that the multi-block model can be used as a valid alternative to the Newmark single-block approach when the actual kinematics of the different parts of a sliding soil mass under earthquake loading are likely to be important.

The proposed method should in particular prove useful in the study of landslides that are expected to break up into few coherent blocks during seismic motion. Moreover, it gives the possibility of analyzing motion in slopes with a negative critical acceleration k_C (i.e. "statically" unstable), which otherwise could not be examined.

This study has also pointed out that the determination of k_C has a critical importance for estimating the displacement by the standard Newmark method, which means that particular attention must be devoted to modelling the failure mechanism.

A limitation of the multi-block approach is that the critical acceleration must be the same for each block, while in reality different parts of a landslide do not usually start moving at the same time.

Finally, for all methods using rigid blocks, further research is required in order to account for pore pressure build-up in the soil due to cyclic earthquake loading.

Acknowledgments

The cooperation of Studio Geotecnico Italiano is gratefully acknowledged; the figures were prepared by S. Bacci. This work has been supported by the European Community, under EPOCH Research Program n° CT91 - 0046.

References

[1] Hansen A. and Franks C.A.M. (1991). *Characterisation and mapping of earthquake triggered landslides for seismic zonation.* Proceedings Fourth International Conference on Seismic Zonation, Stanford, vol. 1, p.149-195.

[2] Ishihara K. and Hsu H.L. (1986). Considerations for landslides in natural slopes triggered by earthquakes. *Proceedings Japan Society of Civil Engineering (Geotechnical Engineering)*, n° 376/III-6, p. 1-16.

[3] Crespellani T., Ghinelli A., Madiai C. and Vannucchi G. (1990). *Analisi di stabilità dei pendii naturali in condizioni sismiche.* Florence, Italy (in Italian).

[4] Matasovic N. (1991). *Selection of method for seismic slope stability analysis.* Proceedings Second International Conference on Recent Advances in Geotechnical Earthquake Engineering and Soil Dynamics, St.Louis, paper n° 7.20, p. 1057-1062 .

[5] Toki K., Miura F. and Oguni Y. (1985). Dynamic slope stability analysis with a non-linear finite element method. *Earthquake Engineering and Structural Dynamics*, vol. 13, p. 151-171.

[6] Succarieh M.D., Lin J.I. and Elgamal A.W.M. (1991). *Computation of Dynamic large displacements in earth systems.* Proceedings Second International Conference on Recent Advances in Geotechnical Earthquake Engineering and Soil Dynamics, St.Louis, paper n° 7.9, p. 1000-1012 .

[7] Newmark N.M. (1965). Effects of earthquakes on dams and embankments. *Geotechnique*, vol. 15, n° 2, p. 137-160.

[8] Chang C.J., Chen W.F. and Yao J.T.P. (1984). Seismic displacements in slopes by limit analysis. *Journal of Geotechnical Engineering, ASCE*, vol. 110, n° 7, p. 860-874.

[9] Sarma S.K. (1979). Stability analysis of embankments and slopes. *Journal of the Geotechnical Engineering Division, ASCE*, vol. 105, n° 12, p. 1511-1524.

[10] Andrighetto R. (1991). *Metodi semplificati per la modellazione della risposta sismica dei pendii.* Thesis for Civ. Eng. Degree, Technical University of Milan (in Italian).

[11] Franklin A.C. and Chang F.K. (1977). Earthquake resistance of earth and rock-fill dams. *Soils and Pavements Laboratory, U.S. Army Engineer Waterways Experiment Station*, Report 5, Vicksburg.

[12] Cividini A., Pergalani F. and Petrini V. (1991). La risposta dei versanti ad azioni sismiche attraverso un modello semplificato. *Ingegneria Sismica*, n° 3, p. 28-44 (in Italian).

[13] Idriss I.M. (1985). *Evaluating seismic risk in engineering practice.* Proceedings 11th international conference on soil mechanics and foundation engineering, San Francisco, vol. 1, p.255-320.

[14] Battistella C. and Faccioli E. (1991). *Studi di pericolosità sismica ed instabilità dei pendii in zone appenniniche.* Atti dellle Conferenze di Geotecnica di Torino, XV ciclo, vol. 15.02 (in Italian).

DETERMINATION OF ARTIFICIAL BEDROCK MOTIONS FOR DYNAMIC SLOPE STABILITY ANALYSIS

Giovanni BOSCO* and Caterina VIGGIANO**

Abstract: The evaluation of the stability conditions of natural slopes, cuts and earth embankments located in seismic regions requires the performance of several different analyses that pose difficult questions. Among these, one that certainly deserves further attention is the estimation and representation of ground shaking that can be expected to occur during the design life of the slope. In this paper a procedure developed to determine artificial acceleration time histories representative of bedorock motions induced by past events is presented. The procedure has been used to determine artificial motions for the 1980 Irpinia event. The artificial and real accelerograms have been integrated twice to simulate a slope stability analysis. The results of this analysis are also presented and discussed in the paper.

Sommario: Per valutare le condizioni di stabilità di un pendio naturale, di uno scavo o di un rilevato di terra ubicati in una zona ritenuta sismica, è necessario sviluppare una serie di analisi che pongono differenti tipi di problemi. Fra questi, un aspetto che merita tuttora notevole attenzione è quello della valutazione e rappresentazione delle sollecitazioni che possono essere indotte dal sisma in un prefissato arco di tempo. Con riferimento a tale problema nel presente lavoro si illustra una procedura per la determinazione di accelerogrammi di progetto artificiali. Tale procedura è stata impiegata per la determinazione di accelerogrammi artificiali relativi al terremoto del 1980 verificatosi in Irpinia. Gli accelerogrammi artificiali così ricostruiti, e quelli registrati nel corso della scossa, sono stati integrati nel tempo per simulare un procedimento di analisi utilizzato nelle verifiche di stabilità di pendii in presenza di sisma. I risultati di tali integrazioni sono riportati e discussi.

1 Introduction

The potential for damages and loss of lives of strong or even moderately strong earthquakes has long been recognized all over the world. However, due to the many and complex aspects involved in the study of this natural phenomenon, it was not untill recently that researchers could shed some light on the difficult task of understanding and modeling some of the processes causing the occurence of earthquakes as well as their effects on the response of soils and structures.

----

* Research Engineer Università Degli Studi Della Basilicata, Potenza

** Engineer Università Degli Studi Della Basilicata, Potenza

Nevertheless, despite the extensive studies and research developed in the field of earthquake engineering several problems are still far from being completely solved in all their aspects. Among these the estimation of the seismic actions that may be induced in natural slopes, cuts, and earth embankments, in the occurrence of an earthquake, certainly deserves further investigation.

The evaluation of the safety factor of a slope with respect to its stability against sliding under earthquake loading conditions requires a rather complex analysis that has to be developed through a sequence of different steps. These include a study on the seismicity of the area to locate sources of earthquake that may affect the stability of the slope, the definition of appropriate loading time histories to represent bedrock motions expected underneath the slope, the detrmination of stresses and strains induced in the slope and, eventually, an evaluation of the resulting permenent displacements also on the basis of appropriate laboratory test data on representative soil samples.

However while most of these aspects have been examined in depth, the estimation of the chacarteristics of ground motions induced by earthquakes still arises several difficult questions. In fact procedures and criteria to obtain undisturbed samples, representative of the different soil types that can be encountered in natural slopes, as well as to perform laboratory tests under dynamic loading conditions have been well established and are now widely used. At the same time several numerical models to represent the response of soils under pulsating loading conditions as well as numerical techniques to perform dynamic analyses of soil deposits, slopes and embankment dams subjected to ground shaking have been developed and are today readily available for use in engineering practice. On the contrary the problem of selecting or reconstructing a time history of loads representative of the ground motion induced by an earthquake at a given site is still a matter of research.

In this respect it should be osbserved that to determine ground shaking at any given site two different types of problems must be faced. Firstly the intensity and characteristics of the motion that can reasonably be expected within a given period of time must be estimated. Secondly appropriate acceleration, velocity or displacement time histories representative of the estimated ground motion must be selected to express numerically the transient loads that may be induced by the earthquake.

Thus to provide an improved basis for evaluating the stability of slopes in seismic regions, the problem of determining time histories representative of the motions induced by an earthquake at a given site, where records of past events are not available, has been investigated. On the basis of the results of this study, a procedure that can be used in practice to reconstruct artificial accelerograms relative to a given earthquake has been developed. This procedure, described herein, can be applied, in principle, to any event for which records of ground shaking are available and allows to determine acceleration time histories representative of bedrock motions for any given distance from the earthquake source.

Up to this stage of the research, this procedure has been tested using the available records of the 6.5 magnitude Irpinia earthquake that struck Southern Italy in November 1980. In this respect it should be pointed out that for the geological and seismic characteristics of the region, this earthquake has been considered as well representative of the tectonical setting of the area. Also the registered magnitude is probably the highest that can be expected for events generated by sources located in this part of the country.

To check the reliability and quality of the results obtained with respect to the stability analysis of slopes in seismic areas, some artificial accelerograms have been reconstructed for sites where acceleration time histories were recorded during the 1980 Irpinia earthquake. Subsequently, the artificial and real acceleration time histories have been integrated twice, with respect to time, in order to simulate the application of the Newmark approach [1] to evaluate the stability of a slope. The results of this double integration process, carried out assuming different values of the yield acceleration, to scan the different parts of the acceleration records, are also presented and discussed in the paper.

2 Determination of artificial bedrock motions

The procedure presented herein has been developed on the basis of some requirements set forth in the attempt to minimize loss of information about the real event and to keep as much as possible unaltered its fundamental characteristics. Hence to reconstruct artificial motions resembling as closely as possible those induced by the 1980 Irpinia earthquake the following parameters were selected to describe the original event: maximum horizontal acceleration, response spectra, duration, shape of the acceleration time history. The latter was defined by means of an envelope curve encompassing the entire record.

All these parameters were obtained from acceleration time histories on outcropping bedrock, back calculated using the accelerograms recorded at ground surface during the 1980 event. It was considered, in fact, that accelerograms can provide an insight into the fundamental characteristics of a motion. Moreover they are widely used when performing dynamic analyses of slopes or structures in seismic regions. Of course the information contained in such records can be considered well representative of the real acceleration time history only if the presence of the instrument in the ground does not alter significantly ground shaking and the differences between real and recorded motions, resulting from the processes of recording, digitizing and correcting the raw data, are minimal and can, therefore, be neglected.

Motions on outcropping bedrock have been preferred in order to obtain consistent data regardless of the position of the recording station and also because accelerations recorded at ground surface bear the modifications due to the propagation of the seismic waves through the compressible soil layers overlaying bedrock. Thus, acceleration time histories for use in the analysis were obtained after deconvolution of surface motions filtering out any information that was not related directly to the inherent characteristics of the source earthquake.

It is worth mentioning that at this stage of the research deconvolution has been carried out using a numerical procedure developed for horizontally layered profiles [2]. Also the soil profiles at the different sites of the recording stations were reconstructed on the basis of the available data on the geology of the region affected by the earthquake. These data were derived from the literature since results of detailed soil explorations are not yet available. Due to the lack of dynamic laboratory tests, data for soils encountered at the sites selected were chosen from typical relationships for shear modulus and damping factor versus strain, presented in the literature for different types of soils [3].

Once the bedrock motions induced by the earthquake had been computed, the parameters selected to describe the source event could be readily obtained. The definition of appropriate correlations to determine any of these parameters as function of the distance from the earthquake source was then attempted. It should be pointed out, however, that while sets of numerical data, such as values of accelerations and relative distances from source, can be easily correlated, the definition of relationships to determine a response spectrum or the shape of an accelerogram for a given value of the distance from the earthquake source is not a simple and straightforward process. In fact a curve cannot be represented uniquely by a single numerical value, no matter if the peak of the curve, the average or the slope along a well defined interval are used.

Therefore in order to define appropriate correlations to reconstruct the curves of the response spectra and of the envelope of the acceleration time histories, given the distance from the earthquake source, a technique that can be applied to both types of cuves was specifically developed. The proposed technique involves, firstly, a discretization of the curves in several small intervals as shown schematically in fig.1. Each interval is then assigned a single numerical value equal to the area of the interval. Since each curve representing a response spectrum or the shape of an accelerogram pertains to a single record and thus to a well defined value of the distance from the earthquake source, the areas of the different intervals can be plotted against the relative values of distance from the source of motion.

As a result two sets of curves relating the distance from the earthquake source with the area of the first, second, third ... interval for the response spectra and shape of accelerograms were constructed. As an example the cor-

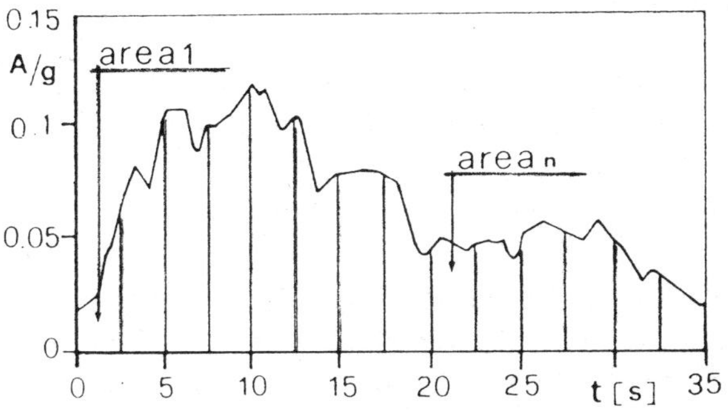

Fig. 1: Description of procedure for discretizing curves in finite intervals to obtain correlations with distance from source.

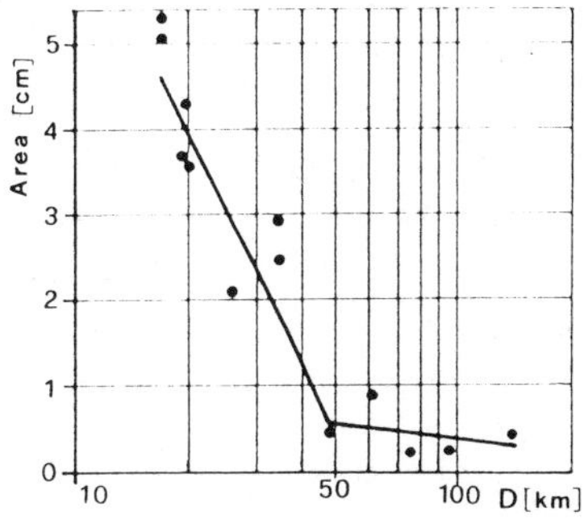

Fig. 2: Area 1 of pseudovelocity response spectra vs distance from source.

relations shown in figg. 2 and 3 allow to determine the values of area 1, as a function of distance from the earthquake source for pseudovelocity response spectra and envelope curves of accelerograms respectively. Thus, for any given value of distance from source, the area of the different intervals underneath the curve being reconstructed can easily be read off from plots as those shown in figg. 2 and 3. Since the width of each interval along the time (fig. 2) or frequency (fig. 3) axis is kept constant, the value of the area defines uniquely a point of the curve that has to be reconstructed.

Once correlations as those shown have been established, these can readily be used, together with the curve traced in fig. 4, to obtain all the parameters required to describe the bedrock motions that might have been induced by the earthquake at any given distance from the source. The curve shown in fig.4 was determined as the best fit of peak horizontal accelerations values relative to motions on outcropping rock and bedrock motions obtained after deconvolution of ground surface records. For comparison an average curve presented in the literature for 6.5 magnitude events is also shown in the same figure [4].

Using the parameters selected to describe the bedrock motions induced by the earthquake at any given distance from the source, synthetic accelerograms, consistent with the real event, were then generated. This was accomplished using a numerical procedure that requires knowledge of the maximum acceleration, pseudovelocity response spectrum and shape of the acceleration time history of the motion to be generated[5]. The procedure allows to reconstruct an infinite number of slightly different accelerograms for the same set of parameters previously defined. These artificial records differ, in fact, only for the phase angle used in generating the synthetic motions. In order to explore the effects of a random choice of the phase angle, 8 accelerograms were generated for any given value of the distance from the earthquake source.

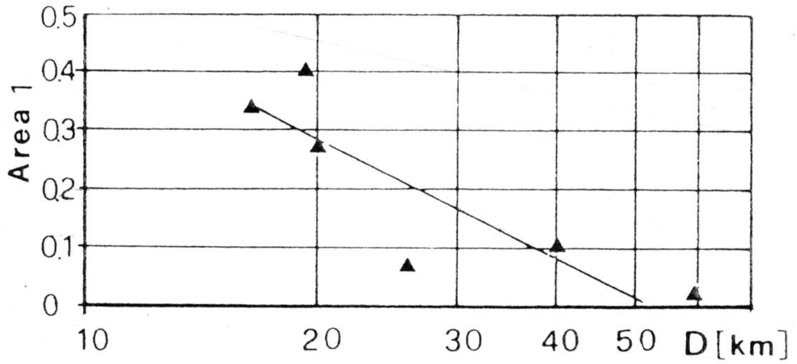

Fig. 3: Area 1 of envelope curves vs distance from source.

3 Comparison of permanent displacements induced by artificial and real motions

The effectiveness and applicabilty of the proposed procedure to reconstruct artificial acceleration time histories, consistent with a given earthquake, for use in stability analyses of slopes located in seismic regions has been investigated in different ways.

As a preliminary check response and Fourier spectra, envelope curves, and values of the Arias intensity of the real and artificial motions were calculated. Envelope curves obtained for the recording station Calitri using the real acceleration time history and that obtained using the procedure presented herein are shown in fig. 5, while values of the Arias intensity are compared in fig. 6.

As can be seen the two types of motions seem to bear several similarities when comparing the parameters shown in these figures. With few exceptions a similar trend was usually observed when comparing envelope curves as well as response spectra for other recording stations. On the basis of these results it would seem, therefore, that the proposed procedure can provide a valuable tool for reconstructing artificial acceleration time histories closely resembling those induced by the source event.

For a better estimation of the effects of using the artificial motions determined in this study to evaluate the stability conditions of a slope, the reconstructed and the real records were integrated twice with respect to time. This was meant to compare the different motions in terms of the permanent downhill displacements that would be calculated for a generic slope following the approach proposed by Newmark. In practice it was implicitly assumed that the artificial and real records represented the average inertia forces induced by the earthquake on the potentially unstable portion of the slope. Clearly in a real case bedrock motions must first be propagated through the compressible soil layers and eventually within the slope before displacements can be calculated.

Considering the scope of the present research it was not deemed important, at this point of the study, to make any definite assumption on the geometry of the slope nor on the type of soils within the slope and their geotechnical properties. Thus the potential for instability was described by means of a single dimensionless parameter, namely the yield acceleration Kc expressed in units of g, the acceleration of gravity.

Double integration of real and artificial acceleration time histories was carried out assuming different values of yield acceleration. As shown schematically in fig. 7 this allowed to estimate the effects of including progressively larger portions of the records.

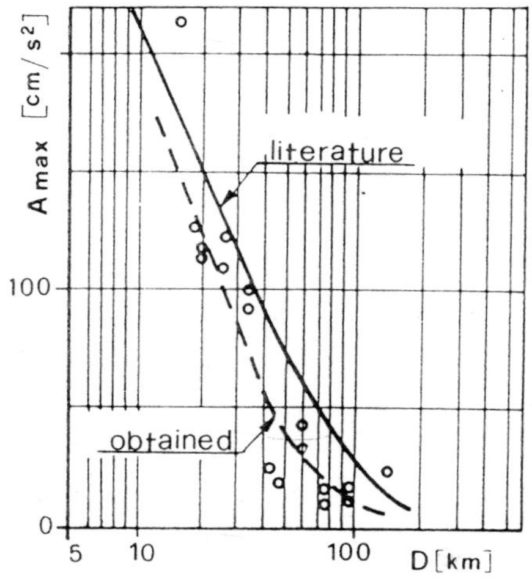

Fig. 4: Peak horizontal accelerations on bedrock vs distance from source.

Of the three components of the motions recorded at the different stations, only the accelerations in the two horizontal directions, North-South and East-West, were used in the analysis. Thus for any given value of yield acceleration 4 values of displacement were obtained for each station, relative to the positive and the negative parts of the two components of the real motions. To obtain a single displacement value and render comparison easier, the maximum value of the two displacements, calculated with double integration of the positive and negative parts of the records, was first selected for each horizontal component. These two values were then squared, added, and the square root

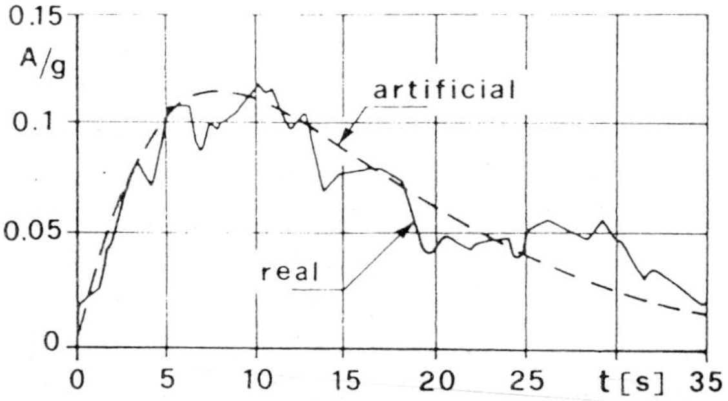

Fig. 5: Envelope curves of real and artificial motions for station at Calitri.

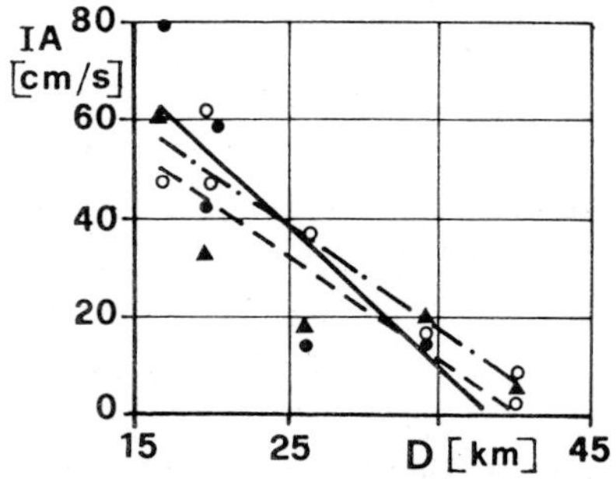

Fig. 6: Arias intensity of real and artificial motions.

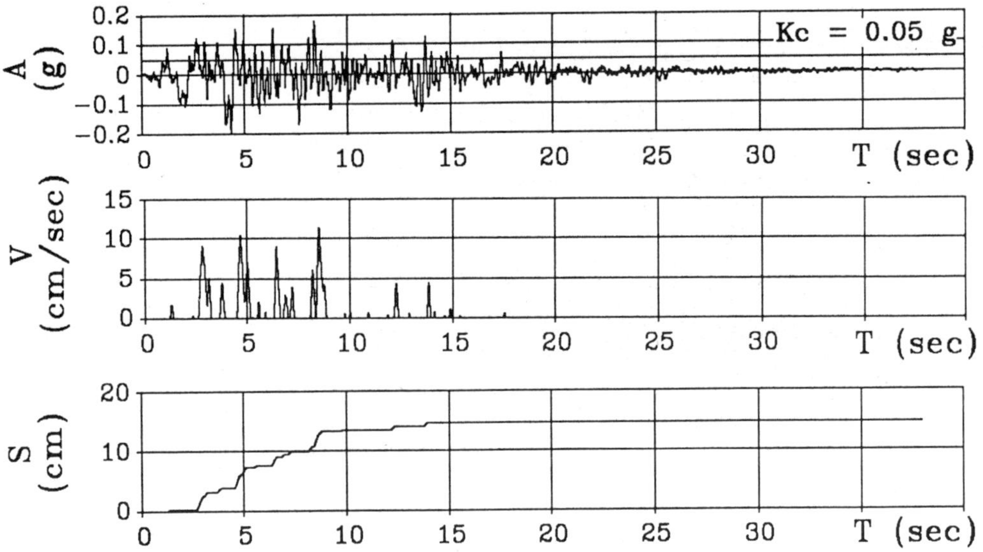

Fig. 7: Schematic description of double integration procedure used for both artificial and real acceleration time histories: accelerations, velocities, cumulative displacements vs time.

of the sum taken as the final value of cumulative displacement. A similar procedure was followed when integrating the artificial accelerograms with the difference that in this case the same accelerogram was assumed to represent ground motion in the two directions, North–South and East-West. Hence in this case the displacement was obtained as the square root of twice the maximum of the two values relative to the negative and positive portion of the artificial motions.

Values of cumulative displacements calculated using the acceleration time histories recorded at the stations triggered during the 1980 Irpinia earthquake, are plotted in fig. 8 versus the yield acceleration. A comparison between the results obtained using the real and artificial motions relative to the recording stations at Bagnoli and Brienza are presented in fig. 9. In this figure the dotted lines give the displacements calculated for any of the 8 artificial time histories reconstructed in the study. As can be seen the results determined from the 8 artificial motions, that differ only for the phase angle, tend to fall within a relatively narrow band. The differences between any of these motions seem rather limited and can be considered, therefore, negligible for all practical purposes.

It is worth noting that the average curves of the 8 artificial accelerograms show reasonably close agreement with the curves obtained from the recorded motions. However a definite trend cannot be established since in one case (Brienza) the use of the artificial motions leads to an overestimation of the total displacement, while in the other case (Bagnoli) displacements are always

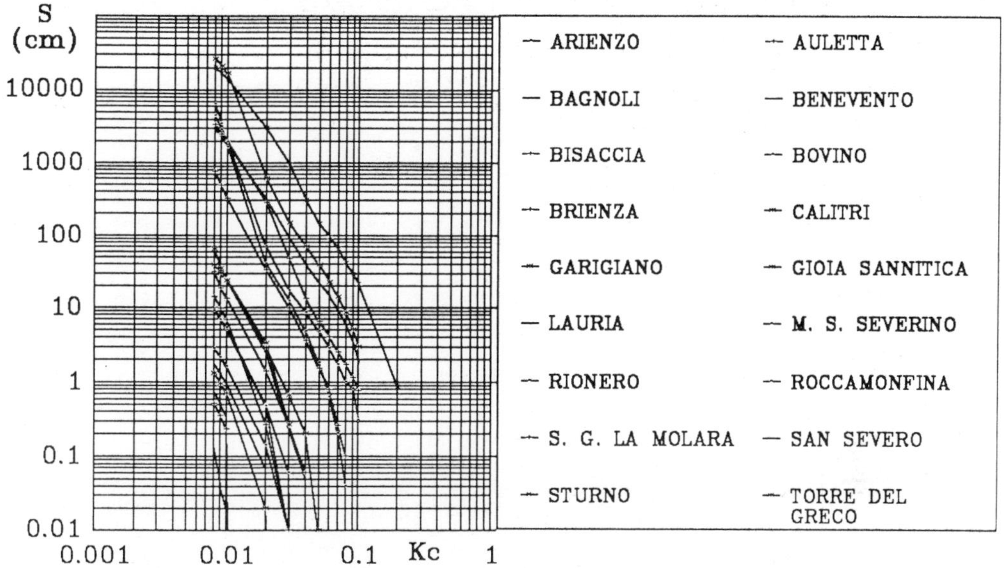

Fig. 8: Cumulative displacements vs yield accelerations obtained from accelerograms recorded at different stations triggered by the Irpinia earthquake.

underestimated with respect to the data obtained using the real motion.

In order to provide a better estimation of the differences between the values of the displacements calculated using the real and the artificial motions the curves shown in figg. 10 and 11 were constructed. In fig.10 the ratio of the values of displacements obtained with the real (Sr) and the artificial motion (Sa) is plotted for the recording station at Bagnoli versus the displacements calculated from the artificial acceleration time history (Sa). As can be seen, for relatively large displacements, greater than about 10 cm, the real and artificial motions yield results that show a relatively good agreement. Instead for lower values of the calculated displacement the artificial records lead to an underestimation of the total displacement.

Again it would seem that in the range of interest for practical purposes the procedure presented herein may be applicable to determine artificial motions for use in slope stability analyses. Similar results were obtained for other sites though in some cases the differences between the real and artificial motions tend to be somewhat larger.

Finally on the basis of the data shown in fig. 11, obtained averaging all the calculated displacements, it can be observed that for ratios of the yield acceleration to the maximum acceleration of the record in the range 0.5 to 1.0 the artificial accelerograms resemble closely the recorded motions. Therefore it would seem that, on the average, the use of the reconstructed motions leads to

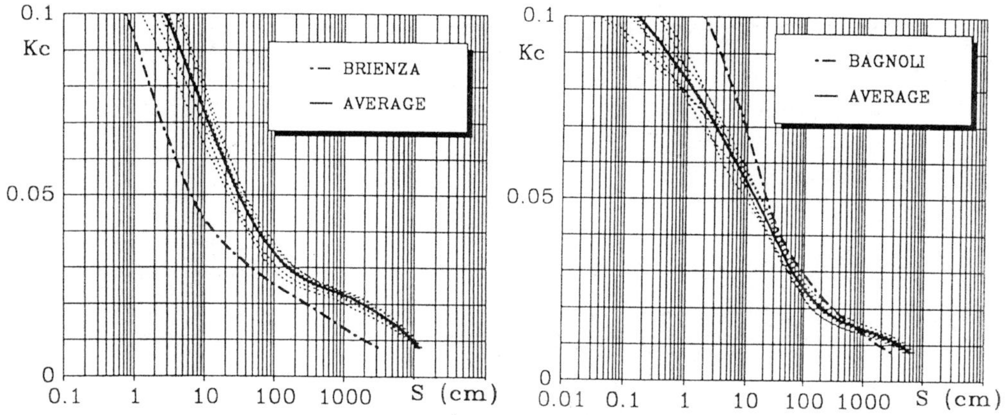

Fig. 9: Cumulative displacements vs yield accelerations obtained from real and reconstructed accelerograms for the stations at Brienza and at Bagnoli.

a conservative estimation of the displacement of a slope. Clearly at present this statement holds only with reference to the 1980 Irpinia earthquake.

4 Conclusion

The evaluation of the stability conditions of natural slopes, cuts and earth structures located in seismic areas depends on the results of several different types of analyses that must be performed correctly in order to attain a reliable estimation of the degree of safety of the slope. Among these analyses one requiring particular attention, due to the difficult questions that it poses, is the estimation of the characteristics of ground shaking to which the slope may be subjected in the occurrence of an earthquake.

In the attempt to overcome the difficulties related to this kind of analyses, the available records of ground shaking induced by the 1980 Irpinia earthquake have been analised. A procedure has then been developed to reconstruct artificial acceleration time histories that could be considered representative of the motions recorded during the event. For this purpose a technique that allows to correlate the fundamental characteristics of an earthquake, such as shape of the recorded motions and response spectra, with the distance from the source of the event has also been defined. With these correlations the parameters required to reconstruct synthetic accelerograms can be determined for any given distance from the earthquake source.

At this stage of the research this procedure has been applied to obtain artificial bedrock motions representative of the 1980 Irpinia earthquake. Real and artificial motions have been compared using response spectra, Arias intensity and envelope curves determined for both records. The reasonable agree-

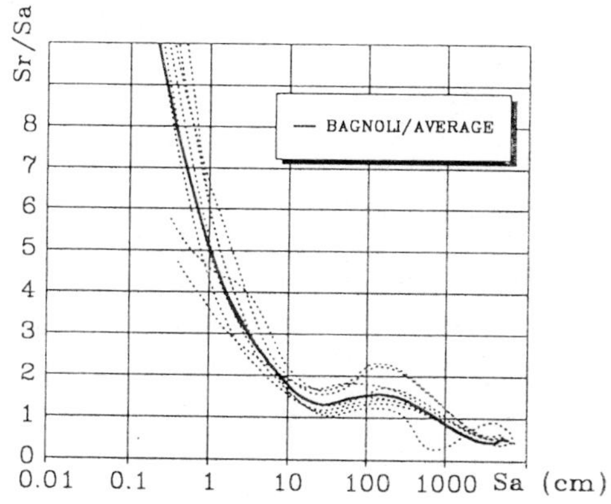

Fig. 10: Ratio of values of cumulative displacements obtained from real and artificial motions vs cumulative displacements from real motion relative to the station at Bagnoli.

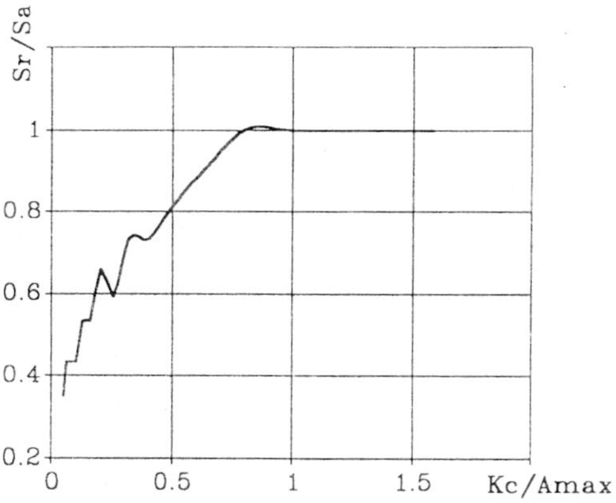

Fig. 11: Ratio of values of cumulative displacements obtained from real and artificial motions vs ratio of yield accelerations to maximum record acceleration.

ment between the results yielded by the two sets of motions seems to indicate the potential applicability of the proposed procedure for use in design studies.

To check if this procedure could be applied also to reconstruct artificial motions for use in stability analyses of slopes located in seismic regions, the artificial and real accelerograms were integrated numerically twice with respect to time. This integration process was meant to simulate the application of the Newmark approach under the assumption that the inertia forces induced by the earthquake on the potentially sliding mass were equal to the artificial or real records. Thus integration was carried out taking into account several different values of the yield acceleration, to represent, schematically, slopes having different values of the safety factor with respect to static loading conditions.

Comparison of the cumulative displacements calculated using the motions recorded at six stations and the artificial accelerograms, determined for the same distance from source as that of the stations, shows that the results obtained in the two cases are generally close or in good agreement. The greatest differences between values of displacement relative to the two sets of motions, real and artificial, are recorded for the highest values of yield acceleration, thus when the calculated displacements become relatively small.

On the contrary for increasing values of displacements the agreement tends to improve. It would thus seem that for all practical purposes the procedure described herein may prove effective in determining artificial acceleration time histories for use in slope stability analyses.

However further research is desirable to achieve a better understanding of the results obtained so far and also to improve the proposed procedure.

In this respect it is worth mentioning some of the points that could still be developed. Firstly it should be observed that in the analysis presented herein only horizontal motions were used. Secondly the propagation of seismic waves was studied with reference to one dimensional conditions. Also soil profiles and material properties were estimated on the basis of available data obtained from the literature. Finally it is worth mentioning that a better estimation of the proposed relationships, to determine the characteristics of ground motions as a function of distance from source, might probably be achieved if other factors, such as nature, depth to surface and morphology of bedrock, besides distance from source only, could be taken into account in the study.

Acknowledgments

The studies previously described were performed with partial support of funds provided by the Italian C.N.R. Consiglio Nazionale Delle Ricerche.

References

[1] Newmark N. M. (1965). Effects of earthquakes on dams and embankments. *Geotechnique,* vol. 15, n. 2, p. 139 - 173.

[2] Schnabel P.B. and Lysmer J. (1972). *SHAKE - A computer program for earthquake response analysis of horizontally layered sites.* Report n. EERC 72-12, Earthquake Engineering Research Center, Berkeley.

[3] Seed H. B. and Idriss I. M. (1970). *Soil moduli and damping factors for dynamic response analyses.* Report n. EERC 70-10, Earthquake Engineering Research Center, Berkeley.

[4] Seed H. B., Murarka R., Lysmer J. and Idriss I. M. (1969). Relationships of maximum accelerations, maximum velocity, distance from source and local site conditions for moderately strong earthquakes. *Bulletin of the Seismological Society of America,* vol. 68, p. 677 - 690.

[5] Gasparini D. (1976). *SIMQKE: a program for artificial motion generation, user's manual and documentation.* Dpt. of Civil Engineering, M.I.T., Cambridge, Massachussets.

ON SOME ASPECTS OF THE NUMERICAL EVALUATION OF PERMANENT DISPLACEMENTS

A. CIVIDINI * and F. PERGALANI **

Abstract: The computation of displacements, through suitable numerical procedures, can be useful in assessing slope stability and in estimating the damage associated with the occurance of slides. The stress analysis of geotechnical systems under dynamic conditions is however a complex problem due to the nonlinear material behaviour and to the multiphase nature of soil and rock media. In the framework of a research project on stability in seismic conditions, the slope stability under earthquake loading will be investigated by means of simplified approaches and subsequently by means of complex nonlinear procedures.

The paper includes a summary of the approaches for stability problems and a brief outline of some constitutive models used in the numerical analyses of dynamic problems.

Riassunto: Nel campo dell'ingegneria delle terre e delle rocce, può essere necessario, per verificare la stabilità dei versanti naturali ed artificiali, determinare gli spostamenti indotti da un intervento o da una azione sismica. Il comportamento dei versanti è assai complesso e dipende da svariati fattori, quali la natura non lineare e dipendente dal tempo dello scheletro solido, la natura plurifase di terre e rocce, etc. Nell'ambito di un progetto di ricerca sulla stabilità dei versanti in condizioni sismiche, è stato avviato uno studio, avente quale obiettivo il confronto tra i risultati di analisi di tipo semplificato ed analisi in cui è tenuto in conto il comportamento non lineare delle terre e delle rocce. Nel presente lavoro vengono indicate le procedure per l'analisi numerica e le principali caratteristiche del legame costitutivo, che potrà essere impiegato nell'analisi.

1. Introduction

In the field of soil and rock engineering, it is often necessary to assess stability, both in static and dynamic conditions, concerning slopes, open cuts, embankments, earth retaining structures, etc. [28,34,23]. Their overall behaviour is characterized by many complex aspects such as: non linear and time-dependent behaviour of the

----

* Politecnico (Technical University) di Milano, Italy.
** Istituto per la Geofisica della Litosfera CNR, Milano, Italy

skeleton; multi-phase nature of soil and rock and the consequent influence of the degree of saturation; the coupling between the flow process of the pore fluid and the deformation of the skeleton in cohesive soils; the interaction between soil or rock and deformable supporting structures.

The stability problems in situation characterized by relatively simple geometry, boundary conditions and material behaviour can be solved by means of analytical or closed form solutions, while numerical methods are nowadays frequently adopted in engineering practice for the solution of complex boundary value problems. In principle, they offer the possibility of handling problems with complex geometry and loading histories (as in the dynamic conditions), involving layered and non homogeneous soil mass, and of using complex constitutive laws. In addition, they permit to model construction and/or excavation sequences. In fact stress and strain variations in the soil and rock masses have been induced by: changes in the geometry caused by erosion process, excavation and construction sequences (underground openings, installation of temporary and permanent supports, stabilization procedures, etc.); changes in underground water conditions (including the effects of rainfall infiltration); changes (reduction) in material mechanical properties; etc. Therefore the in situ stress state can be modelled through the numerical analysis (pre-earthquake analysis). In addition numerical procedures offer the possibility of analizing seepage flows and deformation processes for two-phase media (i.e. consolidation processes). In fact, the analysis of seepage flows is a necessary step for assessing, for instance, stability of slopes and of earth and retaining structures, and for studying the efficiency of drainage systems. The consolidation process, which represents a typical part of the behaviour of cohesive soils, is of interest not only in the evaluation of foundation settlements or in the evaluation of land subsidence due to water removal, but also in the post-earthquake analysis.

Geological and geotechnical information from the site [1] is of fundamental importance in defining the scheme to be adopted in the simulation of soil and rock masses. Fine graded soil deposits or soft and highly weathered rocks can be approximated as a continuum and their behaviour can be modelled through standard finite element and boundary element procedures [e.g.11,57,6]. In rock mass containing joints or faults it is also necessary to model the interaction along the discontinuities. Indeed mass deformability, strength and permeability properties depend upon the characteristics of the joint sets, that in the numerical analysis can be modelled by means of the so called "joint" elements. After reaching the failure state, blocks of rock or of stiff soil can undergo relative movements. This behaviour can be modelled by means of the so called "distinct" elements [15].

Under static conditions the approaches currently used for solving stability analysis problems can be grouped into two main categories. The first one includes procedures for the evaluation of collapse load (i.e. Limit Analysis and "limiting equilibrium" methods); the second one collects approaches for the evaluation of the critical conditions, considering progressive deformation and failure (such as the stress-strain analyses by means of the finite element or the discrete element methods). For earthquake stability analyses similar partitioning can be adopted and the approaches are usually referred to, respectively, as pseudostatic and dynamic ones.

In the following only some of the relevant aspects of the methods belonging to the two categories are briefly mentioned and, subsequently, some considerations are recalled on the constitutive laws to account for non linear material behaviour in the framework of the finite element analysis.

2. Pseudo-static methods

In evaluating the effects of an earthquake loads [50,51] the soil or rock mass is considered as a rigid-plastic medium and the inertia force is considered as an equivalent, permanent horizontal force (expressed as the product of the seismic coefficient K and the weight of the sliding mass, usually applied at the center of gravity). For an assumed failure mechanism, the problem reduces to the calculation of the horizontal force coefficient K that leads to the collapse of the slope. Then the location of the failure surface can be changed and, perhaps with the aid of an automatic search algorithm, the critical failure mechanism can be identified.

Limit analysis solutions for the collapse conditions of rigid-plastic medium (stable in the Drucker sense) can be obtained by means of static approach (i.e. ensuring that the equilibrated stress state in the soil or rock mass does not violate the failure condition, so obtaining a lower bound to the sought collapse load) and of kinematic approach (so obtaining an upper bound). The "exact" collapse load is uniquely determined when both static and kinematic conditions apply. However from an engineering point of view, often it is enough to use the two approaches and to obtain a range within which the collapse load value is included.

In the framework of the finite element method [e.g.56] the first (static) approach makes use of the so called "equilibrium" elements (the nodal variables are the stress components; so that equilibrium equations on infinitesimal volume and the interelement stress continuity are always satisfied). For the second (kinematic) approach the standard "compatible" elements are used. The displacements are nodal variables, the strain-displacement compatibility equations are verified, but the equilibrium equations (and the interelement stress continuity) are imposed only in the "average" or in a weak form.

The solution strategies commonly used in investigating the slope stability, the so-called "methods of slices" (also referred to as "limiting equilibrium" methods) can not be regard either as static or as kinematic approaches [50]. Consequently it is not possible to judge of the approximation to the "exact" solution. In practice however they represent a simple and powerful tool for geotechnical analyses and they are routinely used. For instance, an integrated system for the stability analysis for different kind of rock slopes has been proposed by Aydan et al. [5]. Various methods, based on the limiting equilibrium concept, can be selected and the most critical situation detected.

3. Dynamic analysis

The effects of deformation and of progressive deformation and failure in soil and rock masses, can be considered through various solution strategies. In general terms the strategies can be grouped into two categories. The first one includes the procedures for the analysis of linear problem (based on the finite element method [e.g.11,60,57] or on the boundary integral equation method [e.g.6]. The second one considers two procedures for non linear analysis, namely the distinct element method [15] and the finite element method [e.g.60,57,11].

3.1 LINEAR PROBLEMS

Linear procedures are mainly adopted in solving wave propagation problems, in the evaluation of amplification effects in presence of non homogeneous materials, or irregular boundaries, etc. Finite element method and boundary integral equation method have advantages and disadvantages when applied to the analysis of the seismic response of a non homogeneous soil deposit [e.g. 6,11]. On the one hand, the boundary integral equation method is very efficient when dealing with homogeneous region and linear elastic behaviour, since it requires only the discretization of the boundaries. On the other hand, finite element procedures can easily deal with non homogeneous and complex constitutive behaviour, but they require considerable effort in preparing input data and a lot of computational time if the discretization extends to the far field (alternatively, it is necessary to adopt appropriate adsorbing boundaries to avoid wave reflection induced by rigid boundaries). In addition, the finite element mesh has to be chosen carefully in order to avoid spurious reflections due to element size changes (or to the presence of discontinuities) and the frequency of the applicable (external) loads is limited by the cutoff frequency of the mesh (i.e. the maximum frequency of a wave that can travel through the mesh). Thus, the coupling of the two methods can provide a powerful tool for the stress analysis of problems with a complex geometry, not only in static condition [32] but also under dynamic loading.

3.2 NONLINEAR PROBLEMS

Various aspects make the dynamic problem complex: the non linear behaviour of the soil skeleton; the interaction between the seepage of the fluid and the deformation of the soil skeleton [61]; the nonlinear behaviour of the fluid phase (e.g. for nearly saturated porous media [21], etc. The numerical procedures that can be used for dealing with nonlinear problems are here grouped into two categories: the distinct element and the finite element methods. The first one is used in one-phase analyses, the second one can be used in the analysis of two-phase media, considering non linear material behaviour. In the following some aspects relatively to the distinct element method, to the finite element analysis of two-phase media and finally to the constitutive laws that can be used in the finite element analysis under static and cyclic loading conditions will be summarized.

3.2.1 Distinct element methods

In this numerical solution procedure [15] the equilibrium contact forces and displacements of a stressed assembly of blocks are found through a series of calculations tracing the movements of each individual element. These movements are the result of propagation through the medium of disturbances originating at the boundaries. It is a cyclic computational process, each cycle consisting of two steps: (1) Newton's second law gives the motion of a block resulting from the forces acting on it; (2) the force-displacement law is used to find contact forces from displacements.

For instance, the distinct element method can be applied in the estimation of the design load for a rock shed [39], considering the collision between a falling rock and a cushion and evaluating the effect of the arrangement of crushed material in the cushion on the energy absorption.

The well-known Newmark's method [43] for the evaluation of permanent displacement of a rigid block can be seen as a particular case of the mentioned procedure [e.g.2,24]. Given the time history of the ground excitation (acceleration, velocity and displacement components), the motion of the block resting on a sloping surface is computed. When the shear force at the interface between the base and the block reaches the frictional resistance, a relative displacement is possible. Then, if the relative velocity reduces to zero, the relative movement stops, until a possible subsequent increase in the shear force.

Experimental observations [e.g.7,8,9,14,33,34] indicate that the frictional resistance at the interface is displacement-dependent. Therefore such dependence can be taken into account in the numerical analysis [e.g.23,10] considering two failure envelopes (cf. fig. 1a): the first one is bilinear and represents the peak strength of the jointed mass [44,9]. The second one represents the residual shear resistance that develops for relative displacements greater then a threshold value, s-r, as shown in fig. 1b. Of course, more complex interface behaviours can be easily introduced in the analysis.

The above mentioned procedure allows to consider, for preliminary analyses, the influence of the geotechnical parameters (cf. fig. 2) and of the variability of the ground motion. For instance, the correlation between permanent displacements and some of the traditional ground motion parameters, used to represent the earthquake ground motion, such as peak ground acceleration, record duration, root mean square acceleration, Housner spectral intensity, Arias intensity [19,3] can be investigated [23].

3.2.2 Two-phase analysis

In a saturated, deformable porous medium the dissipation of the excess pore water pressure induced, for instance, by ground motion or by the construction of shallow foundations is associated to changes in the total and effective stresses and, consequently, to the deformation of the porous medium.

In quasi-static case (or consolidation analysis) this phenomenon can be tackled analytically mainly in one-dimensional conditions, adopting the well known Terzaghi's theory [58]. In dynamic case an analytical solution was obtained by Simons et al. [54] for the one-dimensional problem too.

Thus the use of computer oriented solution techniques permits the solution also of two- and three-dimensional problems through the theory proposed by Biot [12,13] and firstly applied to solve consolidation problems in the framework of the finite element method by Sandhu [48]. This analysis is referred to as "coupled" since the problems of seepage and deformation of the porous medium are solved simultaneously. On the contrary, the term "uncoupled" identifies the consolidation analysis based on Terzaghi-Rendulic approximation [e.g.52] that ignores the total stress redistribution within the soil mass taking place during consolidation, and reduces the problem to an ordinary diffusion (e.g. heat transfer) analysis.

Nowadays the finite element method is frequently adopted for solving consolidation problems through the coupled approach, considering as nodal variables the pressure of the pore fluid and the displacements of the soil skeleton. In particular this permits, on the one hand, the description of the non linear and time-dependent (viscous) skeleton behaviour by means of the elasto-visco-plastic constitutive laws adopted in one-phase analyses [22] and it also permits, on the other hand, an account of the presence of water through the same procedure used in the so called "fixed" mesh solution of the seepage flow.

For the general case, which includes the inertial effects [61], a variety of finite element approximations can be adopted. In particular, two finite element formulations and various time integration schemes were presented and discussed in Simons et al. [54]. The first one is a displacement formulation that considers as nodal

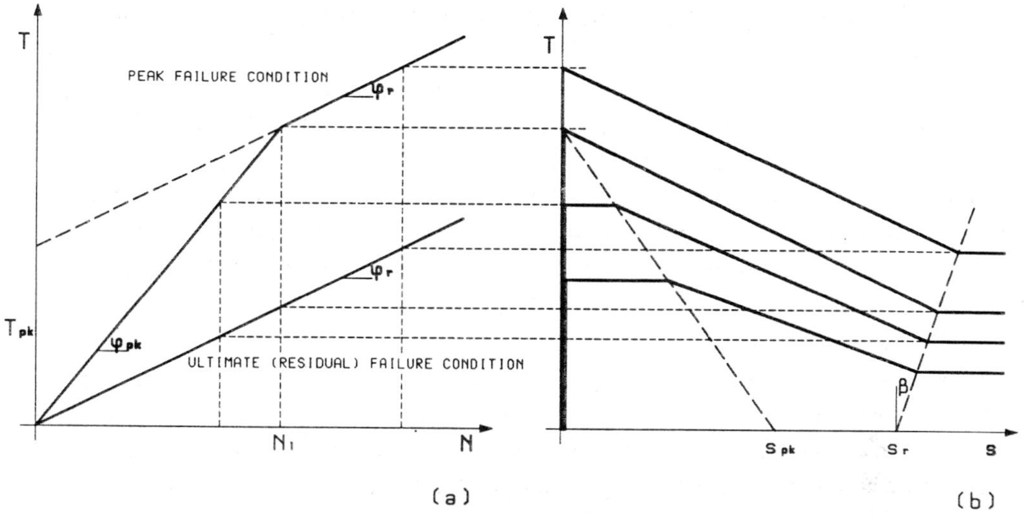

Fig. 1 Constitutive behaviour for the interface between the base and a rigid block. (a) Peak and residual effective strength envelopes. (b) Relation between shear resistance, T, and shear displacement, s, for different normal loads, N.

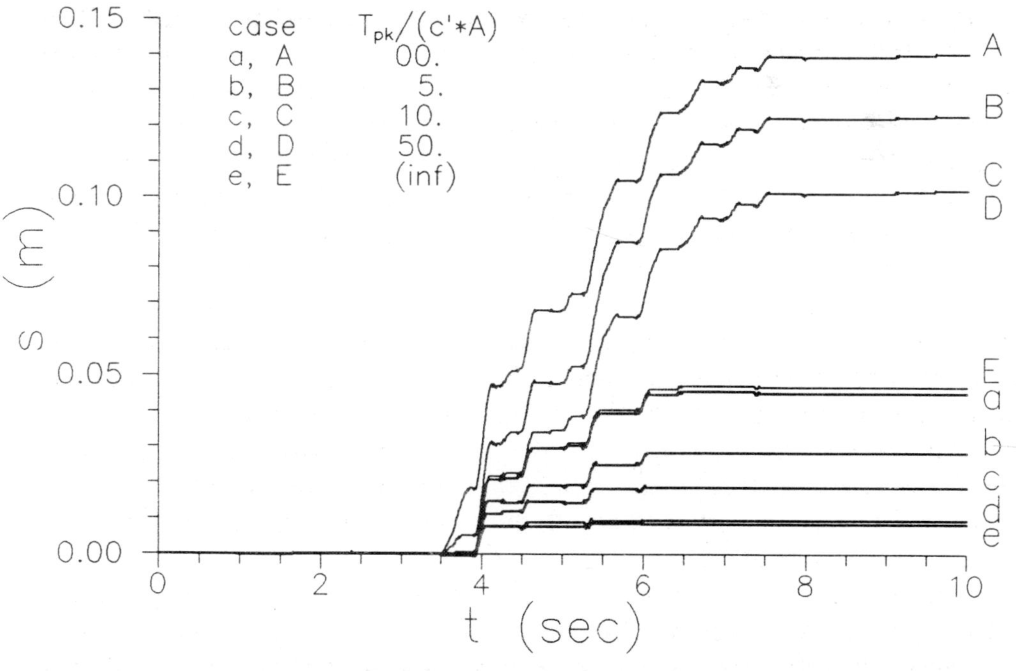

Fig.2. Influence of the shear resistence T-pk on the computed displacements (cases a-e: dry block; cases A-E: saturated mass and constant, i.e. static, water pressure).

variables the displacements of the pore fluid and the displacements of the soil skeleton. The second one, based on the approximation that the change in relative velocity can be neglected, is a mixed formulation that adopts as nodal variables pore pressures and displacements as in the analysis of quasi-static (consolidation) phenomenon.

3.2.3 Constitutive laws

Experimental observations [e.g.55,37,17,29,46] show that the soil behaviour is highly nonlinear in both static and cyclic loading conditions. Consequently, the variety of constitutive models used in soil and rock mechanics reflects the complexity of the investigated behaviours.

A complete discussion of the problems associated with the formulation of constitutive laws falls outside the limit of this paper [62,41,18]. The mere aims of the comments here presented is to recall the main features of some of the most popular material models used in soil and rock mechanics in view of the engineering applications of these models [20]. The discussion is limited to the stress-strain laws in small strain conditions and unless otherwise specified, effective stresses (in Terzaghi's sense [59]) will be considered.

Some well known constitutive laws used in geotechnical engineering belong to the class of hypoelastic models, like the hyperbolic laws proposed by Duncan & Chang [27]. They are frequently adopted for representing the experimental behaviour observed in cyclic triaxial and shear tests and they are used in the numerical evaluation of the dynamic response of horizontally layered deposits of soil, subjected to upward propagating shear waves. Their use in stability analyses is not straightforward, because in the slopes the stress and strain states that develop during earthquake are complex, while the constitutive models were formulated in simple conditions, considering only one stress (strain) component.

The elasto-plastic models appear more promising for the dynamic analysis and it is perhaps worthwhile recalling a few aspects of this class of constitutive laws. In general terms an elasto-plastic law consists of three main parts: a) the yield condition, represented by a surface in the stress space, defining the stress states at which plastic deformation may develop; b) the hardening law, that governs the possible changes in shape, size and position of the yield surface; c) the flow rule, governing the increment of the plastic strains.

Particularly important elasto-plastic materials are those stable in Drucker sense [38]. They are characterized by a convex yield condition, associated flow rule and positive hardening or perfectly plastic behaviour. In this case it is possible to show that an incremental elasto-plastic problem has an unique solution, that the fundamental theorems (static and kinematic) of limit analysis hold. A complete discussion on the existence of the direct and inverse incremental stress-strain laws for rocks can be found in Maier & Hueckel [38].

For sake of simplicity the constitutive models are here grouped in the following four categories: a) perfectly plastic models; b) cap and critical state models; c) bounding surface models; d) hierarchical models.

a) Perfectly plastic models. The well known Mohr-Coulomb and Drucker-Prager models and the more recently proposed Matsuoka-Nakai [40] and Lade [36] models belong to this first group. They are characterized by different shapes of the yield surfaces. Drucker-Prager yield criterion is represented in the principal stress space by a cone, having circular cross section and axis coinciding with the octahedral axis. This criterion can be easily implemented in computer programs since it is expressed by a unique quadratic equation. This is perhaps the main reason why Drucker-Prager yield surface is so frequently adopted in elasto-plastic finite element analyses in geomechanics (static conditions). Its simplicity, however, is also the cause of a limited accuracy in the description of the actual behaviour of geomaterials. In particular, this condition is independent of the third stress invariant (in fact, it has a circular cross section).

A more accurate yield criterion, from the experimental view point, is the classical Mohr-Coulomb condition, represented in the principal stress space by a pyramid with hexagonal non regular cross section. Note that Mohr-Coulomb and Drucker-Prager yield conditions for frictionless materials reduce respectively to Tresca and to Huber-Hencky-von Mises criteria, introduced in 1860 and 1913 as failure conditions for metals. However, these conditions are also used in geotechnical engineering for the so called total stress analysis of saturated soils in undrained conditions. In such a case the yield surface is expressed in terms of total, instead of effective, stresses.

The existence of edges between adjacent sides of Mohr-Coulomb pyramid (i.e. of discontinuities in the yield surface gradient) introduces some difficulty in the numerical analysis and non negligible changes of the elasto-plastic routine are necessary. A yield criterion that does not present the mentioned drawback, but still takes into account the influence of the third stress invariant, is the one proposed by Matsuoka and Nakai. Finally the yield surface proposed by Lade takes into account that the shear resistance of geomaterials does not increase linearly with increasing octahedral stress [e.g.35].

b) Cap and critical state models. The development of the cap models [49] was suggested by the experimental observation that during compression tests on normally consolidated soils the non reversible deformation develops continuously starting from the very early stages of the test (no appreciable initial elastic range exists). Failure is produced when the non linear strains reach particularly large values. In other words, for these materials a clear distinction exists between the stress conditions at which the non linear behaviour initiates and the ones corresponding to failure.

Constitutive models able to reproduce this behaviour are characterized by a hardening yield surface or yield locus (referred to as cap), that in the stress space represents the boundary of the current elastic domain, and by a fixed failure surface that, as previously observed, is the limit for the admissible stress states. The hardening cap and the fixed failure surface allow, respectively, for the non linear behaviour of soil under increasing volumetric stress and for its failure under increasing shear stresses. Note that the hardening cap is limited only to the part of the stress space contained within the failure surface.

The so called critical state models [53] can be viewed as an extension of the previous cap models. They have been originally proposed for modelling the behaviour of cohesive and non cohesive soils. More recently they have been extended also to rocks and rock like materials [31] and to the prediction of the behaviour of soils under cyclic loading (densification model [16]; reflecting surface model [45]).

A main feature of the critical state models, which is not shared by the cap models, is that they are able to account for the softening behaviour of stiff soils and rocks. This is obtained by extending the yield surface in the zone above the failure surface, which is also referred to as critical state surface.

For the original version of the critical state model, known as Cam Clay model [53], it has been observed that the deviatoric stress level at which plastic strains initiate could be overestimated for small values of the volumetric stress. In order to limit this drawback the ellipsoid can be substituted by a cone in the vicinity of the origin. Note, however, that such a type of provision might lead to discontinuities in the gradient of the yield function.

c) Bounding surface models. The bounding surface models [e.g.25,41,47,4] are in principle critical state models characterized by a non linear stress-strain relationship even when the stress path is contained within the yield surface, while no conceptual differences exist between the behaviours of these two classes of models for stress states belonging to the yield or failure surfaces.

The bounding surface models intend to eliminate the effects of the sharp transiction between elastic and plastic regimes that takes place in critical state and cap models when the stress path reaches the yield surface. In addition, they can also provide a better approximation of the material behaviour under cyclic loading, since non linear deformation can develop even when the variation of the stress level does not reach the yield surface.

A bounding surface model is characterized, in addition to the yield and failure surfaces, by the following main ingredients: (1) a region, referred to as elastic nucleus, within which the material behaviour is linearly elastic. This nucleus is contained within the yield surface and might collapse to a point, or pole, when no linear elastic range exists. (2) a rule that permits to define, on the basis of the actual stress point within the yield surface, an image point on the yield surface and a corresponding point on the elastic nucleus. (3) A law governing the smooth transition between purely elastic and elasto-plastic behaviour on the basis of the relative position of the actual stress point and of its images on the yield surface and on the elastic nucleus.

d) Hierarchical models. Finally the so called hierarchical constitutive models can be mentioned [26]. These are characterized by a yield surface expressed as a polynomial function of the invariants of the stress and plastic strain tensors and of the mixed invariants of the same tensors. The term hierarchical is adopted since these models allow for a progressive degree of complexity by increasing the number of terms in the polynomial expression. No distinction is made between yield and failure surface, but only an unique yield function is given that leads to the failure surface when an ultimate condition is reached. Recently, the hierarchical models have been extended to describe the behaviour of interfaces subjected to static and cyclic loading conditions [42].

Finally, a shear-hysteretic elasto-plastic model was recently proposed by Elgamal [30]. The model is capable of generating hysteretic Masing-type behaviour and the formulation was presented for a one-dimensional stress-strain condition, but according to the Author the formulation can be extended to a general stress and strain states. Therefore a procedure having the same type is presently considered for the development of the nonlinear constitutive model to be used in the finite element analysis (the discussion will be presented in a subsequent communication).

4. Concluding remarks

It is often necessary to analyze slope stability under static and dynamic conditions. The geotechnical problem is very complex. It can be treated under simplifying hypothesis (for instance condidering the motion of a single rigid block) or alternatively through numerical procedures, so that all the aspects relevant to the geotechnical problem can be included. Even though the matter here illustrated covers only a limited portion of the problems related to the numerical analysis applied to slope stability under dynamic condition, it is hoped that this discussion can represent a useful basis for the analysis of the complex problems presented at the colloquium.

Acknowledgements The financial support of the Italian Ministry of the University and

Research (CNR - GNDT) is gratefully acknowledged.

References

[1] AGI - Associazione Geotecnica Italiana (1977). Raccomandazioni sulla programmazione ed esecuzione delle indagini geotecniche.

[2] Ambraseys N.N., Menu J.M. (1988). Earthquake-induced ground displacements. Earth.Engin. Struc.Dynam.

[3] Arias A. (1970). A measure of earthquake intensity, Seismic design for nuclear power plants. R.J.Hansen ed., MIT.

[4] Aubry D., Hujeux J.C., Lassoudiere F., Meimon Y. (1982). A double model with multiple mechanisms for cyclic soil behaviour. Proc.1st NUMOG, Zurich.

[5] Aydan O., Ichikawa Y., Shimizu Y., Murata K. (1991). An integrated system for the stability of rock slopes, Proc. 7th IACMAG, Cairns.

[6] Banerjee P.K., Butterfield R. (1981). Boundary elment methods in engineering sciences. McGraw-Hill, London.

[7] Barbero M., Barla G., Scavia C., Zaninetti C. (1989). Comportamento meccanico di giunti in roccia in condizioni dinamiche. Gruppo Naz. Studi Ingegneria Geotecnica 88/89.

[8] Barton N. (1976). The shear strength of rock and rock joints. Int.J. Rock Mech.Min.Sci. Geomech.Abstr.

[9] Barton N. (1988). Some aspect of rock joint behaviour under dynamic conditions. Proc. 2nd Conf. "Meccanica e ingegneria delle rocce", Torino.

[10] Barton N., Bandis S., Baktar K. (1985). Strength, deformation and conductivity coupling of rock joints. Int.J. Rock Mech.Min.Sci. Geomech.Abstr.

[11] Bathe K.J. (1982) Finite Element Procedures in Engineering Analysis. Prentice-Hill.

[12] Biot M.A. (1941). General theory of three-dimensional consolidation. J.Appl.Phys.

[13] Biot M.A. (1956). The theory of propagation of elastic waves in a fluid saturated porous solid. J.Acou.Soc.Am., 168-191.

[14] Brown E.T., Hudson J.A. (1974). Fatigue failure characteristic of some models of jointed rock. Earth.Engin. Struc.Dynam.

[15] Cundall P.A. (1971). A computer model for simulating progressive large scale movement in blocky rock system. Symp.ISRM, Nancy.

[16] Carter J.P., Booker J.R., Wroth C.P. (1982). A critical state model for cyclic loading. In Soil Mechanics - Transient & Cyclic Loads (Eds. G.N.Pande & O.C.Zienkiewicz).

[17] Cavounidis S., Sotiropoulos E. (1980). Hypothesis for progressive failure in a marl. ASCE GT, 106.

[18] Chen W.F., Saleeb A.F. (1982). Constitutive equations for engineering materials, Vol.1 - Elasticity and Modeling. Wiley, NewYork.

[19] Conte J.P., Pister K.S., Mahin S.A. (1990). Influence of earthquake ground motion

process and structural properties on response characteristics of simple structures. Rep.UCB/EERC-90/09.

[20] Cividini A. (1992). Constitutive behaviour and numerical modelling. In Comprehensive Rock Engin.(Eds.E.T.Brown, J.A.Hudson, C.Fairhurst, E.Hoek) vI.16 Pergamon Press.

[21] Cividini A., Genna F. (1987). Undrained dynamic analysis of partially saturated porous media, TR6, Dept. Struct.Engin., Politecnico di Milano.

[22] Cividini A., Gioda G., Barla G. (1985) A numerical analysis of tunnels in saturated two-phase media. Proc.11th ICSMFE, San Francisco.

[23] Cividini A., Pergalani F., Petrini V. (1991). La risposta dei versanti ad azioni sismiche attraverso un modello semplificato. Ingegneria Sismica, 3.

[24] Crawford A.M., Curran J.H. (1981). The influence of shear velocity on frictional resistance of rock discontinuities. Int.J. Rock Mech. Min.Sci. Geomech.Abstr.

[25] Dafalias Y.F., Herrmann L.R. (1980). A bounding surface soil plasticity model. Proc.Int.Symp. on Soil under Cyclic & Transient Loading, Swansea, 335-345.

[26] Desai C.S., Somasundaram S., Frantziskonis F. (1986). A hierarchical approach for constitutive modelling of geologic materials. Int.J.Num.Analyt. Meth. Geomech.

[27] Duncan J.M., Chang C.Y. (1970). Nonlinear analysis of stress and strain in soils. ASCE SM, 96, 1629-1653.

[28] D'Elia B., Esu F., Pellegrini A., Pescatore T.S. (1985). Some effects on natural slope stability induced by the 1980 italian earthquake. Proc. 11th ICSMFE, San Francisco.

[29] van Eekelen H.A.M., Pott D.M. (1978). The behaviour of Drammen Clay under cyclic loading, Geotechnique.

[30] Elgamal A.W. (1991). Shear hysteretic elasto-plastic earthquake response of soil system. Earth.Engin. Struc.Dynam.

[31] Elliott G.M., Brown E.T. (1986). Further development of a plasticity approach to yield in porous rock. TN, Int.J. Rock Mech. Min.Sci. Geomech.Abstr.

[32] Gioda G., Carini A. (1985). A combined boundary element-finite element analysis of lined opening. TN, Rock Mechanics & Rock Engineering.

[33] Goodman R.E., Ohnishi T. (1973). Undrained shear testing of jointed rock, Rock Mechanics & Rock Engineering.

[34] Goodman R.E., Seed H.B. (1966). Earthquake-induced displacements in sand embankments. ASCE SM, 92.

[35] Hoek E., Brown T. (1980). Empirical strength criterion for rock masses. ASCE GT.

[36] Kim M.K., Lade P.V. (1984). Modelling rock strength in three dimensions. Int.J. Rock Mech. Min.Sci. Geomech.Abstr.

[37] Lemos L.J.(1991). Shear strength of shear surface under fast loading. 10thECSMFE.

[38] Maier G., Hueckel T. (1979). Non associated & coupled flow rules of elastoplasticity for rock-like materials. Int.J. Rock Mech. Min.Sci. Geomech.Abstr.

[39] Masuya H., Kajikawa Y. (1991). Numerical analysis of the collision between a fal-

ling rock and a cushion by distinct element method. Proc.7th IACMAG, Cairns.

[40] Matsuoka H., Nakai T. (1974). Stress-deformation and strength characteristics of soils under three different principal stresses. In Proc. JSCE, 232, 59-70.

[41] Mroz Z., Norris V.A. (1982). Elastoplastic and viscoplastic constitutive models for soils with application to cyclic loading. In Soil Mechanics - Transient & Cyclic Loads.

[42] Navayogarajah N., Desai C.S., Kiousis P.D. (1992). Hierarchical single-surface model for static and cyclic behaviour of interfaces. ASCE EM, 118.

[43] Newmark N.(1965) Effects of earthquake on dams and embankments. Geotechnique

[44] Patton F.D. (1966). Multiple modes of shear failure in rock. Proc.1st ICRM, Lisbon.

[45] Pietruszczak S., Poorooshasb H.B. (1985). On modelling of cyclic behaviour of soils. Developments in Soil Mech. & Found.Engin.

[46] Prater E.G. (1980). Cyclic shear resistance of noncohesive soil, ASCE GT, 106.

[47] Prevost J.H. (1978), Plasticity theory for soil stress-strain behaviour. ASCE EM.

[48] Sandhu R.S. (1968). Fluid flow in saturated porous elastic media. PhD Thesis, Univers. of California at Berkeley.

[49] Sandler I.S., DiMaggio F.L., Baladi G.Y. (1976). Generalized cap model for geologic materials. ASCE GT, 102.

[50] Sarma S.K. (1979). Stability analysis of embankments and slopes. ASCE GT, 105.

[51] Sarma S.K. (1981). Seismic displacement analysis of earth dams. ASCE GT, 107.

[52] Schiffman R.L., Chen A. T-F., Jordan J.C. (1969). An analysis of consolidation theories. ASCE SM, 95.

[53] Schofield A.N., Wroth C.P. (1968). Critical state soil mechanics. McGraw-Hill.

[54] Simon B.R., Wu J.S-S., Zienkiewicz O.C., Paul D.K. (1986). Evaluation of u-w and u-p finite element methods for the dynamic response of saturated porous media using one-dimensional models. Int.J.Num.Analyt. Meth. Geomech.

[55] Skempton A.W. (1977). Residual strengh of clay in landslides, folded strata & the laboratory. Geotechnique.

[56] Sloan S.W., Assadi A. (1991). The stability of a square tunnel in a purely cohesive material whose strength increases linearly with depth, Proc.7th IACMAG, Cairns.

[57] Smith I.M., Griffiths D.V. (1987). Programming the finite element method, Wiley,

[58] Terzaghi K. (1923). Die Berechnung der Durchlassigkeitsziffer des Tones aus dem Verlaug der Hydrodynam. Spannungserscheinungen, Ak.Wis.Wien math-nat.KlAbt.IIa 132.

[59] Terzaghi K. (1943). Theoretical Soil Mechanics, Wiley, NewYork.

[60] Zienkiewicz O.C. (1986). The Finite Element Method. McGraw-Hill, London.

[61] Zienkiewicz O.C., Bettess P. (1982). Soil and saturated media under transient, dynamic conditions; general formulation and the validity of various simplifying assumption. In Soil Mechanics - Transient & Cyclic Loads (Eds. G.N.Pande & O.C.Zienkiewicz).

[62] Zienkiewicz O.C., Leung K.H., Pastor M. (1985). Simple model for transient soil loading in earthquake analysis. Int.J. Num.Analyt. Meth. Geomech.

AMPLIFICATION EFFECTS OF WAVE PROPAGATION IN SLOPES

Enrico CONTE* and Giovanni DENTE*

ABSTRACT: Referring to a slope under plane strain conditions, the induced displacement field is determined for incidence of harmonic SH, SV, P and Rayleigh waves. Computations are made using a numerical procedure based on the Boundary Element Method.

RIASSUNTO: Con riferimento ad un pendio in condizione di deformazione piana, è stato determinato il campo degli spostamenti indotti da onde armoniche del tipo SH, SV, P e Rayleigh, al variare dell'angolo d'inclinazione del pendio. Per il calcolo, ci si è avvalsi di una procedura numerica basata sul Metodo degli Elementi di Contorno.

INTRODUCTION

In recent years, significant developments have been made in analyzing the stability of slopes and earth dams in the presence of earthquake loadings. It has been recognized that the seismic safety of a slope can be more suitably evaluated on the basis of the computed values of permanent soil deformations induced by a given earthquake motion, rather than by means of the safety factor involved in the pseudo-static

---

* Dip. Difesa del Suolo, Università della Calabria, Cosenza, Italy

approach. It has also been ascertained that the magnitude of permanent displacements strongly depends on the characteristics of the seismic excitation used as input. Design methods currently employed for predicting permanent displacements usually assume that the soil moves as a rigid body (Ref.1) and neglect the amplification effects on seismic motion due to the complex wave phenomena that occur in the slope. These phenomena cause significant variations of the ground motion and hence need to be taken into account in the calculations.

This paper deals with the propagation of seismic waves in slopes. In particular, the study centers on the effects of topography on ground motion. A numerical technique based on the Boundary Element Method (BEM) is used for this purpose. The method has certain inherent advantages and is highly suitable for dynamic problems involving semi-infinite domains (Ref.2). In fact, as is known, only the boundary of the domain needs to be discretized and consequently the procedure is less costly from a computational point of view, compared to other numerical techniques (such as FDM or FEM). Moreover, the fundamental solution used in the BEM formulation automatically satisfies Sommerfeld's radiation condition and, therefore it is not necessary to employ fictitious absorbing boundaries. The procedure has been previously applied to study wave diffraction from a topography irregularity of arbitrary shape and has provided accurate results (Refs.3 to 5).

The numerical applications made in the present study concern a cliff with various values of the slope angle under conditions of plane strain. The soil is modeled as a homogeneous isotropic elastic half-space. Harmonic incident SH, SV, P and Rayleigh waves, having wavelengths comparable with the sizes of the topographic irregularity, are considered. Particular attention is devoted to the case of incident Rayleigh waves because, as pointed out (Ref.6) the retrograde motion induced by Rayleigh waves is most effective in causing slope instabilities. The influence of the slope angle and of the incident wave type is investigated in some detail in order to quantify the amplification effects of topography on seismic ground motion in slopes.

WAVE PHENOMENA IN A SLOPE

A brief qualitative description of the complicated wave process occuring in a slope is premised here. When incident waves travelling in the half-space encounter a topographic irregularity, such as a cliff, they produce scattered waves which radiate outwards, both returning into the half-space as body waves, and propagating along the topographic surface, as surface waves. In addition, under plane strain conditions, except in the case of incident SH waves which generate scattered waves of SH type only, due to the presence of the scatterer, the energy associated with the incoming waves is generally converted into other wave forms. This phenomenon is known as mode conversion.

In particular, in the case of SV waves vertically incident at the surface of a cliff, Ohtsuki and Harumi (Ref.7) showed that Rayleigh waves are produced markedly near the slope of the cliff, and are transmitted along the upper and lower ground surfaces. In addition, SV waves are reflected at the surface and radiated downward. Scattered P waves with small amplitudes are also produced. Similar features have been found for incident Rayleigh waves propagating along the upper surface of a cliff (Ref.8).

Wave scattering and mode conversion are responsible for local effects on ground motion, and, therefore, they can play an important role on the dynamic response of slopes.

NUMERICAL RESULTS

In order to analyze the amplification pattern induced by seismic waves in slopes, the simple ground model shown in Figure 1 was considered. To account for the geometrical effects of the cliff, various values of the slope angle were assumed. A vertical slope was also considered. In the examples, the corners of the ground surface were locally smoothed with a curved outline, because their presence can give rise to unreal focusing of energy (Ref.4). The amplitude of displacements occurring at various points located both on

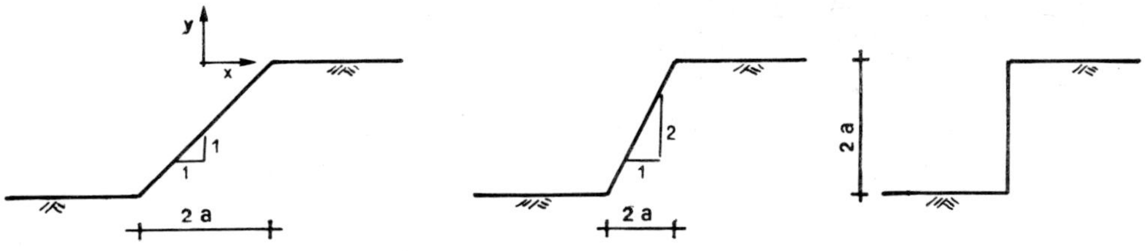

Fig.1: Slope shapes considered in this study

the boundary and along some verticals of the cliff was determined. The calculated values were normalized with regard to unit amplitude of the incident waves. All the results are given for a Poisson's ratio of 1/3. The dimensionless frequency η was assumed equal to 0.5 or 1, where $\eta = 2a/\lambda$, a is the characteristic dimension of the topographic irregularity, and λ is the incident wavelength. Plane SV and P waves travelling vertically in the elastic half space, and Rayleigh waves propagating along the upper surface of the slope are the main ones considered. For completeness, some results concerning the case of incident SH waves are also shown.

The free field amplitudes (in the absence of topographic irregularity) are given in Table I. These values are useful references for quantifying the overall effects of the slope on the ground motion.

wave type	Ux	Uy	Uz
P	0	2	-
SV	2	0	-
R	1	1.56	-
SH	-	-	2

Tab. I. Free-field amplitudes

The numerical results are reported in Figures 2 to 12, where the vertical displacement Uy is depicted by a solid line while the horizontal displacement Ux is drawn by a dotted line. Out-plane displacement is denoted by Uz.

The displacement field induced by vertically incident SV waves is first analyzed. From the results shown in figures 2 to 5, it appears that maximun values of the horizontal displacement occur at the upper surface near the top of the cliff. However, these values are not much greater than those relative to a homogeneous half-space. In all the examples shown, a deamplification effect takes place along the slope. On the lower surface, horizontal displacement amplitude is fairly near that of the free field condition. Moreover, it is interesting to note that although the free field vertical component is zero, vertical displacements arise in the ground. At the foot of the cliff, values of Uy are insignificant, however, they increase rapidly along the slope toward the upper surface, where their amplitude is comparable with that of horizontal displacements. This can be explained by the fact that due to mode conversion phenomena, some of the energy associated with the horizontal component is converted to vertical energy. In the cases examined, these effects, as well as the magnitude of the calculated displacements, seem to be very little influenced by the slope angle and the value of η. Higher frequencies do not give rise to significant changes in maximum amplitude but, probably due to interaction between the direct and diffracted waves (Refs.7, 8), only a more oscillatory pattern is produced (Fig.5).

Figure 6 is relative to the case of incident P waves. As can be observed, P waves generally cause little differences with regard to the free field condition. Moreover, horizontal displacements of small amplitude are also produced. Similar comments can be made on the displacement field induced by incident SH waves (Fig.7). As a partial conclusion, it can be asserted that for incident S and P waves, the amplification ratio reaches a peak value which is about 50 per cent greater than that of a homogeneous half-space.

The wave amplification patterns of Ux and Uy generated by a Rayleigh waves propagating along the upper surface of the cliff from right to left are shown in Figures 8 to 11. The results indicate that large amplification effects of both

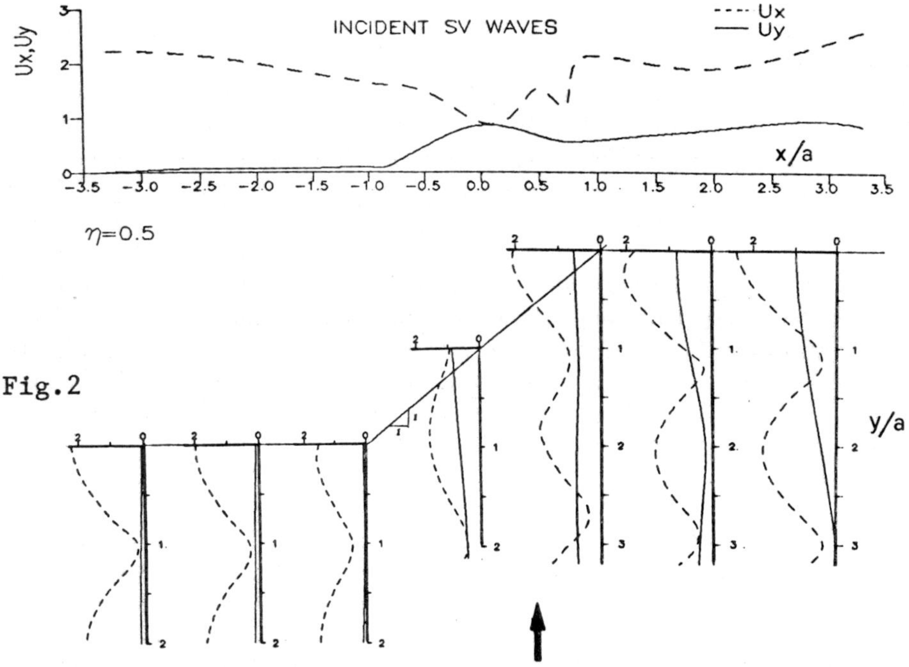

Fig.2

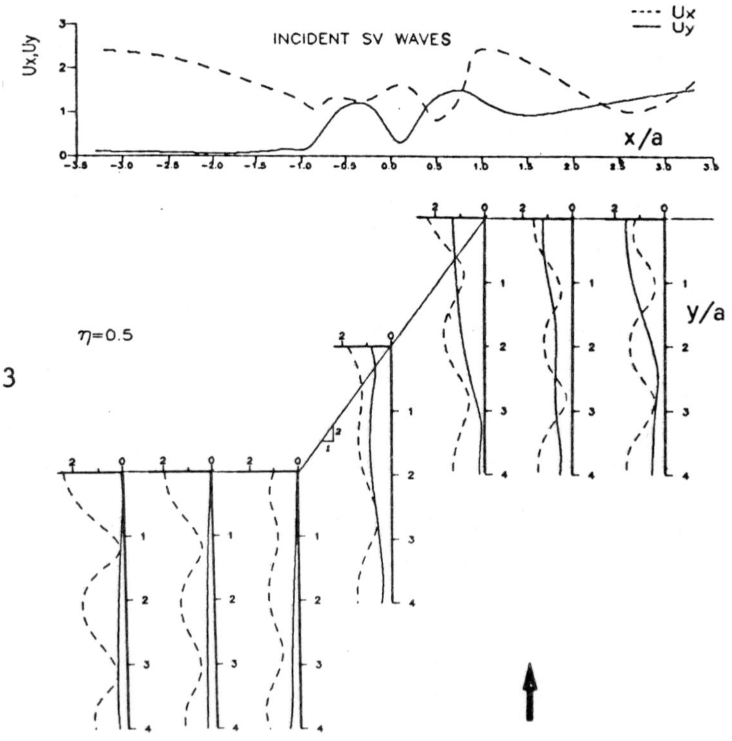

Fig.3

Figs.2-3: Displacement amplitudes for incident SV waves

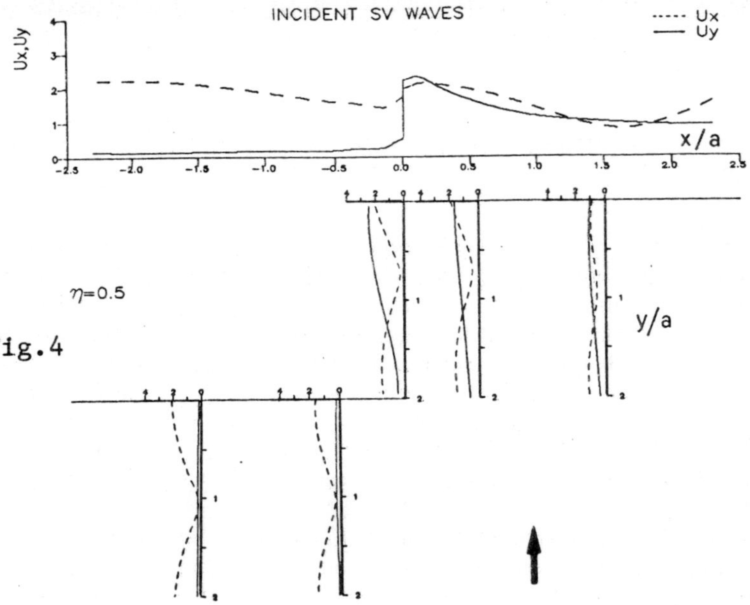

Fig.4

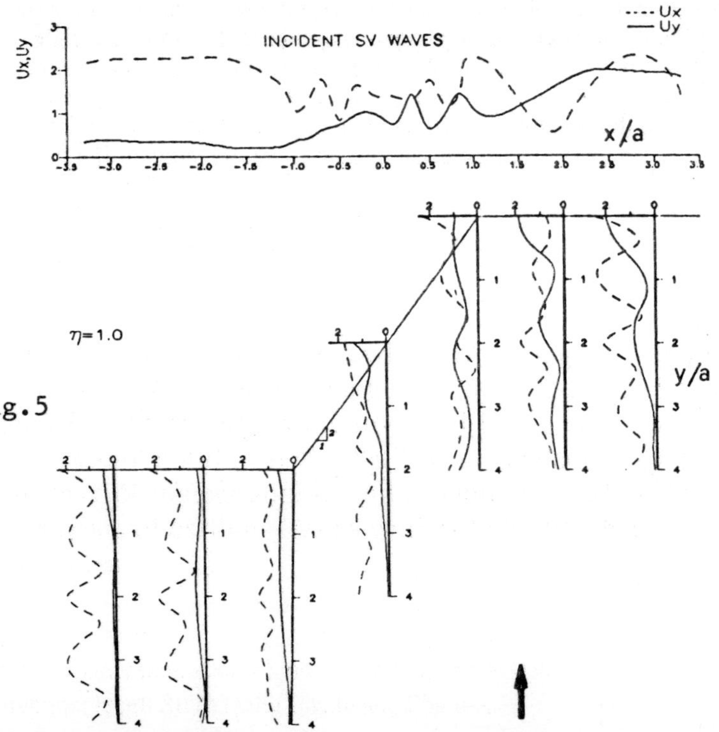

Fig.5

Figs.4-5: Displacement amplitudes for incident SV waves

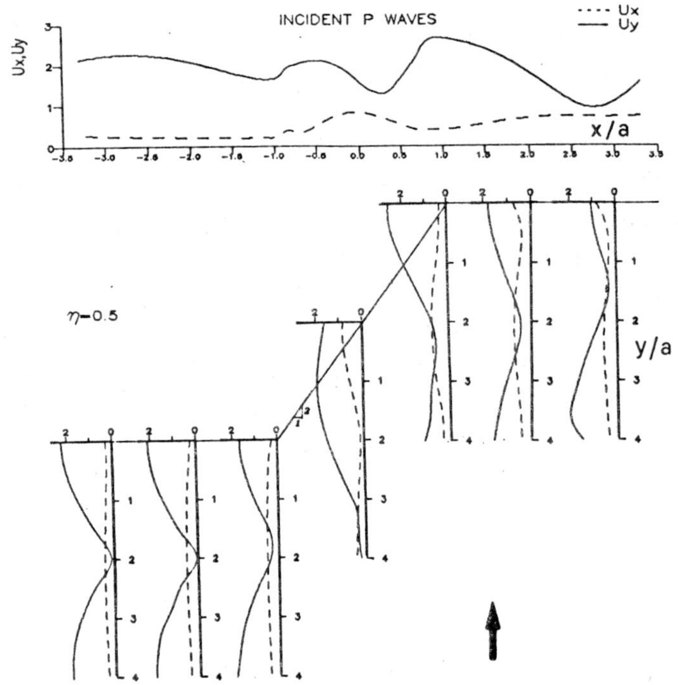

Fig.6: Displacement amplitudes for incident P waves

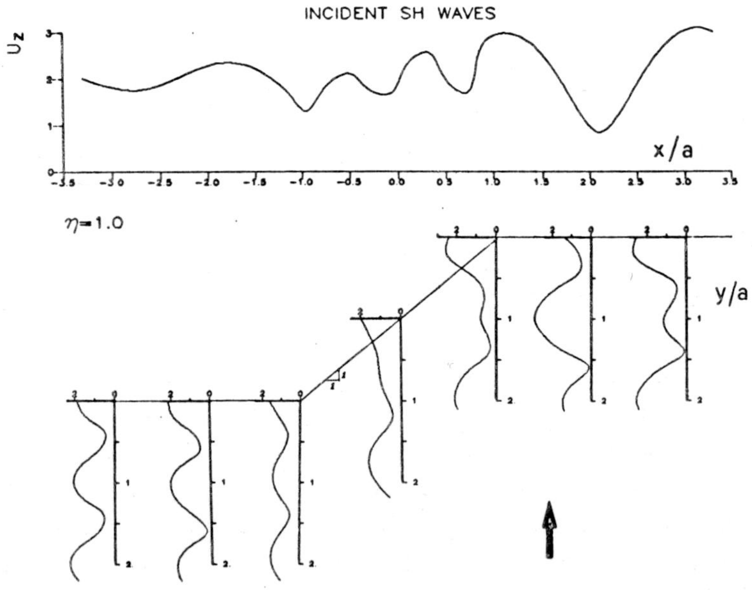

Fig.7: Displacement amplitudes for incident SH waves

vertical and horizontal components occur around the top of the cliff. As expected, being an incident surface wave, displacement amplitudes gradually decay into the interior of the half-space. The foot of the slope and the lower surface, on the contrary, are zones where a significant reduction in wave amplitudes takes place. Moreover, surface displacement distribution exhibits a peak value in proximity to the upper corner of the slope, while it takes a minimum value at the foot of the cliff. These effects become more pronounced as the angle of the slope increases. In particular, for the vertical slope (Figs.10, 11), the upper parts of the slope undergo significant amplification. In this case, the horizontal amplitude is more than 3.5 times as large as that of the incident waves. At the same time, the displacement field is highly deamplificated in the ground below the cliff. Therefore, a sort of shield to the propagation of seismic surface waves is produced in this zone of the slope. Finally, as can be noted, peak values of Ux and Uy attain the same magnitude, especially when the slope angle is highly inclined (Figs.9 to 11).

CONCLUSIONS

Some results are presented for the displacement field generated in slopes by the propagation of P, S and Rayleigh waves. Computations are made using a numerical procedure based on the Boundary Element Method. For the cases examined here, the results show some features which can be summarized as follows. A zone of amplification takes place at the top of the slope while a significant reduction of wave amplitudes can occur at the foot of the cliff, and along its slope. For vertically incident S and P waves, displacement amplitude is generally around 3, whereas the maximum amplitude of free field displacement is 2, which means that the amplification is about 1.5. In addition, for incident SV waves, due to mode conversion phenomena, vertical displacements whose amplitude is not insignificant, are produced at the top of the cliff. Large amplification values can be caused by Rayleigh waves

Fig.8

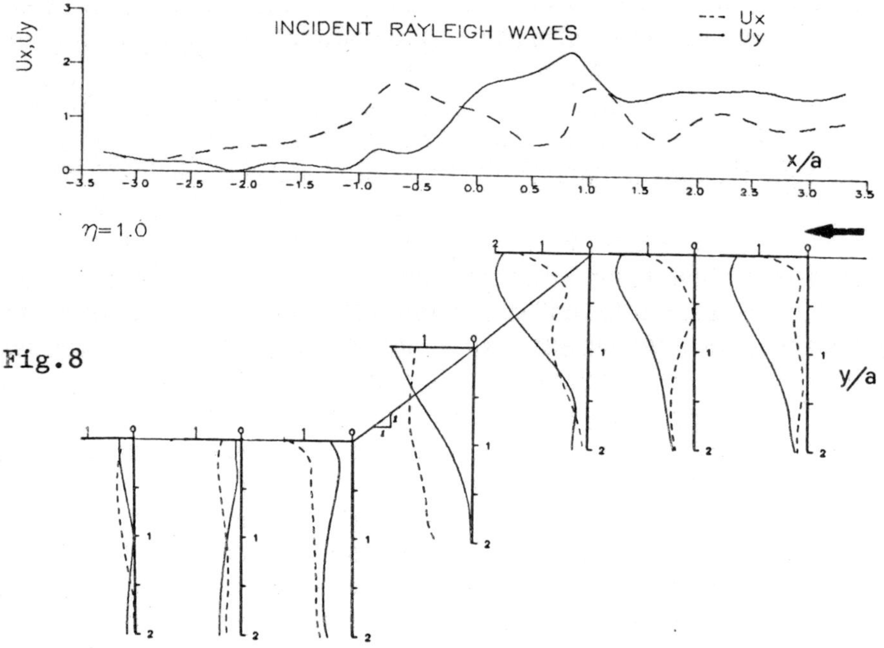

Fig.9

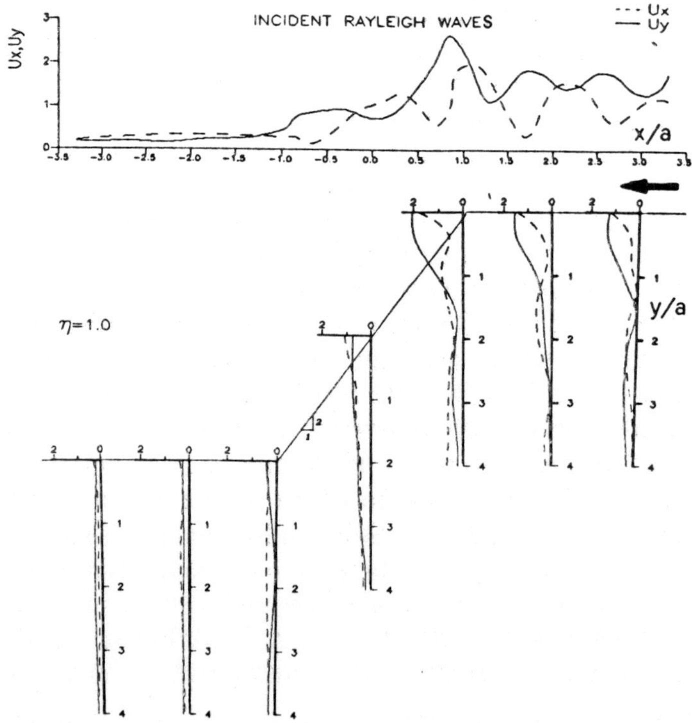

Figs.8-9: Displacement amplitudes for incident Rayleigh waves

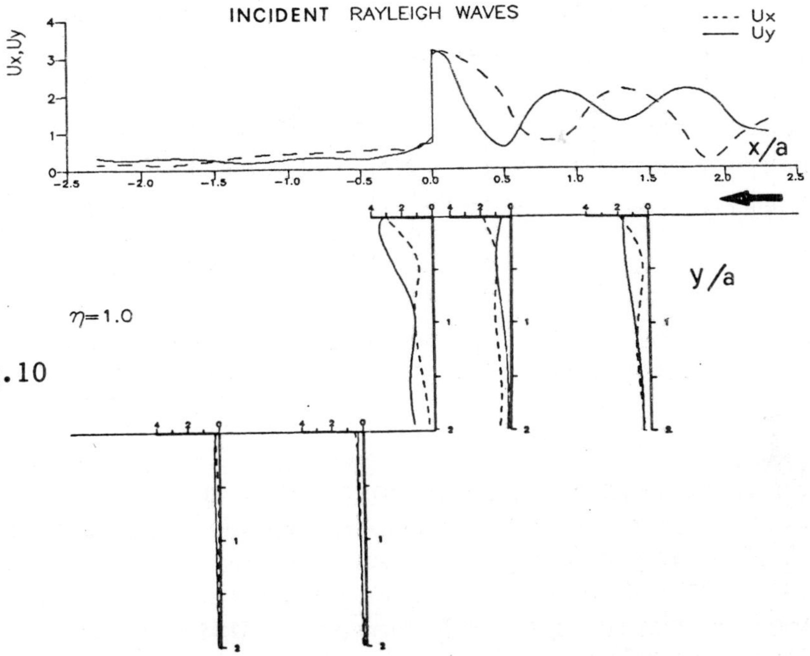

Fig.10

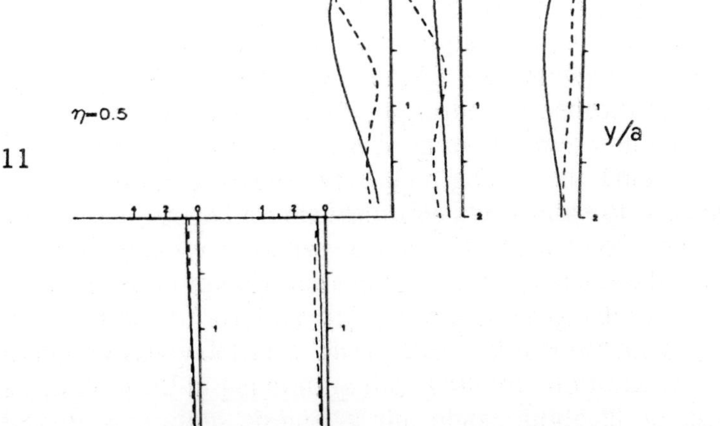

Fig.11

Figs.10-11: Displacement amplitudes for incident Rayleigh waves

propagating from the back of the cliff. In particular, for a vertical slope the horizontal component near the upper corner of the slope reaches a value more than 3.5 times that of the incident waves.

The results obtained in this study can prove useful for the assessment of the seismic response of slopes although they are relative to highly idealized conditions.

REFERENCES

1. Newmark N.M. (1965). "Effects of earthquakes on dams and embankments". Geotechnique, Vol.15, No.2, 139-160.

2. Brebbia C.A., Telles J.C.F. and Wrobel L.C. (1984). "Boundary Element Technique". Spring-Verlag Ed.

3. Conte E., Dente G. and Guerra I. (1988). "Calcolo dell'effetto sismico locale mediante il Metodo degli Elementi di Contorno". Atti 7° Conv. Ann. Gruppo Naz. di Geof. della Terra Solida, 341-354, Roma.

4. Conte E. and Dente G. (1989). "L'indice TEI per la misura degli effetti sismici in presenza di irregolarità topografiche". Atti 4° Conv. Naz. L'Ingegneria Sismica in Italia, Vol.1, 272-279, Milano.

5. Conte E. (1990). "Diffrazione di onde P e SV in presenza di irregolarità topografiche". Atti Riun. Ann. Ricerc. Ing. Geotec., CNR, Roma.

6. Wong H.L. (1982). "Effect of surface topography on the diffraction of P, SV and Rayleigh waves". Bull. Seism. Soc. Am., Vol.72, No.4, 1167-1183.

7. Ohtsuki A. and Harumi K. (1983) "Effect of topography and subsurface inhomogeneities on seismic SV waves". Earthq. Eng. and Struct. Dynam., Vol.11, 441-462.

8. Ohtsuki A., Yamahara H. and Harumi K. (1984) "Effect of topography and subsurface inhomogeneity on seismic Rayleigh waves". Earthq. Eng. and Struct. Dynam., Vol.12, 37-58.

SEISMIC STABILITY ANALYSIS OF SLOPES INCLUDING AMPLIFICATION EFFECTS

Teresa CRESPELLANI[1], Claudia MADIAI[1], Giovanni VANNUCCHI[1]

Abstract: The procedure described herein is an extension of a method previously proposed by the authors, for overcoming some implicit limitations in Newmark's model for the seismic and post-seismic stability analysis of clayey slopes. It is applicable also in complex morphological and stratigraphic conditions. This paper presents further modifications to the procedure, introduced to account for the amplification effects along the slope and for the time variations of the shear strength due to the loading rate and to the cyclic degradation.

Sommario: La procedura descritta e' un'estensione di un metodo, precedentemente proposto dagli Autori allo scopo di rimuovere alcuni limiti impliciti nel modello di Newmark, per l'analisi sismica e post-sismica delle condizioni di stabilita' di pendii argillosi, applicabile anche in presenza di condizioni morfologiche e stratigrafiche complesse. Nel presente lavoro sono presentate le ulteriori modifiche alla procedura, introdotte per tener conto degli effetti di amplificazione lungo il pendio e della variazione nel tempo della resistenza al taglio, dovuta alla velocita' di applicazione dei carichi ciclici e alla degradazione ciclica.

Introduction

A deformation method for stability analysis of natural clayey slopes, during and after seismic events, was recently presented by the authors (Crespellani et al., 1990). The application of this method requires a preliminary definition of design accelerograms. Saturated cohesive soils, subjected to cyclic stress paths in undrained conditions, show a positive pore pressure increment only if shear strain becomes larger than a threshold value, γ_{tp}; this limiting value is a function of the over consolidation ratio, OCR, and of the plasticity index I_P. According to some recent experiences, the pore pressure increment, at the end of the cyclic stress path, u_r, can be computed as a function of γ_{max}/γ_{tp} ratio, where γ_{max} is the maximum shear strain.

The proposed method consists of two main analyses. The first

[1] Dipartimento di Ingegneria Civile - Universita' di Firenze

analysis needs: to determine the threshold shear strain; to compute the maximum shear strain within the slope due to the design earthquake, and to calculate the eventual pore pressure increment. A post earthquake stability analysis, is cautiously performed using a general pore pressure increment, u_r, applied to the total length of the potential critical slip surface. If such analysis leads to a safety factor less than one, it is possible, yet not certain, that after an earthquake collapse conditions can be reached, thereby leading the slope to a landslide. The second analysis consists of computing the displacement of the potential sliding soil mass, during an earthquake. Generalising the criteria on which Newmark's method is based, this procedure can be extended to natural slopes, therefore including conditions such as: irregular ground surface, general stratigraphic conditions, external distributed loads with any inclination, irregular slip surfaces.

The main steps of the analysis are: a) to determine the critical slip surface, if necessary, using the pseudostatic method and the critical seismic coefficient K_c; b) to compute permanent displacement by a numerical integration of the differential equation which describes the relative motion of the sliding mass, using the design accelerogram; c) to compute the total final displacement as a sum of all the incremental movements computed in the preceding step; d) to compare this displacement with that allowable for the slope.

In deformation methods it is plane or logspiral surfaces which must be adopted. In the proposed procedure, if the slope is homogeneous, the logspiral surface which minimizes the seismic coefficient, is determined directly; in the more general case of a non-homogeneous slope, the critical surface, irregular in shape, is determined with the analytical technique for slices as the STABL-PC method (Crespellani et al., 1986; Madiai, 1989) and is close to an equivalent logspiral surface. In spite of the generality of the approach, an assumption was that the seismic coefficient does not vary during the earthquake and that seismic acceleration is the same for a given time in every point of the slope. As shown by numerous experiences, the undrained shear strength of cohesive soils in cyclic conditions (and thus the seismic coefficient) is, at each moment, influenced by the cyclic load path (that is, substantially, the accelerogram). In particular it increases with the load application rate and diminishes progressively with the number of cycles. In consequence, the displacement during the earthquake depends on such factors. Furthermore, the hypothesis of a uniform field of seismic acceleration in the slope (and therefore the assumption of a single input accelerogram) seems to be extremely limiting, especially for slopes of a elevated height and/or constituted of soils not assimilable to infinitely rigid materials. To overcome the limits of the previous procedure, some modifications have been introduced into the analysis method. These allow time variations of the seismic coefficient to be taken into account as well as the effects of the amplification, or the difference in seismic acceleration within the slope at various depths of the ground surface.

1. Time variations of the critical seismic coefficient

As noted, in Newmark's original method, the mass of soil between the slip surface and the topographic surface is represented as a rigid block resting on a vibrating surface subject to the weight and to the relative seismic force of inertia. Such a force varies during the earthquake and determines the slip of the block every time it exceeds the critical threshold value corresponding to the condition of the equilibrium limit. The value of the seismic coefficient is defined as critical when it corresponds to the condition of the equilibrium limit. Having specified the potential critical slip surface and the design accelerogram, the integration of the differential equation of motion is made by considering those time intervals in which the acceleration of the block exceeds the threshold value and moreover, all those intervals in which the relative velocity of the block is greater than zero. The critical seismic coefficient, K_c, is a function of the geometric, stratigraphic and geotechnical characteristics of the slope. In particular, it is a function of the available shear strength in seismic conditions along the potential critical slip surface, τ_f. During an earthquake the undrained shear strength of a cohesive soil undergoes some variations and consequently the critical seismic coefficient endures analogous variations. These are due to the combined opposite effects of two principal causes: a) the loading rate whose growth induces an increase in strength, and b) the cyclic degradation, probably associated with the development of excess pore pressure and/or the collapse of the inherent structure which undergoes a reduction in strength. In practice this may be expressed as:

$$K_c(t) = K_{c,o}\, R(t)\, D(t) \qquad (1)$$

in which: $K_{c,o}$ is the critical seismic coefficient before the earthquake, corresponding to the available shear strength in static conditions along the potential critical slip surface, $R(t)$ is a coefficient of the loading rate and $D(t)$ is an index of cyclic degradation.

For a quantitative evaluation of the coefficient of the loading rate, reference is made to the equation (Poulos, 1988):

$$R(t) = 1 + F_R \log\left(\frac{r(t)}{r_r}\right) \qquad (2)$$

where: F_R is the rate factor (typically in the range 0.05 to 0.2), which is assumed to be 0.1 in the absence of direct measurement, r_r is the reference loading rate which is assumed as equal to the ratio between the static shear strength in undrained conditions and the reference time t=15', $r(t)=\frac{d\tau(t)}{dt}$ is the actual seismic loading rate at the characteristic depth, d_c, of the potential slip surface:

$$d_c = \frac{\Sigma_i\, b_i d_i}{\Sigma_i\, b_i} \tag{3}$$

in which b_i and d_i are the width and average height respectively of the slice i. To calculate the degradation index, D(t), it is necessary to refer to the strain path during the earthquake at the characteristic depth: the figure 1 contains the definition of the cycle n and of the corresponding cyclic shear strain, $\gamma_{c,n}$, for an irregular strain history. As stated, the cyclic degradation is probably associated to the development of excess pore pressure and/or to the collapse of the inherent structure. A positive increase of the pore pressure in cohesive soils subjected to a cyclic load path in undrained conditions occurs only if the shear strain exceeds the threshold value, $\gamma_{t,p}$. Such a threshold value increases with the over-consolidation ratio OCR, and with the plasticity index, I_P, and can be calculated with the equation (Matsui et al., 1980):

$$\gamma_{tp} = A\,(OCR\text{-}1) + B \tag{4}$$

in which A and B are functional parameters of the plasticity index (fig. 2). Taking account of the fact that for low strain cycles the maximum reduction in strength is about 10% (Thiers and Seed, 1968), the index of degradation of the shear strength in the cycle (n+1) along the potential slip surface is calculated thus[2]:

if $\gamma_{c,n+1}>\gamma_{tp}$ or if $\gamma_{c,n+1}<\gamma_{tp}$ and $D_n>0.9$ $\quad D_{n+1}=D_n\,(1+D_n^{\frac{1}{k_n}})^{-k_n}$ (5)

if $\gamma_{c,n+1}<\gamma_{tp}$ and $D_n < 0.9$ $\quad D_{n+1} = D_n$ (6)

[2] Although the index of degradation is sometimes expressed as a function of the increment of pore pressure, analytical expressions containing only the number of cycles are preferred here, for more general application of the procedure.

in which k_n[3] is a parameter of degradation calculated by a hyperbolic equation of the type:

$$k_n = \frac{\gamma_{c,n}}{a + b\ \gamma_{c,n}{}^{c}} \tag{7}$$

As an example and as values for reference, Idriss et al. (1980) indicate the following values for the constants of the equation (7) for normally consolidated marine clay: a=0.4, b=7.3, c=0.61.

2. Amplification effects

The calculation of the slope displacement during an earthquake is usually carried out with the model of a rigid block resting upon a vibrating surface. The input is the motion of the vibrating surface as defined by an accelerogram, a(t). The output is the motion of the block as defined by an accelerogram, $a_b(t)$. If the sliding resistance of the block on the vibrant surface is not exceeded, the result is that $a_b(t)=a(t)$, and there is no relative displacement of the block; in the opposite case the result is that $a_b(t) \neq a(t)$ and relative displacement of the block occurs. In both cases, since the block is considered rigid and each of its points is subject to the same accelerogram $a_b(t)$, the resulting force of inertia is, at all times, applied at the centre of gravity of the mass.

Fig. 1: Definition of the equivalent strain cycle n

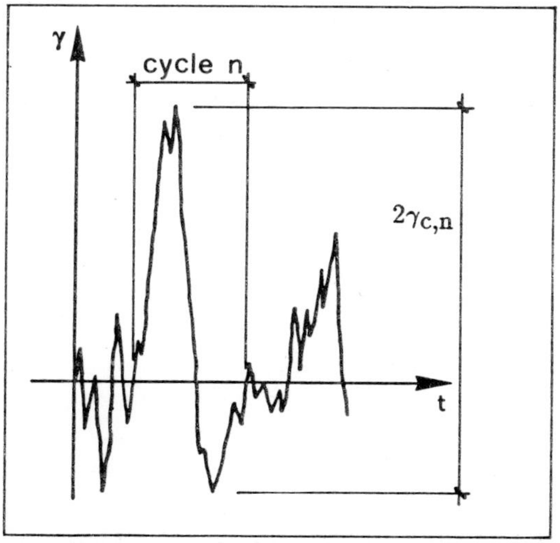

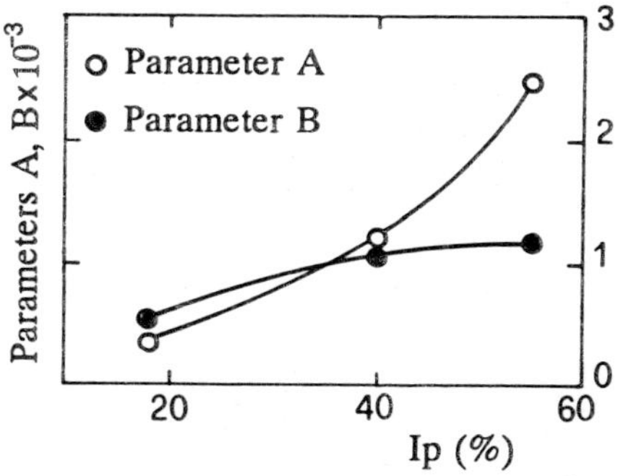

Fig. 2: Parameters A and B versus I_P

[3] The degradation parameter is generally indicated by the letter t; in this paper k_n is used since t is used to indicate the time variable.

In reality the seismic acceleration varies continuously from one point to another of the slope because of several factors (morphological, stratigraphic, geotechnical, etc..), that is, the field of acceleration is not uniform and the resulting seismic force of inertia is not applied to the centre of gravity of the block. The proposed method takes account of the fact that the slope is not a rigid block, but that within it the acceleration varies at any moment from one point to the next. The method provides for: 1) determining the slip surface on the hypothesis of a uniform acceleration at any instant throughout the slope and therefore delimiting the volume of potentially sliding soil; 2) dividing such volume into zones for which a single accelerogram can be assumed. Such zones, of general size and shape can be individualised and delimited in a manner more or less restricted by the geometric, stratigraphic and geotechnic conditions, for determining the accelerograms for each zone[4]; 3) the calculation of the mass and the centre of gravity of each zone; 4) the determination of the representative accelerogram for each zone; 5) the calculation of the seismic force instantaneously associated with and applied to the centre of gravity of each zone as produced in the mass by the seismic acceleration; 6) the determination of the resultant seismic force in each homogeneous zone and its position and direction; 7) the calculation of the cumulated displacement taking account of the time variation of the force of inertia in modulus, direction and position.

---

[4] Two methods are commonly used to determine the seismic motion at any point of a zone: the one-dimensional analysis (1-D) of a column of soil on the hypothesis of an infinite lateral extent, and the two-dimensional FEM analysis (2-D). The first method is much simpler, more rapid and economic, but implies simplified hypotheses about stratigraphic and morphological conditions. Nevertheless, in many cases, and particularly for natural slopes with significant uncertainty about their stratigraphic and geotechnic data, the error margin of amplification effects is acceptable. Elton et al. (1991) have compared results, in terms of shear stresses, of the dynamic analysis of earth dams using SHAKE (one-dimensional) and FLUSH (two-dimensional) methods. They concluded that for a slope of less than 35° the differences are small. According to Stara-Gazetas and Dobry (1991) the comparison of results from "equivalent linear" finite elements and one-dimensional analysis carried out for the lower S. Fernando dam revealed that both methods predict about the same values for peak acceleration, shear stress and strain, but that the detailed characteristics of time histories are remarkably different. Instead, comparison of results from "plasticity" finite elements and the one-dimensional analysis lead to the conclusion that also time histories are in good accord.

3. Illustrative example

Although the method of calculation may be extremely general and applicable to irregular slopes with complex geotechnical and stratigraphic conditions, the illustrative example is deliberately simple in order to follow and check the partial results, and to highlight the differences between the results obtained from different hypotheses. In particular, reference is made to the case of a homogeneous slope of limited height in order to apply also Taylor's method (1948) for static conditions and that of Chang et al. (1984) for the seismic conditions.

In regard to the zones for which it is possible to consider a uniform field of acceleration, reference is made to the subdivision into slices used for the application of the methods of Bishop and of Carter. In the hypothesis that the relation between the depth of bedrock and the height of the slope (depth factor) is sufficiently large not to influence the seismic motion close to the surface, the accelerograms for the various zones are determined by the SHAKE code (one-dimensional analysis), making reference to a single column of homogeneous soil and determining the acceleration time history at the depth of the centre of gravity of each slice measured from the summit.

Geometric and geotechnical data

The assumed geometric and geotechnical data for the slope, which consists, for the most part, of unsaturated material ($\phi_u \neq 0$), are shown in figure 3. Moreover,
shear modulus (fig. 4):

$$G = \frac{G_o}{(1+\alpha\gamma^{\beta})} \quad \text{with } \alpha = 30, \quad \beta = 1.2, \text{ and } \quad G_o = 10\ Z + 100\ (\text{MPa})$$

and damping ratio (fig. 4):

$$D = D_{max} \exp(\lambda \frac{G}{G_o}) \qquad \text{with} \qquad D_{max} = 0.15 \quad \text{and} \quad \lambda = -2.3$$

Seismic data

The design accelerogram at bedrock (fig. 5) is assumed to be equal to the first 20 seconds of the W-E component of the accelerogram recorded at Gubbio (Italy) on 29 April 1984 scaled to $a_{max} = 0.40$ g. The time interval between acceleration values is 0.01 secs. Table 1 shows the values of the safety factor of the slope in static conditions, F_s, and the initial critical seismic coefficient, $K_{c,o}$, calculated by different methods.

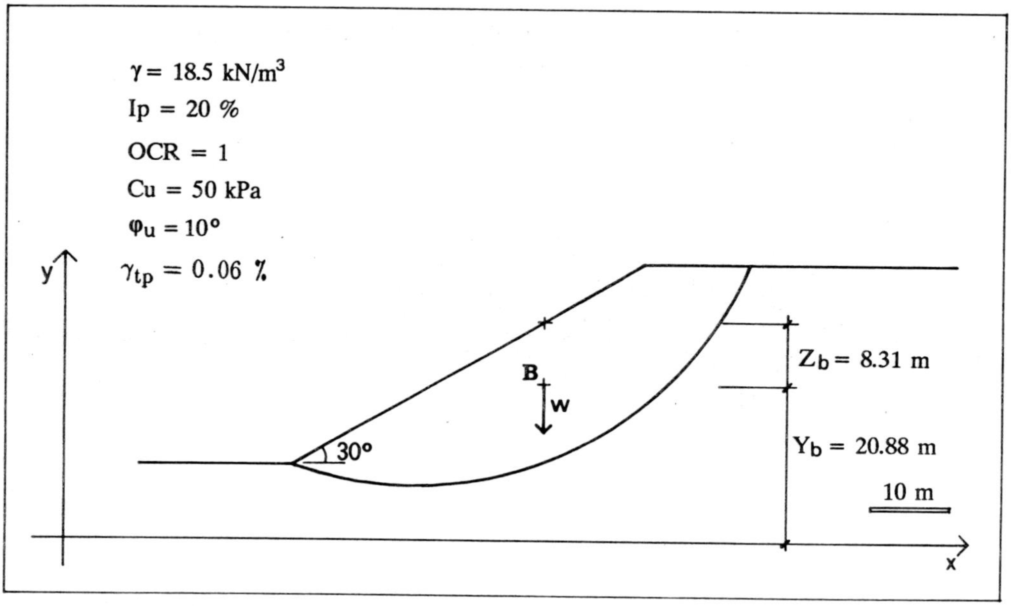

Fig. 3: Geometric and geotechnical data for the illustrative example

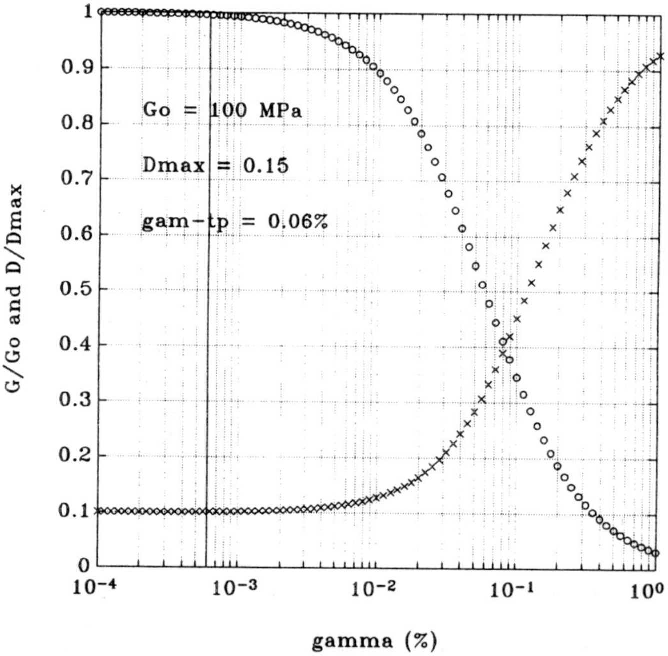

Fig. 4: Dynamic properties for the illustrative example

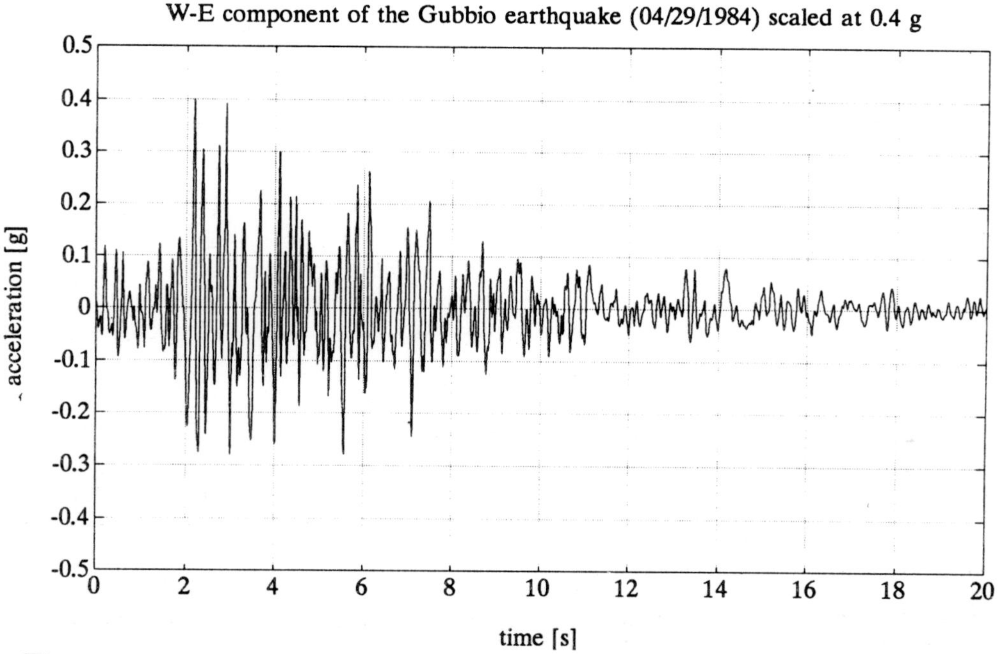

Fig. 5: Design accelerogram at the bedrock for the illustrative example

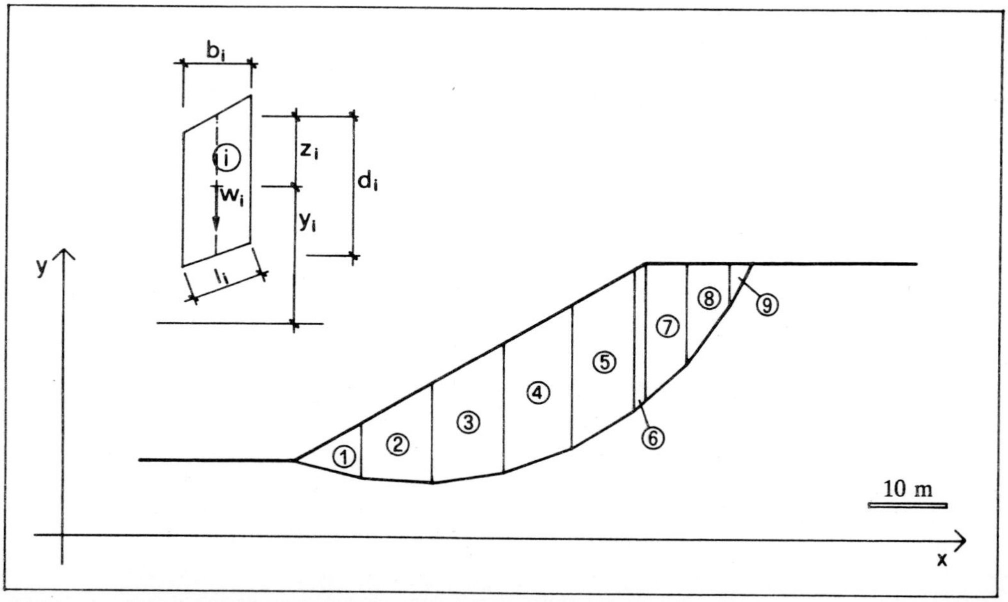

Fig. 6: Zones with uniform field of acceleration (slices) and geometric definitions for the illustrative example

Method	F_s	$K_{c,o}$
Taylor	1.32	----
Bishop	1.20	0.088
Carter	1.12	0.052
Chang et al.	----	0.085

Table 1

In figs. 3 and 6 the circular critical surface is shown as determined by Bishop's pseudostatic method, in fig. 3 the position of the centre of gravity, B, and in fig. 6 the subdivision into 9 zones (with a single acceleration path) of the potentially sliding mass. Referring to the symbols in fig. 6, table 2 shows the numerical values of the weight and the geometrical features of the 9 slices.

i	W_i(kN/m)	Z_i(m)	Y_i(m)	d_i(m)	b_i(m)	l_i(m)
1	648.8657	1.90201	10.75768	3.80403	9.22018	9.5000
2	1855.1520	5.28357	12.77319	10.56714	9.48965	9.5000
3	2698.1450	7.76318	15.74062	15.52637	9.39342	9.5000
4	3056.5220	9.24529	19.54562	18.49059	8.93522	9.5000
5	2910.6310	9.67279	24.04155	19.34558	8.13268	9.5000
6	571.4081	9.48150	27.04865	18.96300	1.62880	2.2052
7	1617.8290	8.11535	28.88465	16.23069	5.38795	7.2947
8	1036.0060	4.97301	32.02699	9.94601	5.63043	9.5000
9	162.1638	1.53004	35.46996	3.06008	2.86450	6.7573

Table 2

Furthermore:
the total weight of the soil in the slip area: $W=\Sigma W_i=14557$ kN/m
the depth of the centre of gravity: $Z_B = 8.307$ m
vertical coordinate of the centre of gravity: $Y_B = 20.883$ m

characteristic depth: $d_C = \dfrac{\Sigma_i\, b_i d_i}{\Sigma_i\, b_i} = 12.97$ m

Using the one-dimensional code SHAKE for seismic analysis of a column of homogeneous soil with design accelerogram at bedrock, the accelerogram at the centre of gravity, Y_B, as well as the stress history, $\tau(t)$, and the shear history, $\gamma(t)$, at the depth, d_C, are calculated. With the techniques described in par. 1 the following have been calculated: a) the stress history, $\tau(t)$, and the coefficient of the loading rate, R(t), (fig. 7a); b) the shear history, $\gamma(t)$, and the index of cyclic degradation, D(t), (fig. 7a); c) the critical seismic coefficient, $K_c(t)=K_{c,o}\ R(t)\ D(t)$, (fig. 7b).

In fig. 8 the accelerogram calculated at the depth of the centre of gravity and the critical seismic coefficient, $K_c(t)$, are shown. To further clarify the calculation procedure any moment of the earthquake can be considered. For example, that corresponding to a time of t = 8.5 sec. This results in:
the instantaneous actual seismic loading rate: $r(t{=}8.5sec)=48.95$ kPa/sec,
the coefficient of the loading rate: $R(t{=}8.5sec)=1.34$,
the actual strain cycle: n=35,
the cyclic shear strain: $\gamma_n=0.015$ %,
the degradation parameter: $k_n=0.0157$,
the degradation index: $D(t{=}8.5sec)=0.924$.

The critical seismic coefficient therefore results in a value of:

$$K_c(t{=}8.5\ sec) = K_{c,o}\ R(t)\ D(t) = 0.088 \cdot 1.34 \cdot 0.924 = 0.109.$$

The seismic acceleration in that instant and the corresponding force of inertia in the 9 zones, at all respective centres of gravity, follow in table 3:

i	a_i(g)	$F_{h,i}$(kN/m)
1	0.1436	93.1648
2	0.1020	189.2070
3	0.0804	216.9794
4	0.0740	226.1673
5	0.0714	207.9035
6	0.0714	40.8151
7	0.0780	126.2311
8	0.1020	105.6623
9	0.1464	23.7400

Table 3

The resulting seismic force is thus: $F_h(t{=}8.5\ sec)=\Sigma F_{h,i}=1229.87$ kN/m,

and is applied at a depth of: $Y_h(t{=}8.5\ sec)=\dfrac{\Sigma F_{h,i} Y_i}{F_h}=20.51$ m.

In fig. 9 the cumulated displacement in the slope calculated as

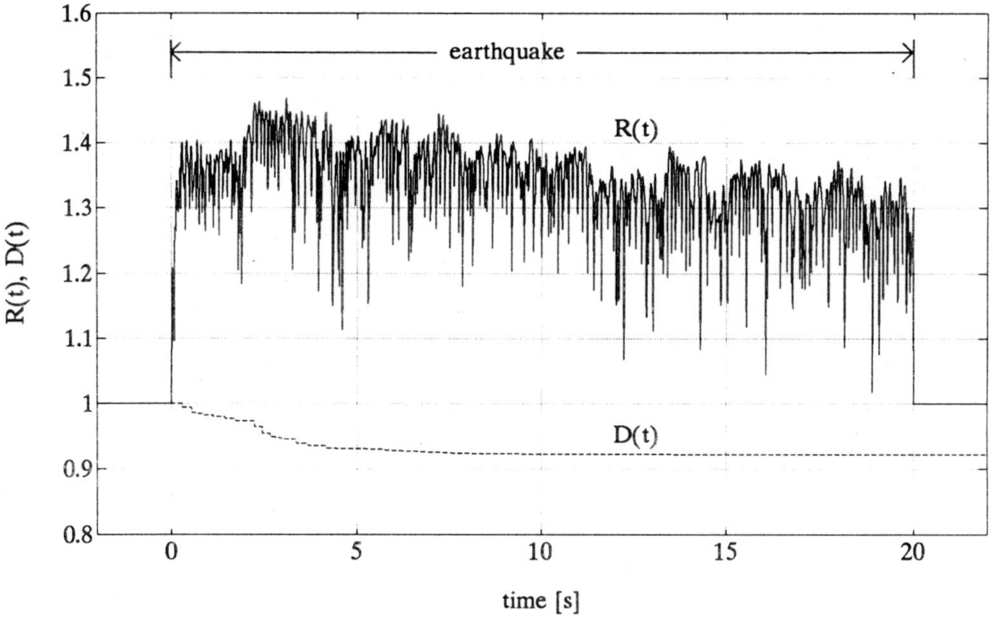

Fig. 7a: Loading rate coefficient, R(t), and cyclic degradation index, D(t), before, during and after the earthquake

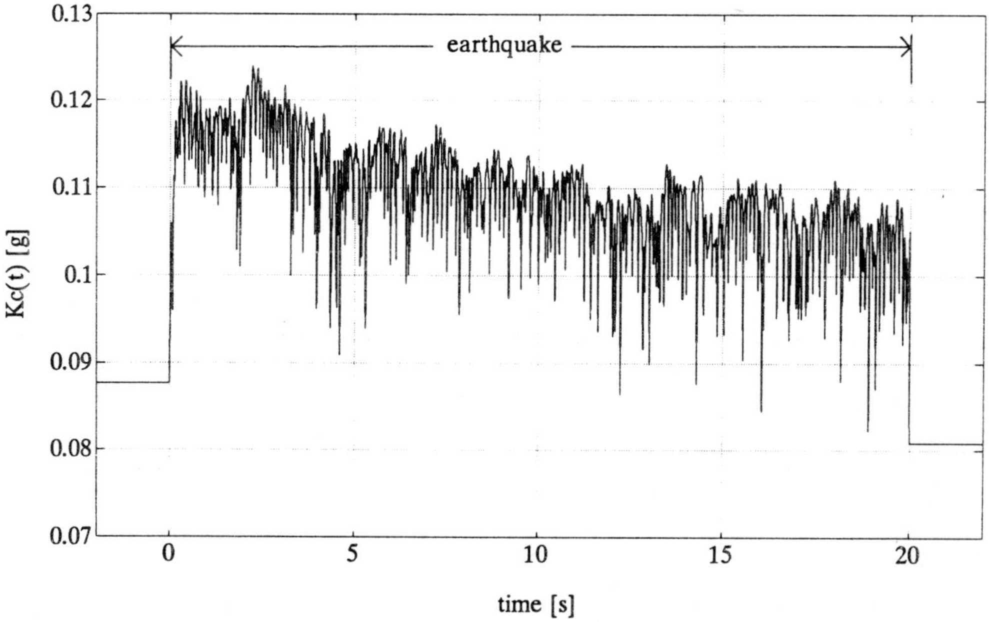

Fig. 7b: Seismic coefficient, $K_C(t)$, before, during and after the earthquake

shown above is compared with that obtained when the effects of the time variation of the critical seismic coefficient are disregarded (assuming $K_C=K_{C,o}$). In order to evaluate the influence of the amplification effects, the cumulated displacement in the slope is calculated assuming a single accelerogram (i.e. that of the bedrock) in the cases of K_C as variable in time and K_C as constant. The results are shown in fig. 10. It is a rather complex computation to take account of the amplification effects in the proposed method, when a different acceleration path for each zone is assumed. Thus a comparison has been made of the displacement obtained by this method with those drawn from a simpler procedure in which a uniform field of acceleration is assumed for the entire mass delimited by the critical surface. Also using the SHAKE code the amplification effects have been calculated only at the depth of the centre of gravity. The accelerogram thus determined (generally differing from that at the bedrock) has been assumed as the design accelerogram for the calculation of the displacement. The case has also been examined with K_C as variable in time and with K_C as constant. The displacements obtained are shown in fig. 11. Table 4 shows the values of the final displacement in the different cases examined.

Field of acceleration	displacement [cm] (for $K_C=K_{C,o}$)	displacement [cm] (for $K_C=K_C(t)$)
Non uniform	15.3	8.9
Uniform with ampl. effects	14.2	8.3
Uniform without ampl. eff.	7.8	4.8

Table 4

From the results obtained it is possible to observe that the amplification effects as well as the cyclic degradation effects, influence the displacement, if considered separately. By contrast, when considered together, we note a compensation effect though to limited extent in this illustrative example. Furthermore the results obtained by the simplified method in which a uniform field of acceleration is assumed (i.e. a single accelerogram for the mass in movement, differing from that of the bedrock), do not differ significantly from those obtained by the method which considers a non-uniform field of acceleration.

Conclusions

The numerical example was introduced only to illustrate the method in detail, and to that end a very simple, schematic example was chosen. In reality using this method it is possible to analyse the actual seismic and post-seismic conditions of very complex morphological, stratigraphic and geotechnical slopes and to consider any surface whatever and different local seismic responses within the mass in movement. In the illustrative example for the calculation of the seismic response, the area was subdivided into zones (slices) delimited by the

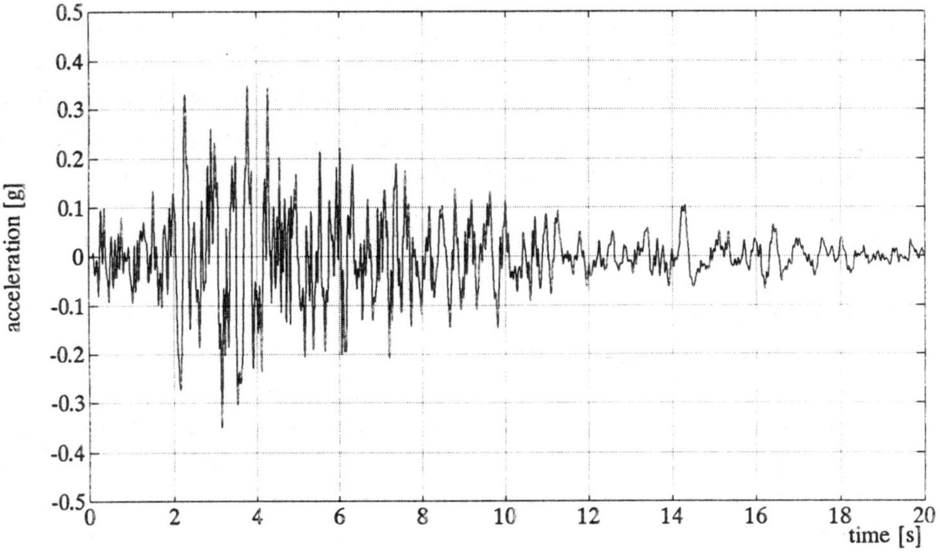

Fig. 8: Computed accelerogram at the centre of gravity for the illustrative example

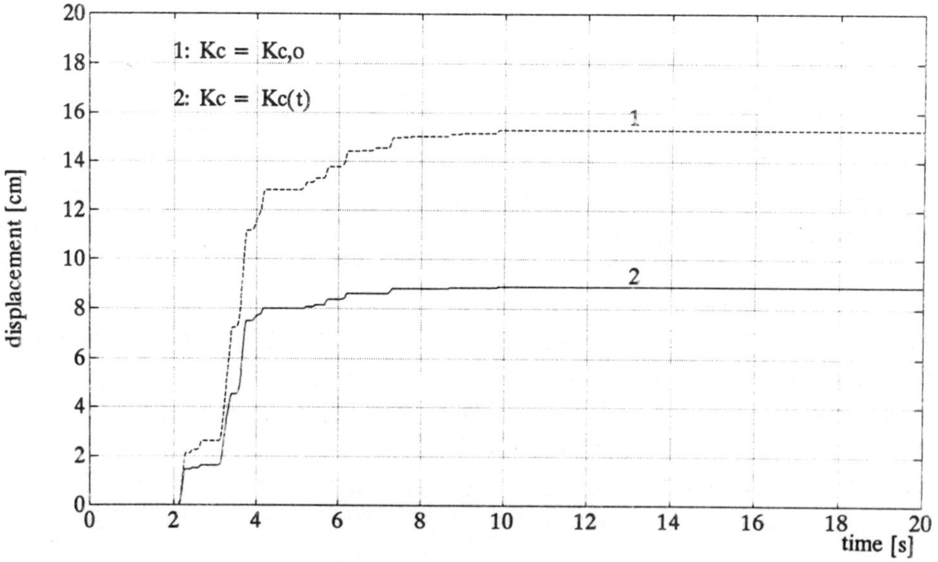

Fig. 9: Computed displacements obtained by assuming a non uniform acceleration field

vertical planes (in these zones the field of acceleration had been considered uniform). But in reality the present procedure allows for zones of any shape whatever to be considered (for example, horizontal layers as proposed by Chen and Liu (1990)). Furthermore, for the sake of simplicity, the SHAKE code has been used to analyse the seismic response but any other code for local seismic response could be used. The proposed procedure is completely general and very flexible.

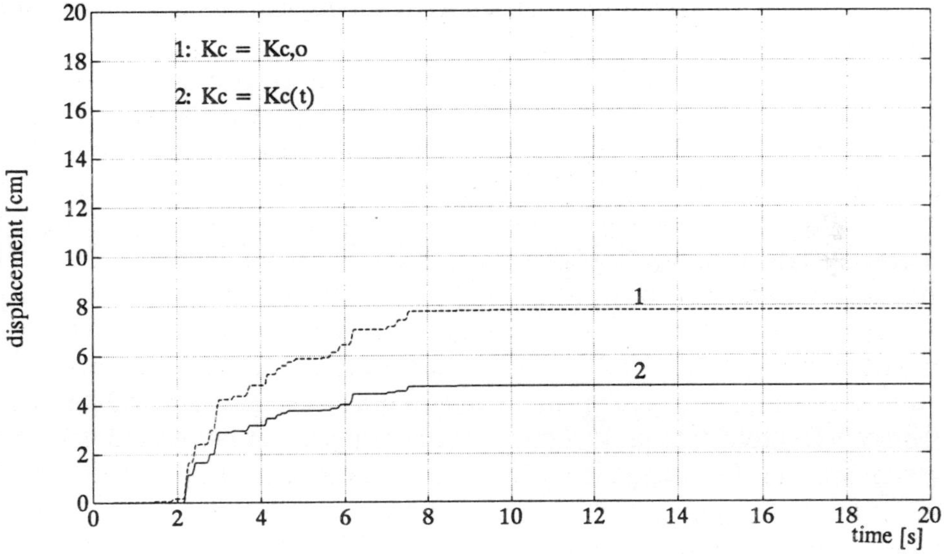

Fig. 10: Computed displacements obtained by assuming a uniform acceleration field (without amplification effects)

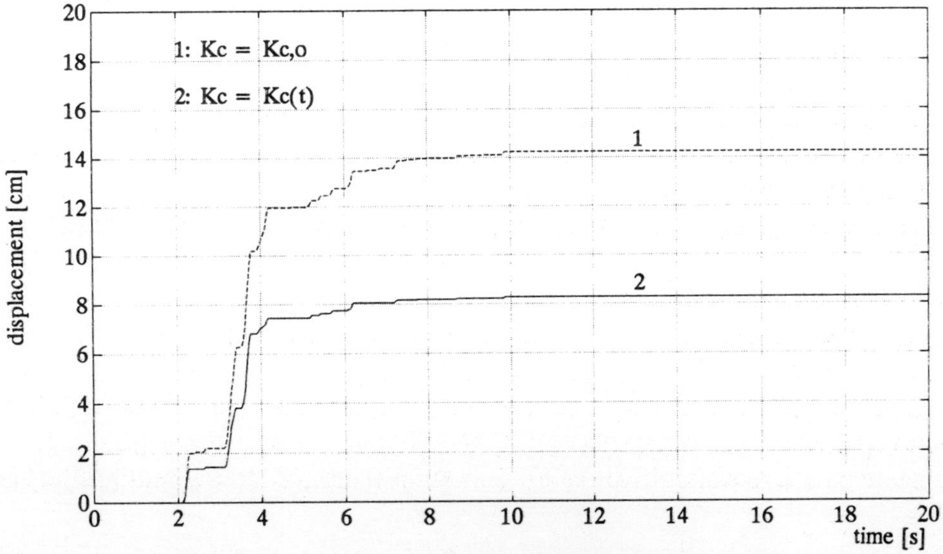

Fig. 11: Computed displacements obtained by assuming a uniform acceleration field (including amplification effects)

REFERENCES

Bishop A.W. (1955) - *The use of the slip circle in the stability analysis of slopes.* Geotechnique, vol. 5, n. 1

Carter R.K. (1971) - *Computer oriented slope stability analysis by method of slices.* MSCE Thesis, Purdue University

Chang G.J., Chen W.F., Yao J.P.T. (1984) - *Seismic displacement in slopes by limit analysis.* Journal of Geotechnical Eng. Div., Proc. ASCE, vol. 110

Chen W.F., Liu X.L. (1990) - *Limit analysis in soil mechanics.* Developments in Geotechnical Engineering, 52, Elsevier

Crespellani T., Ghinelli A., Madiai C., Vannucchi G. (1990) - *Analisi di stabilita' dei pendii naturali in condizioni sismiche.* RIG, XXIV, n. 2

Crespellani T., Ghinelli A., Vannucchi G. (1986) - *Tecniche probabilistiche nell'analisi e nel controllo della stabilita' dei pendii.* XVI° Convegno Naz. di Geotecnica, Bologna

Elton D.J., Shie C.F., Hadj-Hamou T. (1991) - *One- and two-dimensional analysis of earth dams.* Proc. of 2nd Int. Conf. on Recent Advances in Geotechnical Earthquake Engineering and Soil Dynamic, St. Louis, Missouri, Paper n. 7.16

Idriss I.M., Moriwaki Y., Wright S.G., Doyle E.H., Ladd R.S. (1980) - *Behavior of normally consolidated clay under simulated earthquake and ocean wave loading conditions.* Proc. of the Int. Symposium on Soil under Cyclic and Transient Loading, Swansea, vol. 1, A.A. Balkema, Rotterdam

Madiai C. (1989) - *Un codice di calcolo per l'analisi delle deformazioni permanenti nei pendii in condizioni sismiche.* Universita' di Firenze, Atti del Dipartimento di Ingegneria Civile, Sezione Geotecnica n. 1/89

Matsui T., Ohara H., Ito T. (1980) - *Cyclic stress-strain history and shear characteristics of clay.* Journal of Geotechnical Eng. Div., Proc. ASCE, vol. 106, n. GT10, October

Newmark N. W. (1965) - *Effects of earthquakes on dams and embankments.* The fifth Rankine lecture of the British Geotechnical Society, Geotechnique, vol. 15, n. 2

Poulos H.G. (1988) - *Marine geotechnics.* Unwin Hyman Ltd, London

Stara-Gazetas E., Dobry R. (1991) - *Inelastic seismic response of San Fernando and Santa Felicia dams.* Proc. of 2nd Int. Conf. on Recent Advances in Geotechnical Earthquake Engineering and Soil Dynamic, St. Louis, Missouri, Paper n. 7.41

Taylor, D.W. (1948) - *Fundamentals of soil mechanics.* John Wiley and Sons, New York

Thiers G.R., Seed H.B. (1968) - *Cyclic stress-strain characteristics of clay.* Journ. of the Soil Mech. and Found. Div., Proc. ASCE, vol. 94

SEA FLOOR SEDIMENT STABILITY ANALYSIS

L. Dormieux*, J.-L. Auriault**, O. Coussy***

abstract : This paper presents a quantitative study of the coupled phenomena of pore pressure generation and diffusion in a granular seabed subjected to wave loading, from which a method of stability analysis is derived.

résumé : La présente communication décrit une étude quantitative des phénomènes couplés de génération et diffusion de surpressions interstitielles dans un sédiment marin soumis à l'action de la houle, à partir de laquelle on propose une méthode d'analyse de la stabilité.

1 - INTRODUCTION

The ideas developed first in soil dynamics for assessing the liquefaction potential of a saturated soil subjected to a travelling shear wave have been applied in the ten last years to study the response of an underwater bed to the action of water waves. The works of Ishihara and Yamazaki (1984) and then those of Rahman and Jaber (1986) have shown the pivotal role played by the phenomenon of pore pressure generation in the sea bed. These authors proposed simplified methods for evaluating the pore pressures. The purpose of the present paper is to conduct a fuller analysis of the coupled phenomena of pressure generation and diffusion in a bed of saturated granular soil, and to derive a method for computing the stability of this bed subjected to a given wave.

2 - GEOMETRY AND LOADING

We assume in this study that the seabed occupies an homogeneous half space limited on the top by the plane $x_2=0$ (see figure 1). r and ρ^w denote the average mass density of the porous medium and the mass density of the water respectively. Before the wave

* Laboratoire de Mécanique des Solides, Palaiseau, France
** Institut de Mécanique de Grenoble, Grenoble, France
*** L.C.P.C., Paris, France

pressure is applied, the initial stress field $\underline{\underline{\sigma}}^{init}$ within the bed is in equilibrium with the gravity forces and the hydrostatic pressure $\rho^w gd$ acting on the plane $x_2=0$, where d is the local water depth. The initial pore pressure field, denoted u^{init} is assumed to be hydrostatic, that is $u^{init}=\rho^w g(d-x_2)$.

The action of the wave on the seabed is modelled by an overpressure $p(x_1,t)$ with respect to the hydrostatic value $\rho^w gd$. In the following, we will adopt the hypothesis that the pressure of the fluid is continuous through the interface at $x_2=0$. Then, using δu to denote the variation of the pore pressure u in the porous medium with respect to its hydrostatic value, we have :

$$\delta u(x_1,x_2=0,t)=p(x_1,t) \tag{1}$$

To evaluate the overpressure $p(x_1,t)$, the practice is largely to use linear wave theory, which was retained for the present study :

$$p(x_1,t)=p_o \sin(\Lambda x_1-\omega t) \tag{2a}$$

$$p_o=\frac{\rho^w gH}{2\cosh(\Lambda d)} \qquad \Lambda=2\pi/L \qquad \omega=2\pi/T^h \tag{2b}$$

in which H designates the crest to trough height of the waves, L the wave length and T^h the period of the waves. It is assumed that this theory remains valid for a slightly inclined seabed.

3 - THE POROELASTIC PROBLEM

3.1. the poroelastic solution.

$\delta\underline{\underline{\sigma}}$ and $\underline{\underline{\varepsilon}}$ being respectively the variation of the stress tensor with respect to the initial state and the strain tensor, the isotropic poroelastic constitutive law takes the form :

$$\delta\underline{\underline{\sigma}} = \lambda(\mathrm{tr}\underline{\underline{\varepsilon}})1 + 2\mu\underline{\underline{\varepsilon}} - \delta u\ \underline{\underline{1}} \tag{3}$$

The poroelastic solution has been given for an arbitrary overpressure $p(x_1,t)$ on the boundary by Dormieux (1989). The matrix in the base $(\underline{e}_i)$ of the stress solution $\delta\underline{\underline{\sigma}}$ is :

$$\delta\hat{\underline{\underline{\sigma}}}(\underline{x},t) = -\delta u \; \underline{\underline{1}} + x_2 \begin{pmatrix} -\frac{\partial}{\partial x_2}(\delta u) & \frac{\partial}{\partial x_1}(\delta u) & 0 \\ \frac{\partial}{\partial x_1}(\delta u) & \frac{\partial}{\partial x_2}(\delta u) & 0 \\ 0 & 0 & 0 \end{pmatrix} \tag{4}$$

where δu is the harmonic function equal to $p(x_1,t)$ on the boundary

$$\Delta(\delta u) = 0 \quad ; \quad \delta u(x_1,0,t) = p(x_1,t) \tag{5}$$

In the particular case (2) of a sinusoidal pressure, δu is given by

$$\delta u(x_1,x_2,t) = p_o \, \exp(2\pi\tilde{x}_2) \, \sin 2\pi(\tilde{x}_1-\tau) \tag{6}$$

where $\tilde{x}_i=x_i/L$.

3.2. the characteristic times of the problem.

Assuming that the two phases making up the porous medium are incompressible and that the permeability is homogeneous and can be described by an isotropic tensor $\underline{\underline{k}}=k\underline{\underline{1}}$, the diffusion equation satisfied by the volumetric strain takes the classical form :

$$tr\dot{\underline{\underline{\varepsilon}}} = c^m \Delta(tr\underline{\underline{\varepsilon}}) \tag{7}$$

in which the diffusion coefficient c^m is equal to $k(\lambda+2\mu)$.

The wave length L defines the characteristic distance of the problem. Equation (7) allows us to deduce that the quantity $T^c=L^2/c^m$ represents the characteristic time of the fluid mass diffusion in the porous medium stressed by the wave. It is assumed that this result remains valid in the inelastic case.

The loading imposed on the seabed possesses a characteristic time defined by the period T^h of the wave. For usual values of the wave and marine sediments, the ratio $\rho=T^h/T^c$ is of the order of 10^{-3} to

10^{-7}. This remark leads us to make use of the existence of two clearly separated time scales in the problem.

4 - THE INELASTIC CASE

4.1. Consequences of the existence of two time scales.

Let $\mathcal{A}(\underline{x},t)$ be any given quantity of the problem. In relation to the imposed loading, which is periodic with period T^h, it is reasonable to think that $\mathcal{A}$ will be a quasi periodic function of time, oscillating at the period T^h around a mean value whose variations appear only on the scale of the characteristic time of consolidation (see figure 2). The idea of using the double time scale is equivalent to doubling the role played by the argument t according to two dimensionless variables : the "fast" variable $\tau=t/T^h$ and the "slow" one $\theta=t/T^c$. So we will let $\mathcal{A}(\underline{x},t)=\mathcal{A}'(\underline{x},\tau,\theta)$. Intuitively, the variable τ must take into account the rapid fluctuations of the quantity $\mathcal{A}$ within the loading cycle. As far as the variable θ is concerned, it expresses the drift of this quantity on the scale of the consolidation phenomenon. So we will assume that $\mathcal{A}'$ is a periodic function of period 1 with respect to the fast variable τ. Thus the drift $\Delta\mathcal{A}$ of $\mathcal{A}$ within a given cycle of loading can be entirely related to the argument θ

$$\Delta\mathcal{A}= \mathcal{A}(t+T^h)-\mathcal{A}(t) = \mathcal{A}'(\tau,\theta+\rho)-\mathcal{A}'(\tau,\theta) \cong \rho\frac{\partial\mathcal{A}'}{\partial\theta} \tag{8}$$

in which we have used the equality $\mathcal{A}'(\tau,\theta+\rho)=\mathcal{A}'(\tau+1,\theta+\rho)$. The derivative of $\mathcal{A}$ with respect to physical time t can be expressed in partial derivatives of $\mathcal{A}'$ with respect to τ and θ :

$$\frac{\partial\mathcal{A}}{\partial t} = (1/T^h)(\frac{\partial\mathcal{A}'}{\partial\tau} + \rho\frac{\partial\mathcal{A}'}{\partial\theta}) \tag{9}$$

The fluctuations of $\mathcal{A}$, i.e. the term in $\partial/\partial\tau$ will be predominant in the computation of $\partial\mathcal{A}/\partial t$. Heuristically, we therefore assume the existence of an expansion of the field $\mathcal{A}$ by increasing powers of ρ :

$$\mathcal{A}(t) = \sum_{i\geq 0} \rho^i\mathcal{A}^{(i)}(\tau,\theta) \tag{10}$$

in which the functions $\mathcal{A}^{(i)}$ and their derivatives with respect to τ and θ are all of the same order. By limiting ourselves to the first order of ρ, it then follows that :

$$\frac{\partial \mathcal{A}}{\partial t} = (1/T^h)\left(\frac{\partial \mathcal{A}^{(0)}}{\partial \tau} + \rho\left(\frac{\partial \mathcal{A}^{(0)}}{\partial \theta} + \frac{\partial \mathcal{A}^{(1)}}{\partial \tau}\right)\right) \tag{11}$$

4.2. First approach of the inelastic problem

The behaviour of sands subjected to monotonic or cyclic shear stress is marked by the appearance of inelastic volumetric strains. In the following, we try to assess quantitatively the consequences of contractancy in the case of an underwater bed of loose sand subjected to wave action (loading described at section 2). First, the phenomenon of pore pressure increase must be quantified. Thereafter, we must check if the determined pore pressure field is likely to lead to failure of the bed.

In order to take into account the existence of inelastic volumetric strains, we will adopt the following hypothesis

$$\dot{\underline{\underline{\varepsilon}}} = \frac{1}{3}(\dot{\varepsilon}_v^{in})\underline{\underline{1}} + \frac{1+\nu}{E}\dot{\underline{\underline{\sigma}}}' - (\nu/E)(\mathrm{tr}\ \dot{\underline{\underline{\sigma}}}')\ \underline{\underline{1}} \tag{12a}$$

$$\dot{\varepsilon}_v^{in} = \frac{d}{dt}(\varepsilon_v^{in}) \leq 0 \tag{12b}$$

where the inequality (12b) expresses the hypothesis of contractancy for loose sands.

Using (11) in (12b), we get the inequality :

$$\frac{\partial}{\partial \tau}(\varepsilon_v^{in(0)}) + \rho\left(\frac{\partial}{\partial \theta}(\varepsilon_v^{in(0)}) + \frac{\partial}{\partial \tau}(\varepsilon_v^{in(1)})\right) \leq 0 \tag{13}$$

The immediate consequence is that the derivative of $\varepsilon_v^{in(0)}$ with respect to the fast variable τ is negative. Recalling that $\varepsilon_v^{in(0)}$ is a periodic function of τ, one obtains :

$$\frac{\partial}{\partial \tau}(\varepsilon_v^{in(0)}) = 0 \tag{14}$$

In other words, in terms of rate, i.e. derivative with respect to the fast variable τ, the boundary value problem at order zero is poroelastic. This result is particularly useful because the poroelastic problem was given in section 3. More precisely, the stress, strain and pore pressure solutions of the inelastic problem satisfy respectively :

$$\frac{\partial}{\partial\tau}(\delta\underline{\underline{\sigma}}^{(0)})=\frac{\partial}{\partial\tau}(\delta\underline{\underline{\sigma}}^{el})$$
$$\frac{\partial}{\partial\tau}(\underline{\underline{\varepsilon}}^{(0)})=\frac{\partial}{\partial\tau}(\underline{\underline{\varepsilon}}^{el}) \qquad (15)$$
$$\frac{\partial}{\partial\tau}(\delta u^{(0)})=\frac{\partial}{\partial\tau}(\delta u^{el})$$

where the poroelastic solutions $\delta\underline{\underline{\sigma}}^{el}$ and δu^{el} have been defined in (4) and (6). Integrating (15) with respect to τ, one obtains :

$$\delta\underline{\underline{\sigma}}^{(0)}(\tau,\theta)=\delta\underline{\underline{\Sigma}}(\theta)+\delta\underline{\underline{\sigma}}^{el}(\tau) \qquad (16a)$$

$$\underline{\underline{\varepsilon}}^{(0)}(\tau,\theta)=\underline{\underline{E}}(\theta)+\underline{\underline{\varepsilon}}^{el}(\tau) \qquad (16b)$$

$$\delta u^{(0)}(\tau,\theta)=\delta U(\theta)+\delta u^{el}(\tau) \qquad (16c)$$

Thus, the physical quantities that are solutions of the inelastic problem appear as the sum of a cyclic fluctuation term corresponding to the solution of the poroelastic problem, and a drift term which varies at the scale of the characteristic time of the consolidation phenomenon. The quantities $\delta\underline{\underline{\Sigma}}$, $\underline{\underline{E}}$, and δU represent the mean values, over one cycle, of the stresses, the strains and the pore pressure, respectively.

The law which governs the variation of the quantity ε_v^{in} of course still remains to be specified on the basis of the numerous experimental works that have attempted to provide a quantitative description of contractancy.

4.3. Evolution of the inelastic volumetric strain

(14) indicates that $\varepsilon_v^{in(0)}$ varies only at the time scale of the slow variable θ. Eq. (8) relates this variation to the increment of $\varepsilon_v^{in(0)}$ over a loading cycle :

$$\rho\frac{\partial}{\partial\theta}(\varepsilon_v^{in(0)}) = \Delta\varepsilon_v^{in(0)} \tag{17}$$

The experimental data should now be put to use. Let us consider a sample of soil subjected to an initial *effective* stress state defined by the tensor $\underline{\underline{\sigma}}'$ upon which a cyclic deviatoric loading $\underline{\underline{s}}^c(t)$ with zero mean value is superimposed in drained condition. The intensity of this cyclic loading is often measured by the cyclic shear stress ratio $\mathcal{S}'=\tau^c/\sigma'$ where τ^c and σ' denote respectively the amplitude of the variations of the second invariant $J_2^{1/2}$ of $\underline{\underline{s}}^c(t)$ and the mean effective stress. Many investigators have attempted to determine the increment $\Delta\varepsilon_v^{in}$ of inelastic volumetric strain caused by the loading cycle n°N as a function of the contractancy exhibited in the course of the previous N-1 cycles, denoted $\varepsilon_v^{in}(N-1)$ and $\mathcal{S}'$

$$\Delta\varepsilon_v^{in}=\mathcal{F}(\varepsilon_v^{in}(N-1),\mathcal{S}') \tag{18}$$

The calculations will be based on a model proposed by Dormieux (1989) in which :

$$|\Delta\varepsilon_v^{in}|= \mathcal{G}(\varepsilon_v^{in}(N-1),\mathcal{S}') = \frac{\kappa}{\alpha}\,\mathcal{S}'^{\beta}\exp(\frac{\beta-1}{\kappa}\varepsilon_v^{in}) \tag{19}$$

where α, β and κ are 3 material parameters. Usual values of parameters β are of the order of 5.

The structure of the solution given in (16) indicates that the conditions required for use of this type of experimental law are satisfied. More precisely, $\underline{\underline{\sigma}}'$ is here equal to the sum $\underline{\underline{\sigma}}'_{init}+\delta\underline{\underline{\Sigma}}(\theta)+\delta U(\theta)\underline{\underline{1}}$ and the cyclic deviator $\underline{\underline{s}}^c(t)$ is none other than $\delta\underline{\underline{\sigma}}'^{el}$ so that

$$\tau^c=\max\sqrt{J_2(\delta\underline{\underline{\sigma}}^{el})} = p_o(\Lambda x_2)\exp(\Lambda x_2) \tag{20}$$

With (17) and (19), the law of inelastic volumetric strain takes the form :

$$\rho\frac{\partial}{\partial\theta}[\varepsilon_v^{in(0)}](\theta) = \mathcal{G}\left(\varepsilon_v^{in(0)}(\theta),\mathcal{S}'(\theta)\right) \tag{21}$$

4.4 Determination of the mean quantities δU and $\underline{\underline{\Sigma}}$

It may be shown (Dormieux, 1989) that the solution $(\delta U,\underline{\underline{\Sigma}})$ in pore pressure and stresses to the mean-quantities problem is defined as follows :

- δU is solution of the differential equation

$$\frac{\partial}{\partial\theta}(\delta U) = \tilde{\Delta}(\delta U) - K(\frac{\partial}{\partial\theta}\varepsilon_v^{in}) \tag{22}$$

$$\delta U(x_1,0,\theta)=0 \quad \delta U(x_1,x_2,0)=0$$

where K is the drained poroelastic bulk modulus. The physical meaning of (22) is that of a diffusion equation of the pore pressure δU, with a pressure source term equal to the term $-K\partial\varepsilon_v^{in}/\partial\theta$. This last quantity represents the rate of generated pore pressure corresponding to the inelastic strain rate $\partial\varepsilon_v^{in}/\partial\theta$.

- $\delta\underline{\underline{\Sigma}}$ is equal to $\delta\Sigma(\underline{e}_1\otimes\underline{e}_1 + \underline{e}_3\otimes\underline{e}_3)$ where $\delta\Sigma$ is given by

$$-\frac{\partial}{\partial\theta}(\delta\Sigma)=\tilde{\Delta}(\delta U)\frac{1-2\nu}{1-\nu} \qquad \delta\Sigma(x_1,x_2,0)=0 \tag{23}$$

Putting together (22) with the inelastic volumetric strain law (21) and (23), we obtain a system of coupled equations determining the solution of the mean quantities problem, i.e. ε_v^{in} and $(\delta U,\delta\underline{\underline{\Sigma}})$.

The numerical solution can be easily found by discretizing according to the variable θ.

4.5 Numerical results

The geometry is defined by the water depth d. In the computation presented here, the loading is characterized by its period T^h and the ratio H_o/L_o between wave height H_o and wave length L_o in deep water.

Figure 3 presents the inelastic pore pressure δU distribution obtained in the bed for the set of parameters listed in table 1. The distributions are given every 20 cycles.

In this case, we observe a gradual accumulation of the pore pressure. The pressure δU is required to be zero on the boundary $x_2=0$. The very low values of δU computed for high values of $|x_2|$ can be explained by the fact that the ratio $\mathcal{S}'$ which defines the intensity of the loading and controls the contractancy decreases exponentially as a function of $\Lambda|x_2|$, as shown by (20). The pore pressure profile is marked by the existence of a maximum for an ordinate x_2^*, located close to the bed boundary.

material	$c^m=10^{-3}m^2s^{-1}$; $\alpha=.0023$; $\beta=6.644$; $r'=10kN/m^3$; $\varphi=20°$
loading	$H_o/L_o=0.06$; $T^h=10s$
geometry	$d=15m$

table 1

5 - STABILITY ANALYSIS

Let us consider a seabed the upper surface of which makes an angle χ with the horizontal, and is subjected to N loading cycles of the type defined in section 2. The strength of the material is described by a Mohr-Coulomb criterion with friction angle φ.

This section is concerned with the stability of the sea bed, in the sense of the yield design theory (Salençon,1990), *in the course of a given cycle of loading*. It is recalled that the stability in the sense of yield design theory expresses the compatibility between equilibrium and strength, i.e. the existence of a statically admissible stress field which complyies with the strength criterion at all points.
The *effective stress* loading is the sum of two body forces densities : the density $r'\underline{g}$ defined by the buoyant unit weight $r'=r-\rho^w g$, and the density $\rho^w g\underline{i}$ of the so-called *seepage* forces where the "hydraulic gradient" dimensionless vector $\underline{i}$ is classicaly defined by $\rho^w g)\underline{i}=-\underline{grad}\ \delta u$. The effective stress vector on the plane $x_2=0$ is equal to $\underline{0}$.

As δu, the hydraulic gradient $\underline{i}$ is the sum of a cyclic term $\underline{i}^{el}$ corresponding to the case of an elastic behaviour and an accumulating one $\underline{i}^{in}$ which is associated to the inelastic volumetric strains due to contractancy:

$$(\rho^w g)\underline{i}^{el} = -\underline{grad}\,\delta u^{el} \qquad (\rho^w g)\underline{i}^{in} = -\underline{grad}\ \delta U \qquad (24)$$

$(\rho^w g)\underline{i}^{el}$ is a vector of amplitude equal to $\Lambda p_o \exp(\Lambda x_2)$, rotating in the plane $(\underline{e}_1,\underline{e}_2)$ at the angular velocity ω :

$$(\rho^w g)\underline{i}^{el} = -\Lambda p_0 \exp(\Lambda x_2)(\underline{e}_1 \cos\psi + \underline{e}_2 \sin\psi)$$
$$\psi = \Lambda x_1 - \omega t = 2\pi(\tilde{x}_1 - \tau) \qquad (25)$$

On the other hand, the *inelastic* hydraulic gradient depends only on x_2 and on the time θ. It is parallel to $\underline{e}_2$. Figure 4 superimposes the 3 components of the body forces density to which the seabed is subjected.

The yield design theory has been applied in the present case by Dormieux (1989) and leads to the following condition of stability

$$\left[(\rho^w/r')\frac{\sin\varphi}{\sin(\varphi-\chi)}\right].i^{in}(x_2=0) + \left[\frac{1}{\sin(\varphi-\chi)}\right].\Lambda p_o/r'g \le 1 \qquad (26)$$

(26) clearly brings out two parameters quantifying the intensity of the loading that the bed is subjected to. The parameter $Q^1 = \Lambda p_o/r'g$ is a direct measurement of the loading intensity, while the parameter $Q^2 = i^{in}(0)$ represents the inelastic volumetric strains caused by the loading, and therefore measures the loading intensity indirectly in the sense that it quantifies the coupling between loading and behaviour.

The domain K of values of the couples (Q^1,Q^2) that the bed can potentially withstand is the triangle represented in figure 5. In the case where the loading is defined by a block of waves having constant period and Ho/Lo ratio, the loading path in the (Q^1,Q^2) plane is a vertical segment whose origin is on the $Q^2=0$ axis. If this segment leaves the triangle K, it is concluded that the bed is unstable. If it does not, stability is presumed.

Figure 6 presents the variations of $i^{in}(0)$ as a function of the time θ measured here by the number N of cycles. Three different values of the ratio H_o/L_o are considered : 0.05, 0.06 and 0.07. all the other parameters of the problem being given in table 1. Figure 5 shows the loading path corresponding to the 3 situations of figure 6. The abscissa of the loading path increases with H_o/L_o. The length of the path corresponds to the maximum value of $i^{in}(0)$ over the considered duration (400 cycles). For the two considered values of χ, the bed is stable for H_o/L_o=0.05 and unstable for H_o/L_o=0.07. However, for H_o/L_o=0.06, the stability depends on χ : it is ensured for the horizontal seabed but the inclined seabed is not stable. Thus, this diagram shows the high sensitivity of the stability to the H_o/L_o ratio as well as the unfavorable role played by even a small inclination of the seabed.

REFERENCES

Dormieux, L. (1989). Influence de la houle sur la stabilité d'un massif sous-marin. *Thèse de Doctorat de l'E.N.P.C.*

Ishihara, K et Yamazaki, A. (1984). Analysis of wave induced liquefaction in seabed deposits of sand. *Soils and Foundations,* Vol. n°3.

Rahman M.S.et Jaber, C.L. (1986). A simplified drained analysis for wave-induced liquefaction in ocean floor sands. *Soils and Foundations*, Vol. 26, n°3.

Salençon, J. (1990). An introduction to yield design theory and its application to soil mechanics. *European Journal of Mechanics A/Solids*, Vol. 9, n°5.

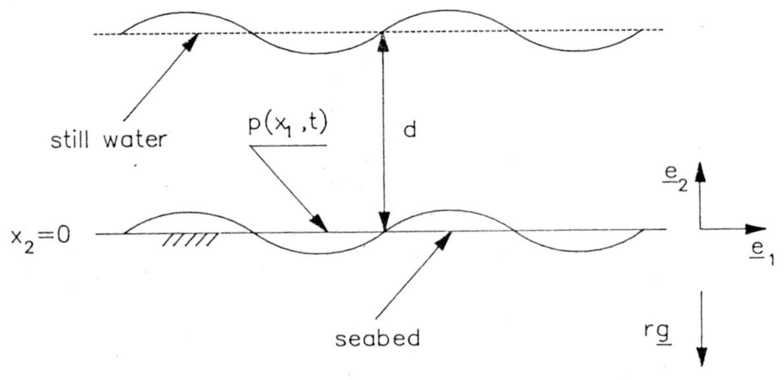

figure 1. Seabed under wave loading

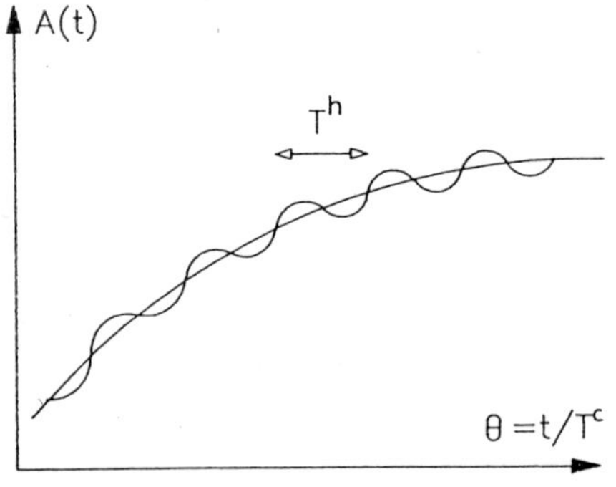

figure 2. Quasi-periodic function A(t)

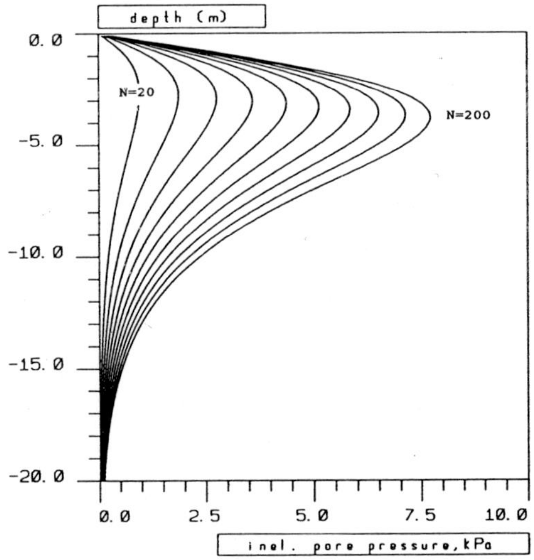

figure 3. inelastic pore pressure for various number of cycles

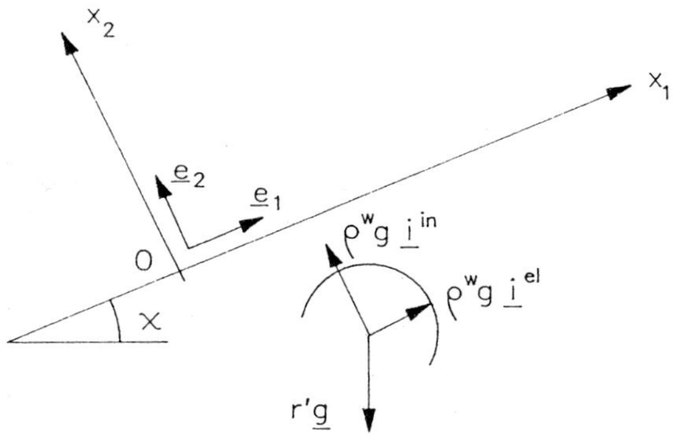

figure 4. The different components of body forces.

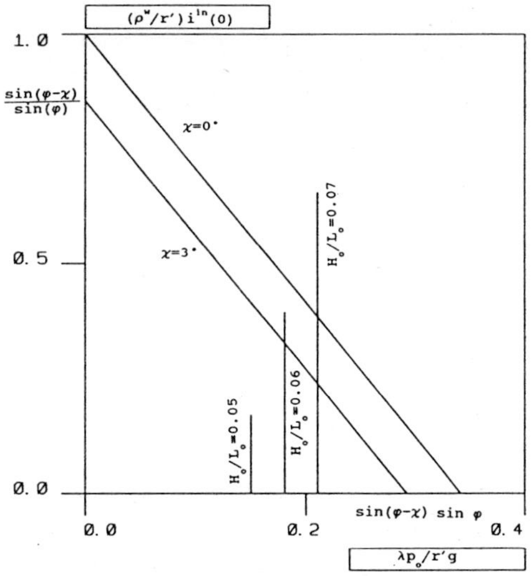

figure 5. Loading paths in plane (Q^1, Q^2)

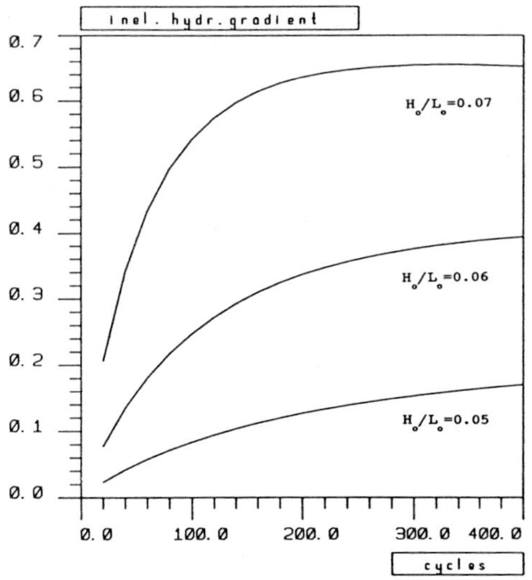

figure 6. Effect of H_o/L_o ratio on the inelastic gradient $i^{in}(0)$

APPLICATION OF THE YIELD DESIGN THEORY TO THE SEISMIC ANALYSIS OF SLOPES

Luc DORMIEUX*, Alain PECKER**, Jean SALENÇON*

Abstract: The present paper describes a method of evaluation of the residual displacements induced by an earthquake in a homogeneous earthdam built in cohesive material. This method constitutes an extension of the yield design theory to the case of a dynamic loading.

Résumé: La présente communication propose une méthode de calcul des déplacements résiduels dus à un séisme dans un barrage homogène constitué d'un matériau purement cohérent. Elle s'appuie sur une extension de la théorie du calcul à la rupture au cas d'une sollicitation dynamique.

1 Introduction

The design of earthdams in earthquake prone areas must account for the inertia forces developing within the soil mass. In the pseudo-static method, this is commonly performed with a classical slope stability analysis in which the static gravity field is modified by introducing horizontal and vertical body forces representing these inertia forces, which are taken constant in time. To overcome these difficulties, the pionnering work of Newmark settled down the idea of a design based on allowable residual (permanent) displacements which the dam undergoes following an earthquake. In the present paper, we propose a method of evaluation of these residual, plane strain, displacements for a homogeneous earthdam built in a purely cohesive material with an undrained shear strength C. This method is related to the one developed by Pecker and Salençon (1991) for the seismic response of shallow foundations.

Under the hypothesis that failure occurs through sliding of a block along a circular surface, the potential failure surface and the critical acceleration, corresponding to the onset of rupture, are determined within the framework of the Yield Design Theory (Salençon, 1983, 1990). Given the acceleration time history at

---

* Laboratoire de Mécanique des Solides - Ecole Polytechnique - Palaiseau - France

** Géodynamique et Structure - Bagneux, France

the base of the dam, the acceleration field within the earth mass is analytically computed; the residual displacements are then evaluated, for the potentially unstable soil mass, when the computed accelerations exceed the critical threshold.

The method presented herein is not, in essence, different from the one originally proposed by Makdisi and Seed (1979). It however presents, over the latter, the advantages of being analytic and of having a sound mechanical basis within the framework of the Yield Design Theory. Besides, it can be easily implemented on a P.C. and requires only a few seconds to yield the answer; it is therefore easily amenable to parametric studies.

The idea of using the yield design theory for the computations of permanent displacements of a slope submitted to a seismic excitation has already been used by Chang et al (1984). Their work, dedicated to a cohesionless material, is however questionable with regards to the use of the kinematic theorem for a material for which the flow rule is clearly not associated. Besides, the dynamic analysis of the slope, which is reflected by a significant variation of the inertia forces over the volume, is not accounted for in their analysis.

2 Determination of the critical failure surface and computation of the yield acceleration

A schematic drawing of the geometry of the problem under study is given on figure 1.

The acceleration field within the soil mass, whose derivation is given in paragraph 3., is taken parallel to the horizontal direction; furthermore, it is admitted that the inertia forces within a mass $\mathfrak{B}$ can be accounted for in the stability analysis by replacing the actual acceleration field $\underline{a}(\underline{x}, t)$ by an average field $\underline{\tilde{a}}(t)$ over the mass $\mathfrak{B}$.

$$\underline{\tilde{a}}(t) = \tilde{a}(t)\, \underline{e}_x = \frac{1}{|\mathfrak{B}|} \int_{(\mathfrak{B})} \underline{a}(\underline{x}, t)\, dV \qquad (1)$$

In accordance with the hypothesis of circular shape failure surfaces, the family of blocks represented on figure 2a are analysed. Each block is limited by a circular arc $\widehat{M_1 M_2}$ located in space by the depth of the exit point $h = y(M_2)$ and, either by the coordinates of its center Ω, or by the angles θ_1 and θ_2 between the horizontal axis and the vectors $\underline{\Omega M}_1$ and $\underline{\Omega M}_2$. In view of the asymmetry of the structure introduced by the upstream water pressure, only the downstream oriented surfaces are analyzed; situations where the water pressure would have a favorable effect on the stability are excluded. In the same way, for practical purposes, it is sufficient to study situations where the acceleration vector $\underline{\tilde{a}}(t)$ is directed upstream ($\tilde{a} > 0$), so that the inertia body forces $- \rho\, \underline{\tilde{a}}$ are directed downstream. For each block $\mathfrak{B}$, the stability analysis aims at determining the upper limit $\tilde{a}^{cr}_{\mathfrak{B}}$ of $\tilde{a}$, beyond which the equilibrium of the block is no more compatible with the strength of the material.

The yield acceleration $\tilde{a}^{cr}$, associated with the so-called "critical block" $\mathfrak{B}^{cr}$ corresponds to the lowest value of $\tilde{a}^{cr}_{\mathfrak{B}}$:

$$\tilde{a}^{cr} = \inf_{\theta_1,\theta_2,h} \tilde{a}^{cr}_{\mathfrak{B}} (\theta_1, \theta_2, h) \tag{2}$$

$\tilde{a}^{cr}_{\mathfrak{B}}$ appears to be the threshold of instability related to a circular rupture of block $\mathfrak{B}$ around Ω. It can therefore be determined through the kinematic approach of the yield design theory.

This approach compares the work $W_{ext}(\mathcal{S}, \tilde{a})$, in the failure mechanism, of the external static forces, noted $\mathcal{S}$, and of the inertia forces $- \rho\, \tilde{a}$, to the maximum resisting work W_{res}. The latter represents the maximum energy the soil can set against the failure mechanism, taking into account the material strength capacities.

The kinematic theorem of the yield design theory states that the inequality $W_{ext}(\mathcal{S}, \tilde{a}) \leq W_{res}$ is a necessary condition for stability. The threshold of instability $\tilde{a}^{cr}_{\mathfrak{B}}$ is the solution of:

$$W_{ext}(\mathcal{S}, \tilde{a}^{cr}_{\mathfrak{B}}) = W_{res} \tag{3}$$

with respect to the unknown $\tilde{a}^{cr}_{\mathfrak{B}}$; equation (3) requires the determination of the two quantities W_{ext} and W_{res}. Let $\underline{\hat{\omega}} = \hat{\omega}\, \underline{e}_z$

($\hat{\omega} > 0$) be the virtual rotational velocity of block $\mathfrak{B}(\theta_1, \theta_2, h)$ around Ω, the work of the external forces in such a rigid body motion writes:

$$\mathcal{W}_{ext}(\mathcal{S}, \tilde{a}) = (\underline{\mathfrak{M}}_a + \underline{\mathfrak{M}}_g + \underline{\mathfrak{M}}_w) \cdot \underline{\hat{\omega}} \tag{4}$$

where $\underline{\mathfrak{M}}_a$, $\underline{\mathfrak{M}}_g$, $\underline{\mathfrak{M}}_w$ represent respectively the moments around Ω of the inertia forces, of the gravity forces and of the hydrostatic water pressure.

These moments have a single component along $\underline{e}_z$

$$\underline{\mathfrak{M}}_a = \rho\, \tilde{a}\, X_a\, \underline{e}_z \;;\; X_a(\theta_1, \theta_2, h) = \int_{(\mathfrak{B})} \underline{\Omega M} \cdot \underline{e}_y \, dV \tag{5a}$$

$$\underline{\mathfrak{M}}_g = \rho\, g\, X_g\, \underline{e}_z \;;\; X_g(\theta_1, \theta_2, h) = \int_{(\mathfrak{B})} \underline{\Omega M} \cdot \underline{e}_x \, dV \tag{5b}$$

$$\underline{\mathfrak{M}}_w = \rho_w\, g\, X_w\, \underline{e}_z \;;\; X_w(\theta_1, \theta_2, h) = \int_{y_W}^{y_{M1}} \underline{\Omega M} \cdot \underline{WM} \, dy \text{ if } y_{M1} > y_W$$

$$X_w = 0 \text{ if } y_{M1} \le y_W \tag{5c}$$

where W represents the intersection of the upstream slope with the free surface of the reservoir (figure 1).

The resisting work in the rotational movement of $\mathfrak{B}$ arises from the velocity discontinuity $[\![\underline{v}]\!]$ along the arc $\widehat{M_1 M_2}$. Since the discontinuity is tangent to the arc $\widehat{M_1 M_2}$, it is associated with a resisting work per unit length equal to $C\,|[\![\underline{v}]\!]|$ whose integral is $\mathcal{W}_{res}$:

$$\mathcal{W}_{res} = \int_{\widehat{M_1 M_2}} C\,|[\![\underline{v}]\!]| \, d\ell = C\, R^2 (\theta_2 - \theta_1)\, \hat{\omega} \tag{6}$$

where R is the radius of the arc $\widehat{M_1 M_2}$. Using (4) and (6), the inequality of the kinematic theorem writes:

$$\rho\, \tilde{a}\, X_a + \rho\, g\, X_g + \rho_w\, g\, X_w \leq C\, R^2\, (\theta_2 - \theta_1) \tag{7}$$

According to (3), the threshold of instability $\tilde{a}^{cr}_{\mathfrak{B}}$ for the set of parameters θ_1, θ_2, h is obtained for the equality in equation (7).

Whenever the 3-parameters set (θ_1, θ_2, h) takes all the permissible values, the set of inequalities (7) defines a family of necessary conditions for stability. As anticipated, the systematic numerical evaluation indicates that the most stringent stability condition is achieved, h being given, when the inertia forces, the gravity force and the water pressure are all driving forces (X_a, X_g, $X_w \geq 0$). The yield acceleration $\tilde{a}^{cr}$ is then given by (2):

$$\tilde{a}^{cr} = \inf_{\theta_1, \theta_2, h} \left\{ \frac{C\, R^2\, (\theta_2 - \theta_1)}{\rho\, X_a} - g \left(\frac{X_g}{X_a} + \frac{\rho_w}{\rho} \frac{X_w}{X_a} \right) \right\} \tag{8}$$

excluding from the search the sets of parameters (θ_1, θ_2, h) for which $X_a \leq 0$. It turns out that the yield acceleration is reached when the exit point of the failure surface is at the downstream toe of the dam (h = H).

In aseismic conditions, i.e. $\tilde{a} = 0$, we define a confidence factor with respect to a circular failure along the arc $\widehat{M_1 M_2}$ as the ratio, when the inertia forces and gravity forces are driving forces ($\rho\, g\, X_g + \rho_w\, g\, X_w > 0$):

$$F\, (\theta_1, \theta_2, h) = \frac{C\, R^2\, (\theta_2 - \theta_1)}{\rho\, g\, X_g + \rho_w\, g\, X_w} \tag{9}$$

With this definition, (8) can be written:

$$\frac{\tilde{a}^{cr}}{g} = \inf_{\theta_1, \theta_2, h} \left\{ \left[\frac{X_g\, (\theta_1, \theta_2, h)}{X_a\, (\theta_1, \theta_2, h)} + \frac{\rho_w}{\rho} \cdot \frac{X_w\, (\theta_1, \theta_2, h)}{X_a\, (\theta_1, \theta_2, h)} \right] \cdot \left(F\, (\theta_1, \theta_2, h) - 1 \right) \right\} \tag{10}$$

For a stable structure in aseismic conditions, the confidence factor $F(\theta_1,\theta_2,h)$ is greater or equal to 1 and $\tilde{a}^{cr}$ is therefore a positive or nul scalar quantity. The 3-parameters set (θ_1,θ_2,h) associated with the minimum of (10) defines the critical block $\mathcal{B}^{cr}$ and the corresponding potential failure surface.

3 Computation of the acceleration field

The computations are based on the shear beam model originally developed by Ambraseys (1960) and Ambraseys-Sarma (1967). It has been extensively used to estimate the lateral seismic response of earthdams, its popularity stemming mainly from its simplicity. A number of rigorous studies have largely corroborated one of the crucial assumptions of the shear beam model, i.e. that strains and displacements are uniformly distributed over the width of the dam.

In the present paper, we restrict our attention to a homogeneous dam with uniform properties (modulus and mass density) over height. More realistic assumptions regarding the variations of the shear modulus G have been studied by Dakoulas and Gazetas (1985) and could be implemented without further difficulties.

Referring to figure 3, the governing equation of motion is derived by considering the dynamic equilibrium of an infinitesimal slice of thickness dy and width b.

$$\rho \frac{\partial^2 \underline{u}^*}{\partial t^2} = \frac{1}{y} \frac{\partial}{\partial y}\left[y\, G \frac{\partial \underline{u}}{\partial y}\right] \qquad (11)$$

where:

$\underline{u}$ horizontal relative displacement with respect to base,

$\underline{u}^*$ absolute displacement = $\underline{u} + u_g \underline{e}_x$

u_g displacement of the base

G shear modulus

Introducing the shear wave velocity ($\rho V_s^2 = G$), (11) can be written

$$\frac{\partial^2 \underline{u}}{\partial y^2} + \frac{1}{y}\frac{\partial \underline{u}}{\partial y} - \frac{1}{V_s^2}\frac{\partial^2 \underline{u}}{\partial t^2} = \frac{1}{V_s^2}\frac{d^2 u_g}{d t^2}\underline{e}_x \qquad (12)$$

Equation (12) is solved using modal superposition; hence $\underline{u}(y,t)$ is obtained as a summation of discretely infinite number of displacement histories corresponding to each of the natural frequencies of the dam

$$\underline{u}(y,t) = \left(\sum_{n=1}^{\infty} P_n U_n(y) D_n(t) \right) \underline{e}_x \tag{13}$$

where $D_n(t)$ is the well-known Duhamel integral.

The absolute acceleration is given by:

$$\underline{a}(y,t) = \underline{e}_x \left[\sum_{n=1}^{\infty} P_n U_n(y) D''_n(t) \right] + \ddot{u}_g \left[1 - \sum_{n=1}^{\infty} P_n U_n(y) \right] \underline{e}_x \tag{14}$$

with

$$I_n = \omega_n \int_0^t \ddot{u}_g(\tau)\, e^{-\beta\omega_n(t-\tau)} \sin[\omega_n^*(t-\tau)]\, d\tau$$

$$J_n = \omega_n \int_0^t \ddot{u}_g(\tau)\, e^{-\beta\omega_n(t-\tau)} \cos[\omega_n^*(t-\tau)]\, d\tau$$

$$D''_n(t) = \frac{d^2}{dt^2} D_n(t) = \frac{1 - 2\beta^2}{(1-\beta^2)^{1/2}} I_n + 2\beta J_n \tag{15}$$

$$P_n = \frac{2}{q_n} \frac{1}{J_1(q_n)} \tag{16a}$$

$$U_n(y) = J_0\left(q_n \frac{y}{H}\right) \tag{16b}$$

where q_n is the nth root of $J_0(.) = 0$; J_0 and J_1 are Bessel's functions of order 0 and 1; β is the material damping ratio.

ω_n is the undamped circular frequency $= q_n \frac{V_s}{H}$ and ω_n^* the damped circular frequency.

A realistic evaluation of the acceleration field $\underline{a}$ (y,t) usually requires that at least 10 modes be included in the analysis.

Referring to figure 4, the average acceleration $\tilde{a}(t)$ over a given depth is then computed as the acceleration developing the same inertia forces within the potentially sliding block (eq. 1). Considering the simplified geometry of figure 4, which is a valid approximation for a circular arc, it can readily be shown that:

$$\tilde{a}\,(t) = \frac{2}{Y^2} \int_0^Y a\,(y,t)\; y\; dy \tag{17}$$

A further refinement is introduced in the computations by computing the strain compatible properties for the dike. Following Makdisi and Seed (1979), the average shear strain through the dam is computed as, retaining only the first mode:

$$(\gamma_{av}) \simeq 0.2 \frac{H}{V_s^2} \operatorname*{Max}_t \,[D_1''\,(t)] \tag{18}$$

from which the compatible damping ratio β and shear modulus G (or shear wave velocity V_s) can be evaluated. A few iterations are required to achieve convergence and $\tilde{a}$ (t) can then be computed from (14) to (17) with the appropriate value of V_s and β.

4 Evaluation of the permanent displacements

In the preceding paragraph, the time history of the average acceleration $\tilde{a}$ (t) has been derived for any potentially unstable block. At the critical time t^{cr} when $\tilde{a}(t)$ reaches the yield acceleration $\tilde{a}^{cr}$, a rupture along the arc $\widehat{M_1M_2}$ is initiated (figure 5). The evaluation of the permanent, irrecoverable, displacements taking place from that time is based on two hypotheses: the first one deals with the kinematic of the rupture and the second one has to do with the material behavior.

Let (Ω, $\underline{e}_x$, $\underline{e}_y$, $\underline{e}_z$) be the coordinates axes attached to the dam, translated with respect to a fixed reference according to the acceleration time history $\tilde{a}$ (t). It is assumed that, beyond the critical time t^{cr}, the kinematic of the block $\mathfrak{B}^{cr}$, relative to the reference $(\Omega, \underline{e}_x, \underline{e}_y, \underline{e}_z)$ is a rigid body rotation around the axis Ω_z, whose angular velocity is noted $\underline{\omega}(t) = \omega(t)\, \underline{e}_z$. The circular arc $\widehat{M_1M_2}$ along which sliding takes place, is a surface of discontinuity for the velocity field within the dike. In compliance with an elastoplastic constitutive behavior, the second hypothesis states that, as soon as sliding occurs, the stress state along the failure surface is located on the boundary of the elastic domain of the material, defined by (3). In other words, equality is ensured in (3) if $\omega \neq 0$

$$\omega > 0 \quad \Rightarrow \quad \tau = C \tag{19a}$$

$$\omega < 0 \quad \Rightarrow \quad \tau = -C \tag{19b}$$

To compute the time history of motion, i.e. $\omega(t)$, it is convenient to apply the principle of virtual work. It implies that, in a Galilean referential, for any rigid body, virtual motion $\underline{\hat{u}}$ of the block $\mathfrak{B}^{cr}$, the work $\mathfrak{A}(\underline{\hat{u}})$ of the acceleration momentum is equal to the work of the external forces acting on the block. The principle of virtual work is now applied to the rotational velocity field in the relative motion of $\mathfrak{B}^{cr}$ with reference to the dike which is taken as a particular virtual motion.

The absolute acceleration $\underline{\gamma}(t)$ of block $\mathfrak{B}^{cr}$ writes down:

$$\underline{\gamma}(t) = \underline{\tilde{a}}(t) + \underline{\omega} \wedge (\underline{\omega} \wedge \underline{\Omega M}) + \dot{\underline{\omega}} \wedge \underline{\Omega M} \tag{20}$$

and the work of the acceleration momentum $\rho\, \underline{\gamma}$ in the relative motion $\underline{U}^r$ is given by:

$$\mathfrak{A}(\underline{U}^r) = \underline{\omega} \cdot \int_{(\mathfrak{B}^{cr})} (\underline{\Omega M} \wedge \rho\, \underline{\tilde{a}})\, dV + \int_{(\mathfrak{B}^{cr})} (\rho\, [\underline{\omega} \wedge \underline{\Omega M}].[\dot{\underline{\omega}} \wedge \underline{\Omega M}])dV \tag{21}$$

or

$$\mathfrak{A}(\underline{U}^r) = \omega(t)\,(J\, \dot{\omega}(t) - \rho\, \tilde{a}(t)\, X_a) \tag{22}$$

where J is the mass moment of inertia of block $\mathfrak{B}^{cr}$ around the axis Ω_z.

The work of the gravity forces, of the water pressure and of the contact forces along the arc $\widehat{M_1M_2}$ in the relative rotational motion of block $\mathfrak{B}^{cr}$ is equal to the scalar product: $\underline{\omega}(t) \cdot (\underline{\mathfrak{M}}_g + \underline{\mathfrak{M}}_w + \underline{\mathfrak{M}}_\sigma)$.

Finally, the principle of virtual work yields the following equality:

$$J \dot{\omega}(t) = \rho \tilde{a}(t) X_a + \rho g X_g + \rho_w g X_w - R \int_{\widehat{M_1M_2}} \tau(t) \, d\ell \qquad (23)$$

or, from equation (8):

$$J \dot{\omega}(t) = \rho X_a (\tilde{a}(t) - \tilde{a}^{cr}) + C R^2 (\theta_2 - \theta_1) - R \int_{\widehat{M_1M_2}} \tau(t) \, d\ell$$

(24)

The situation $\omega < 0$ would correspond to an upstream ("up the slope") movement of the block, in which the gravity forces and the upstream water pressure would be resisting forces. Although that situation could be treated without any additional difficulty, it is not considered in the following and the angular velocity is consequently greater or equal to 0. Introducing (19a) in (24), it follows that:

$$\omega > 0 \Rightarrow J \dot{\omega}(t) = \rho X_a (\tilde{a}(t) - \tilde{a}^{cr}) \qquad (25)$$

According to the sign of $(\tilde{a}(t) - \tilde{a}^{cr})$, the block movement is either accelerated or decelerated. If $\omega = 0$, the result depends on the value of $\tilde{a}(t)$.

$$\omega = 0: \quad \begin{cases} \text{if } \tilde{a}(t) \geq a^{cr}, \ J \dot{\omega}(t) = \rho X_a (\tilde{a}(t) - \tilde{a}^{cr}) \\ \text{if } \tilde{a}(t) < a^{cr}, \ \dot{\omega}(t) = 0 \end{cases} \qquad (26)$$

Equations (25) and (26) are used to numerically integrate the time history of motion (24) with a time discretization of the input motion.

5 Examples

An interesting question to answer is: given a sufficiently conservative static design of the dam (with respect to the two loading parameters: the dead load and the upstream water pressure), is it possible to prevent excessive seismic permanent displacements from occurring? The results presented hereafter throw some light on this question although without bringing a definite answer.

The idea consists in plotting the horizontal displacement of point A (figure 1) as a function of the static confidence factor F. More precisely, given the geometry of the dike and the static loads, the confidence factor is varied by altering the undrained shear strength of the material. For a given seismic input motion, the maximum amplitude of the horizontal displacement $\xi_x(A)$ is computed for each shear strength value. The variations of $\xi_x(A)$, as a function of F, are given on figure 6 for the data gathered in table I. The seismic input motion is the N21E component of the Kern County California earthquake of July 1952 (magnitude M = 7.6), recorded at Taft and scaled to a maximum acceleration of 0.3 g (duration 30 s).

Geometry	Material properties
$H = 45$ m $d = 10$ m $\alpha = 20°$ $h_w = 45$ m	$V_s = 285\ m.s^{-1}$ $\beta = 0.2$ $\gamma_r = 3\ 10^{-4}$

TABLE I

The graph exhibits a hyperbolic shape which could be anticipated. The displacements take large values, associated with an almost complete failure of the dam, when F tends to 1. On the other hand, the displacements are equal to zero for a confidence factor above 1.7 and do not exceed 5 cm when the confidence factor is larger than 1.5. Care must however be taken not to generalize these results to other geometries.

Figure 7 gives the variation of the horizontal displacement at point A versus the water height in the reservoir. The displacement amplitude is insensitive to the water height as long as the latter remains smaller than 2/3 of the total dam height; it then increases to reach a value equal to 3 times the initial value. The existence of this threshold can be explained by looking at the geometry of the critical block $\mathcal{B}^{cr}$: for sufficiently low water levels in the reservoir, the location of the exit point M_1 of the potentially unstable block is located above the reservoir free surface so that the water pressure does not contribute to the motion.

References

N.N. Ambraseys (1960). On the shear response of a two dimensional truncated wedge subjected to an arbitrary disturbance. *Bulletin of Seismological Society of America*, **50**, n° 45.

N.N. Ambraseys, S.K. Sarma (1967). The response of earthdams to strong earthquakes. *Geotechnique* n° 17, (3).

C.J. Chang, W.F. Chen, J.T.P. Yao (1984). Seismic displacements in slopes by limit analysis. *Geotech. Eng. Division, ASCE*, **110**, GT7.

P. Dakoulas, G. Gazetas (1985). A class of inhomogeneous shear models for seismic response of dams and embankments. *Soil Dynamics and Earthquake Engineering*, **4**, n° 4.

A.M. Makdisi, H.B. Seed (1979). Simplified procedure for evaluating embankment response. *J. Geotech. Eng. Division - ASCE* **105**, GT12.

A. Pecker, J. Salençon (1991). *Seismic bearing capacity of shallow strip foundations on clay soils*. CEE Mexico International Workshop.

J. Salençon (1983). *Calcul à la rupture et analyse limite*. Presses de l'Ecole Nationale des Ponts et Chaussées.

J. Salençon (1990). An introduction to the yield design theory and its application to soil mechanics. *European Journal of Mechanics, A/Solids*, 9, n° 5.

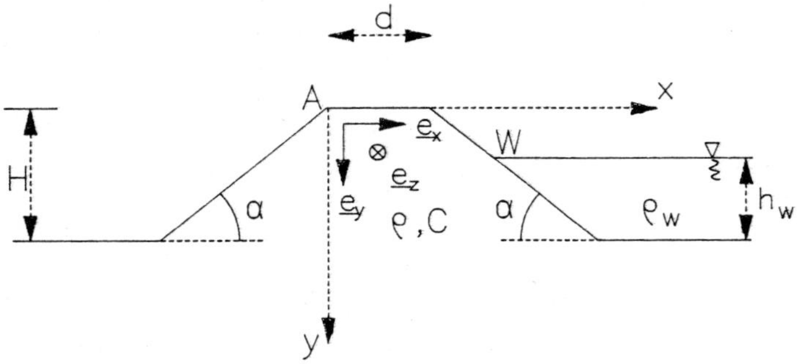

Fig. 1: Geometry and notations

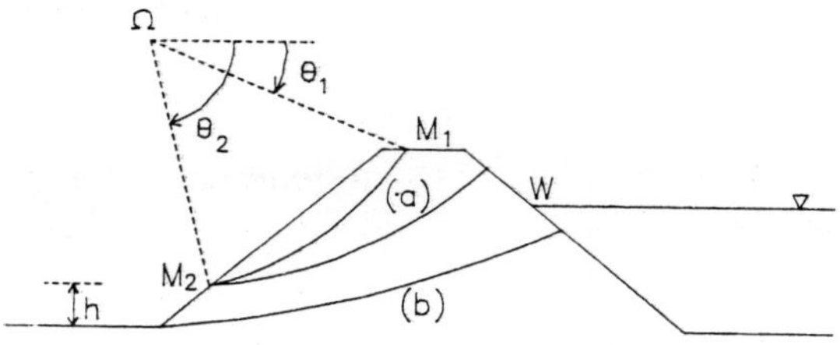

Fig. 2a: The different components of loading

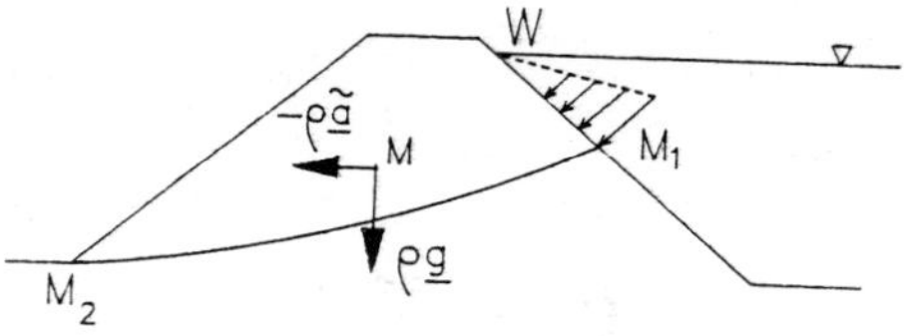

Fig. 2b: Circular failure surfaces

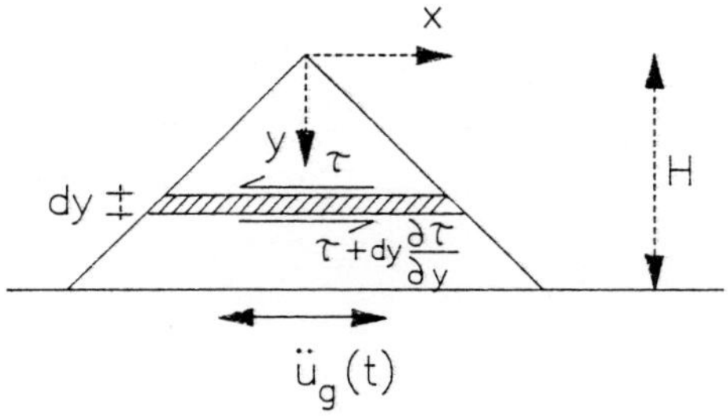

Fig. 3: Dam cross section - Shear beam model

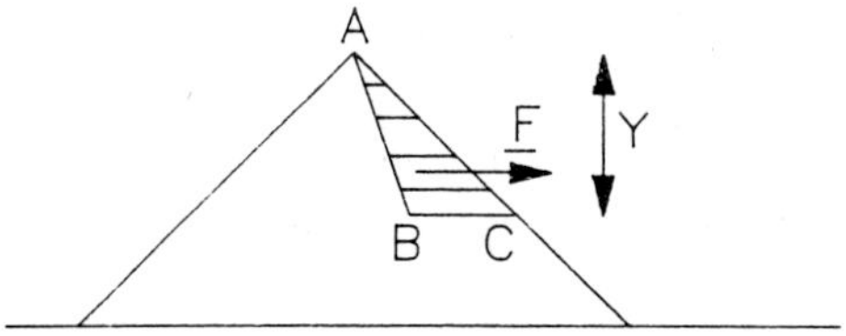

$$\underline{F} = \rho\, \tilde{\underline{a}}(t)\, V_{ABC} = \int_{[ABC]} \rho \underline{a}(y,t)\, dV$$

Fig. 4: Computation of average accelerations

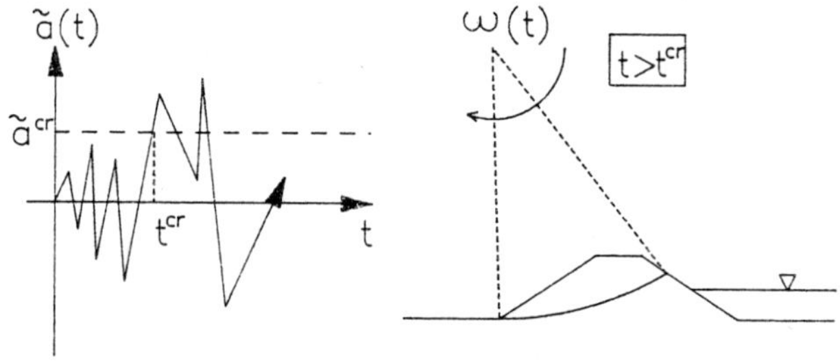

Fig. 5: Beginning of sliding

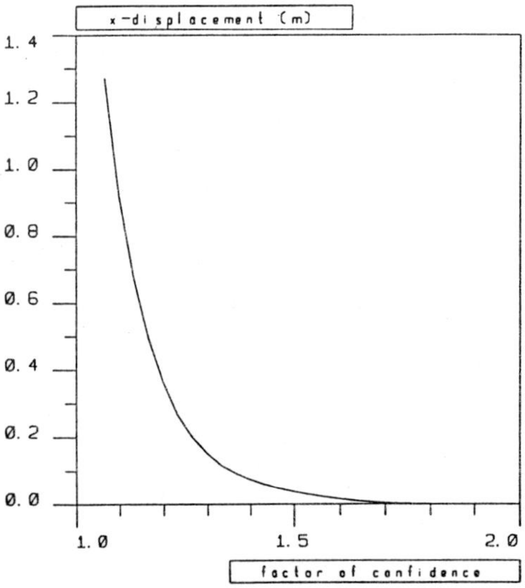

Fig. 6: Variation of ξ_x (A) with the static factor of confidence

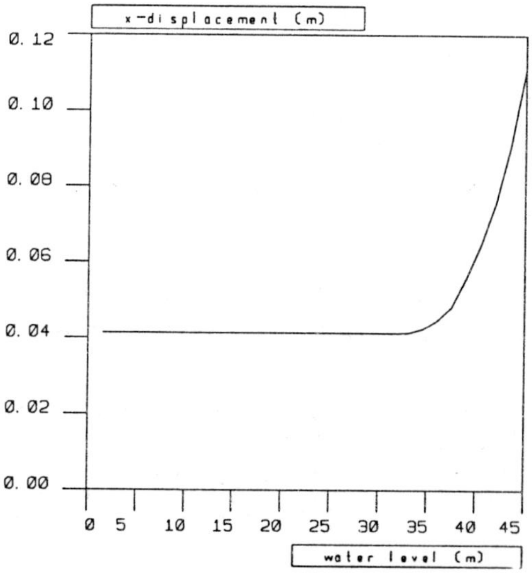

Fig. 7: Variation of ξ_x (A) with the water level in the reservoir

EXPERIENCES ON THE SEISMIC RESPONSE OF TWO EMBANKMENT DAMS

Giovanni LA BARBERA*, Pasquale PALUMBO*,
Giovanni RUGGERI*, Guido MAZZA'**, F. BETTINALI**

Abstract: In this paper various models and computational techniques which have been developed and applied to analyse the seismic response of two embankment dams are briefly described and compared. Comparisons of results, computational efforts and difficulties, relevant to the different approaches are also presented.

Riassunto: In questa nota sono brevemente illustrati e messi a confronto i diversi modelli e le procedure computazionali sviluppate e applicate per le analisi dinamiche di due dighe in terra. Sono inoltre illustrati i confronti tra i risultati, le difficolta' e l'impegno computazionale dei diversi approcci e modelli.

1.0 Foreword

At present the approaches to dynamic analysis of earth dams and embankments can be divided in two main categories. The former gathers the methods facing the problem as total stress. This is a common procedure in geotechnical engineering applications as it allows to tackle complex problems quite easily. The contribution of the liquid phase to balance equations is not explicitly taken into account, thus implying a certain number of hypotheses and simplifications. The latter gathers the most rigourous methods usually associated to the coupled theory of solid-liquid phases. By this method all field equations (equilibrium, continuity, congruence) and boundary conditions are simultaneously satisfied. This paper describes two procedures belonging to the former of the afore-mentioned categories and one procedure belonging to the latter. All the carried out analyses and some representative results obtained by means of the three procedures are here reported.

2.0 Procedures for the evaluation of dynamic safety conditions belonging to the first category

In the following two procedures to evaluate the dynamic safety conditions of a slope will be described. For both procedures the dynamic response analysis of the slope is performed by adopting the equivalent linear approach. The main hypothesis which underlay the equivalent linear method are explained in [1], [2], [3].

* ISMES S.p.A. - BERGAMO
** ENEL - Direzione studi e ricerche/CRIS

A seismic analysis of an embankment dam or a natural slope, presupposes the stress state and state parameters inside the embankment body to be known in the static conditions before the seismic event. In this analysis the self weight, hydrostatic pressures on the upstream slope face and the buoyancy effect on the submerged zone must be considered. This analysis can be done adopting different constitutive laws which define the material behaviour. A linear or non linear elastic or an elasto-plastic behaviour can be choosen. The choice depend on many different factors which are varying for each case. The dynamic response analysis adopting the equivalent linear approach give us only the spatial distribution and the time-hystories of the accelerations induced by the earthquake inside the slope. Since the model is elastic, it cannot give us informations regarding important safety indexes, like permanent displacements or safety factors with respect to local failures. In order to evaluate these safety indexes some additional computations are needed. Two different approaches will subsequently be described; they have been succesfully applied in order to evaluate the dynamic safety conditions of two different embankment dams.

2.1 EVALUATION OF DYNAMIC SAFETY CONDITIONS BY ANALISING LOCAL SAFETY FACTORS DISTRIBUTION.

The safety conditions under sismic action is evaluated comparing in all points of the dam body the shearing stresses caused by the loads with the materials shear resistance. The shear stresses induced by the loads have been determined combining the static stresses with those caused by the earthquake. It is a common practice to combine the static shear stress with the "effective" dynamic shear stress. The latter is defined as the maximum shear stress induced in a certain point by the seismic action, multiplied by a reducing factor f = 0,65. The reducing factor takes into consideration the fact that the maximum value of the dynamic stress is reached in a certain very short time interval only and furthermore that it generally occurs at times varying at the different points of the dam body. The shearing strength available in each point is evaluated by means of the classic Mohr-Coulomb failure criterion. The safety factor at each point is computed as ratio between the material shear resistance and the value of the shearing stress caused by the seismic load. The kind of test takes therefore to the definition of a local safety factor instead of a sliding safety coefficient along some pre-established failure lines. The global analysis of all local safety factors allows to determine the capability of the structure to face the stress state caused by the earthquake, also pointing out the zones that are mostly influenced by the seismic action.

2.1.1 Application of the method to analyze an embankment dam

The procedure outlined at point 2 and 2.1 has been applied to analyze an embankment dam which main characteristics are described in figure 1. The analysis has been developed according to the following main steps:

- static analysis of the embankement main section;
- representation of the design earthquake;
- dynamic analysis of the dam main section;
- evaluation of safety conditions under seismic loads.

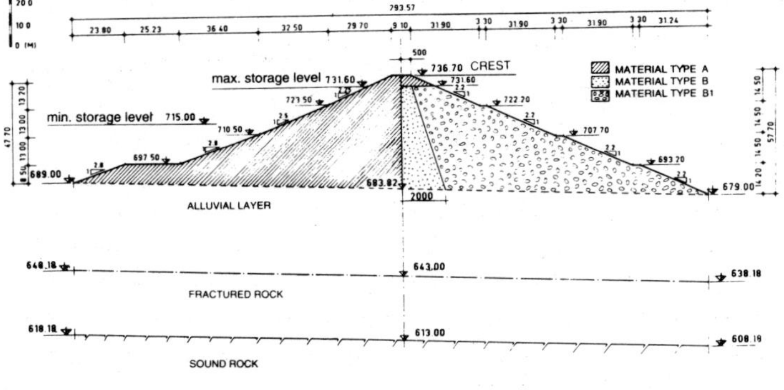

Fig. 1: Main characteristics of the dam

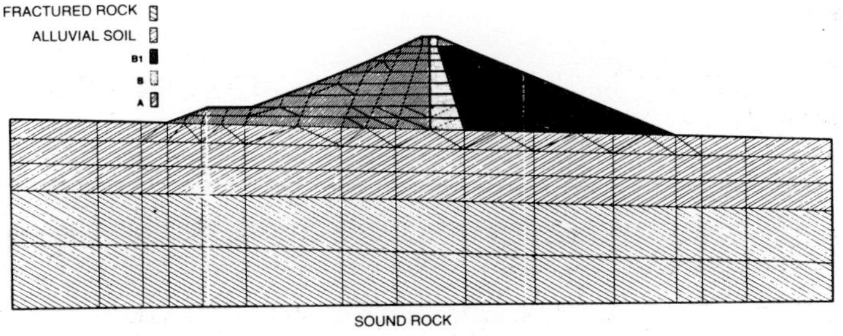

Fig. 2: Numerical model and materials distribution

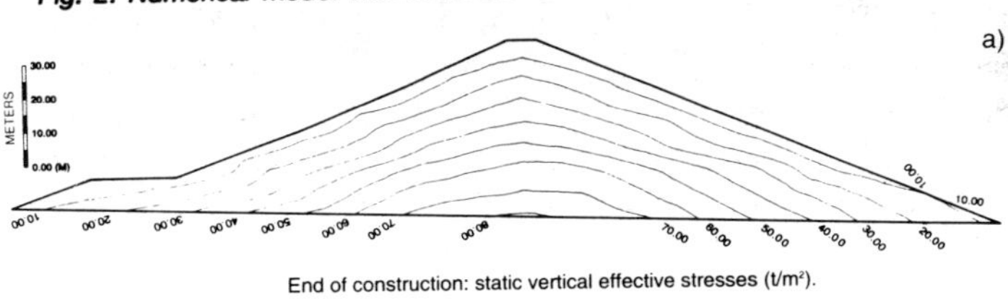

End of construction: static vertical effective stresses (t/m²).

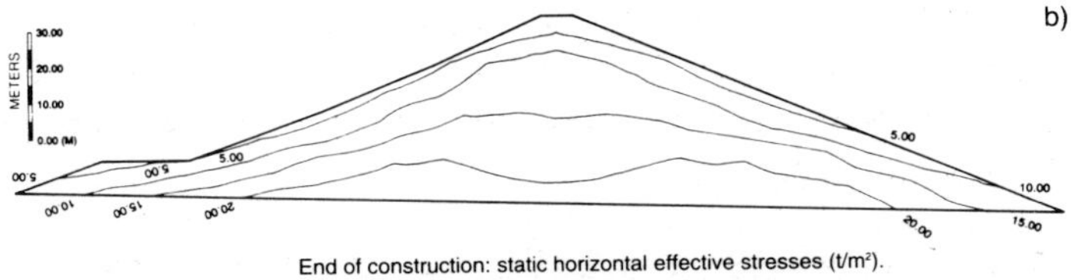

End of construction: static horizontal effective stresses (t/m²).

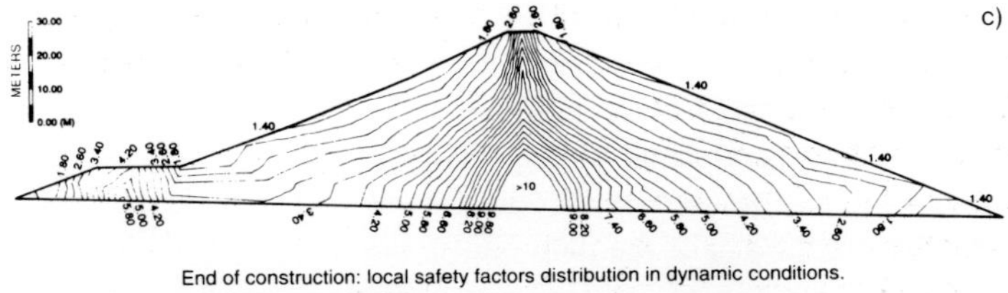

End of construction: local safety factors distribution in dynamic conditions.

Fig. 3: Results of computation at the end of construction

The embankment is 47.70 m high max and it is made of loose alluvium; it is a zoned dam with a damproof blanket. The dam body consists of three zones of materials with different mechanical characteristics and sharing strength. (fig. 1 and 2).

Static analysis

As to the stress-strain behaviour of the materials subjected to the action of static loads, foundation sound and fractured rocks have been characterized according to their homogeneous linear elastic and isotropic constitutive law, while alluvium both in the foundations and in the dam body has been characterized according to Duncan-Chang [4] hyperbolic non linear constitutive law. The adopted computation procedure implies the simulation of the dam progressive construction by horizontal layers. It is therefore possible to consider the non-linearity in the material response and to evaluate that most deformations of each layer, due to the loads of the lower layers, exhausted during laying. After computing the solution at construction completion, the hydrostatic load due to the water storge level is applied. Also for this analysis, the procedure has implied subsequent load increments related to the increasing water levels. Fig. 2 shows the numerical model, with the areas of different materials. For the situation, at the end of construction, figures 3a, 3b, show the contour lines of vertical and horizontal effective stresses.

Seismic input

In order to carry out the dynamic analysis, the spectrum defined by the Italian code [5] for high-seismicity areas (S=12) has been adopted. For the computation of the dynamic time-history response, an artificial accelogram representing the horizontal component of the motion has been obtained with the response spectrum. The obtained accelerogram is shown in figure 4a. It lasts 20 seconds. The same figure reports the comparison between the code spectrum and the artificial accelerogram one, for damping values of 4%, 6% and 8%. The design seismic load is applied to the model base considered as bedrock surface and by diagramming the propagation of the seismic eccitation with the main propagation of the shear waves in vertical direction. The computation has been carried out by using SHAKE program [6] which analyzes the one-dimensional propagation of the shear waves, considering the non-linearity of the material by applying the equivalent linear criterion. Figure 4b shows the accelerogram computed at the bedrock and the corresponding response spectrum for different damping values.

Dynamic analysis

The time-history dynamic analysis has been carried out by employing the finite-element computation program QUADSIM [7] that allows to evaluate the seismic response of two-dimensional earth structures taking into account the non-linear and dissipating behaviour of materials.

Safety evaluation

Figure 3c show the results as contour lines of the local safety factor. It is therefore easy to evaluate the safety conditions of the different parts of the dam subjected to the action of the seismic load and also to obtain the element necessary to have a general picture of the stability conditions of the dam body.

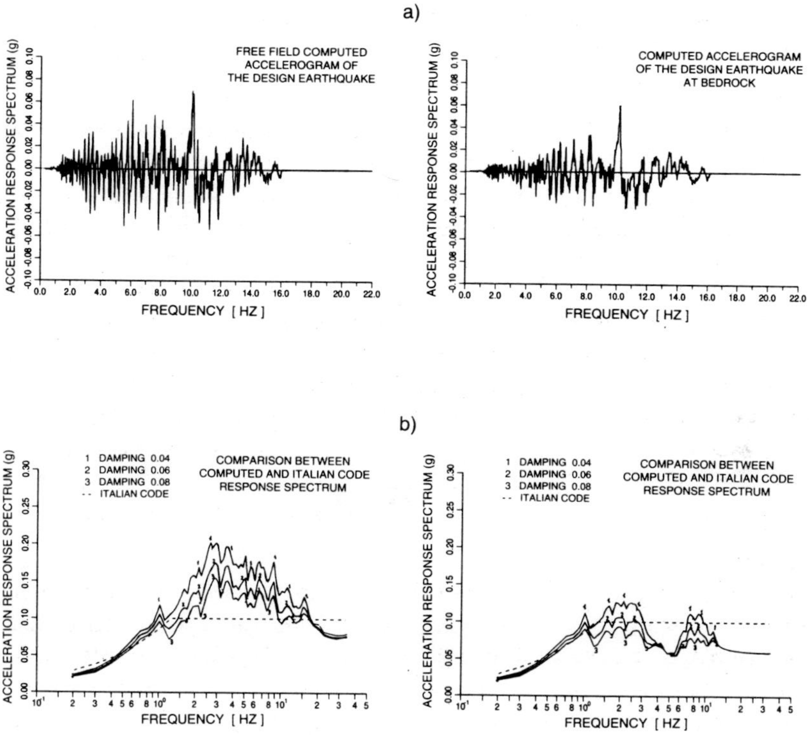

Fig. 4: Free field and bedrock computed accelerograms and response spectrum

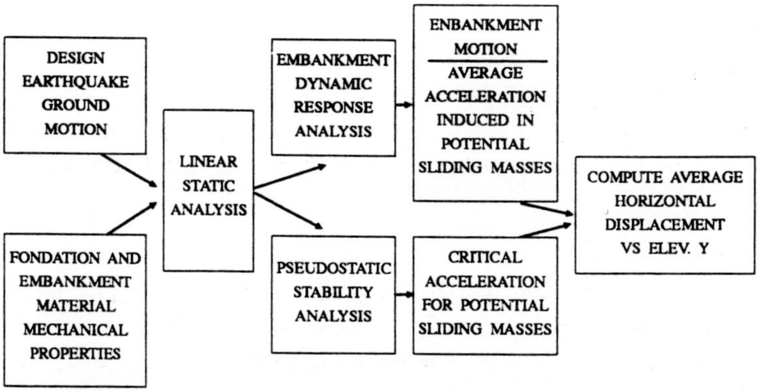

Fig. 5: Flow chart representing the main steps of Newmark type seismic analysis

2.0 EVALUATION OF DYNAMIC SAFETY CONDITIONS ESTIMATING PERMANENT DISPLACEMENTS

Dynamic analysis by adopting the equivalent linear approach is used only to get the distribution of induced acceleration time-hystories inside the dam body. Also, since the model is elastic, it cannot predict directly permanent deformations, further analysis are required as subsequently described. The analytical approach adopted for estimating the permanent deformations is based on the concept proposed by Newmark (1965) in which the displaced part of an embankment is modelled as a rigid block on an horizontal or inclined plane, subjected to induced earthquake motion causing the block to slide on the plane. For a comprehensive explaination the reader is referred to [1], [8], [9], [10], [11]. This method of analysis was pioneered by Newmark (1965). The version adopted for the present analysis was developed by Makdisi and Seed (1978) and differs from the Newmark approach in generating relative displacements by the net acceleration above the sliding surface, whereas Newmark used the net acceleration below the sliding surface.

2.2.1 Application of the method to analise the seismic response of El Infiernillo embankment dam.

The principal steps of the analysis are shown in the flow chart illustrated in figure 5. The analysis were carried out using the following codes: NASTRAN for finite element static and dynamic analysis; PC STABL5M [12] for slope stability analysis and NEWDISP for the evaluation of permanent displacements. For a more comprehensive explaination the reader is referred to [13].

The maximum cross section of the dam with materials distribution is presented in figure 6. The basic description of materials and strength parameters are reported in [13].

Linear Static analysis

The main purposes of the linear static analysis consist in defining the mean effective stress distribution through the embankment. The adopted finite element numerical model is reported in figure 7. An illustrative scheme of the loads considered in the analysis is reported in figure 8. Results of the linear static analysis are reported in figure 9.

Seismic response of the embankment

The dynamic response of the embankment has been computed using an equivalent linear method implemented in a routine which utilizes the NASTRAN multipurpose finite element computer code. In order to perform this analysis it has been choosen NASTRAN code instead of QUADISM and this is due to the fact that with NASTRAN code it is possible to use a wide variety of elements. It is also possible to perform a modal analysis which allows the evaluation of the natural frequencies of the dam, the QUADISM code does not allow it. NASTRAN permits also to analyse, if desired, the dynamic fluid-structure interaction. It should be noted that the linear equivalent method provides conservative values of acceleration due to the fact that it does not consider the soil plasticity effect which imposes a threshold value to the accelerations. Some significant results are reported in figure 11.

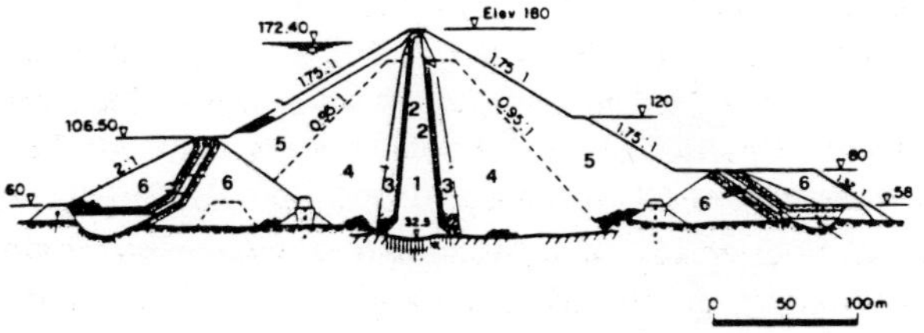

Fig. 6: Maximum cross section with materials distribution

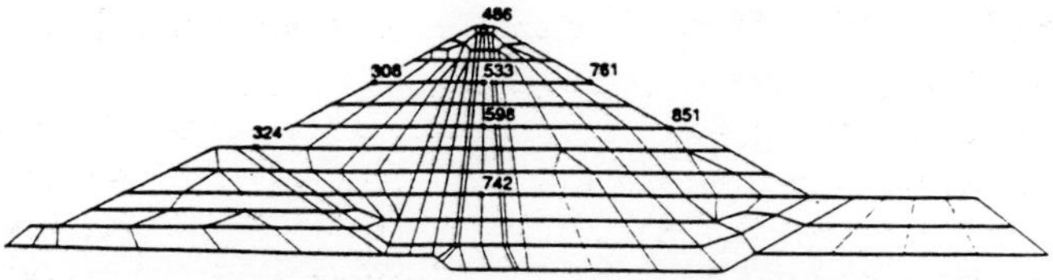

Fig. 7: Adopted finite element model for static and dynamic analysis

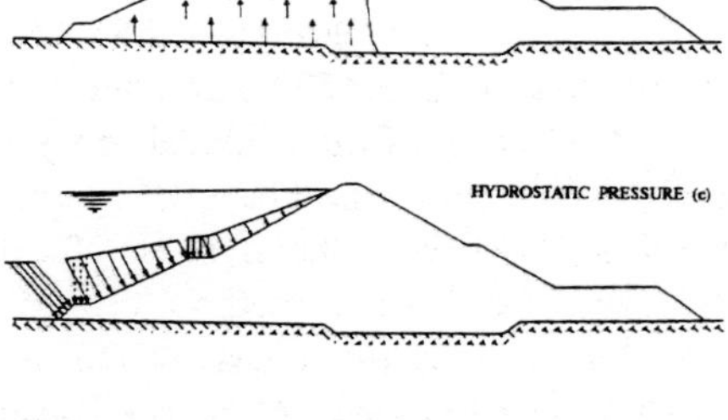

Fig. 8: Loading phases considered for the linear static analysis

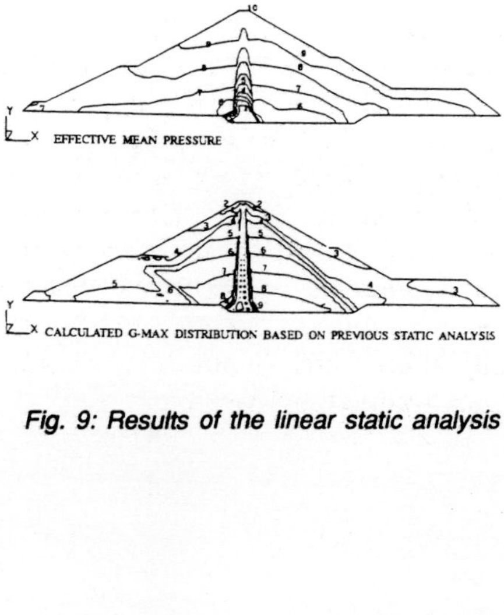

Fig. 9: Results of the linear static analysis

Input ground motion

Two inputs accelerograms have been considered in the present analysis, namely EQ1 and EQ3 (figure 11 reports EQ3 accelerogram and the corresponding results). For the given finite element discretization (fig. 7), the maximum frequency content of the motion which can be accurately reproduced is around 8 Hz. For this reason the given accelerograms have been filtered with a cut-off frequency of 8 Hz.

Evaluation of critical accelerations

For the present analysis the computer code PC STABL5M [12] has been used. Circular sliding surfaces stability conditions have been calculated with the simplified Janbu method. The analysis has been performed on effective stresses basis. Based on the previous static analyses four sliding surfaces have been choosen in the upstream slope and three in the downstream slope. The critical horizontal acceleration has been determined by a pseudostatic analysis performed on each sliding surface; the obtained results are reported in figure 10.

Some significant results using NEWDISP program are reported in figure 11

3.0 A procedure belonging to the second category: coupled effective stress dynamic elasto-plastic analysis

The common engineering practice approaches the analysis of an embankment dam subjected to seismic loads according to the so-called *Total Stress Theory*. It is known that this theory is based on several semplifying assumptions, most of them deriving from the fact that the contribution of the fluid phase to the equilibrium is not explicitely taken into account. A more rigorous analysis may be obtained following the so-called *Coupled Solid Fluid Theory* where both the solid and the fluid phases in the soil are explicitely taken into account in the equilibrium equations, Ref. [14], [15], [16], [17]. The computer program used for the analysis is called OMEGA; this is a FE code developed for the numerical solution of non-linear analysis of both one-phase solid and two-phases solid-fluid medium subjected to static and dynamic loading conditions. The code version used for the seismic analysis presented in this paper has been jointly developed by ISMES S.p.A. (Bergamo, Italy) and CIMNE (Barcellona, Spain). Although this version covers already a wide range of important applications, it has to be considered only as a first release. In fact, its peculiarity is the finite element formulation and the informatic structure, which makes it easy to extend to many other types of engineering problems.

3.1 APPLICATION OF THE METHOD TO ANALISE THE SEISMIC RESPONSE OF EL INFIERNILLO EMBANKMENT DAM

For the analysis of El Infiernillo Embankment dam, two different methods have been developed. This is the second contribution reporting the main results obtained according to the complete coupled-theory, restricted to the case of low excitation (earthquake accelerogram EQ1.DAT). The aim is to propose a useful comparison, both in terms of results and computational costs, between the two approaches [18]. The different material property zones and the mesh employed for the analysis are reported in Figs. 6 and 7. The

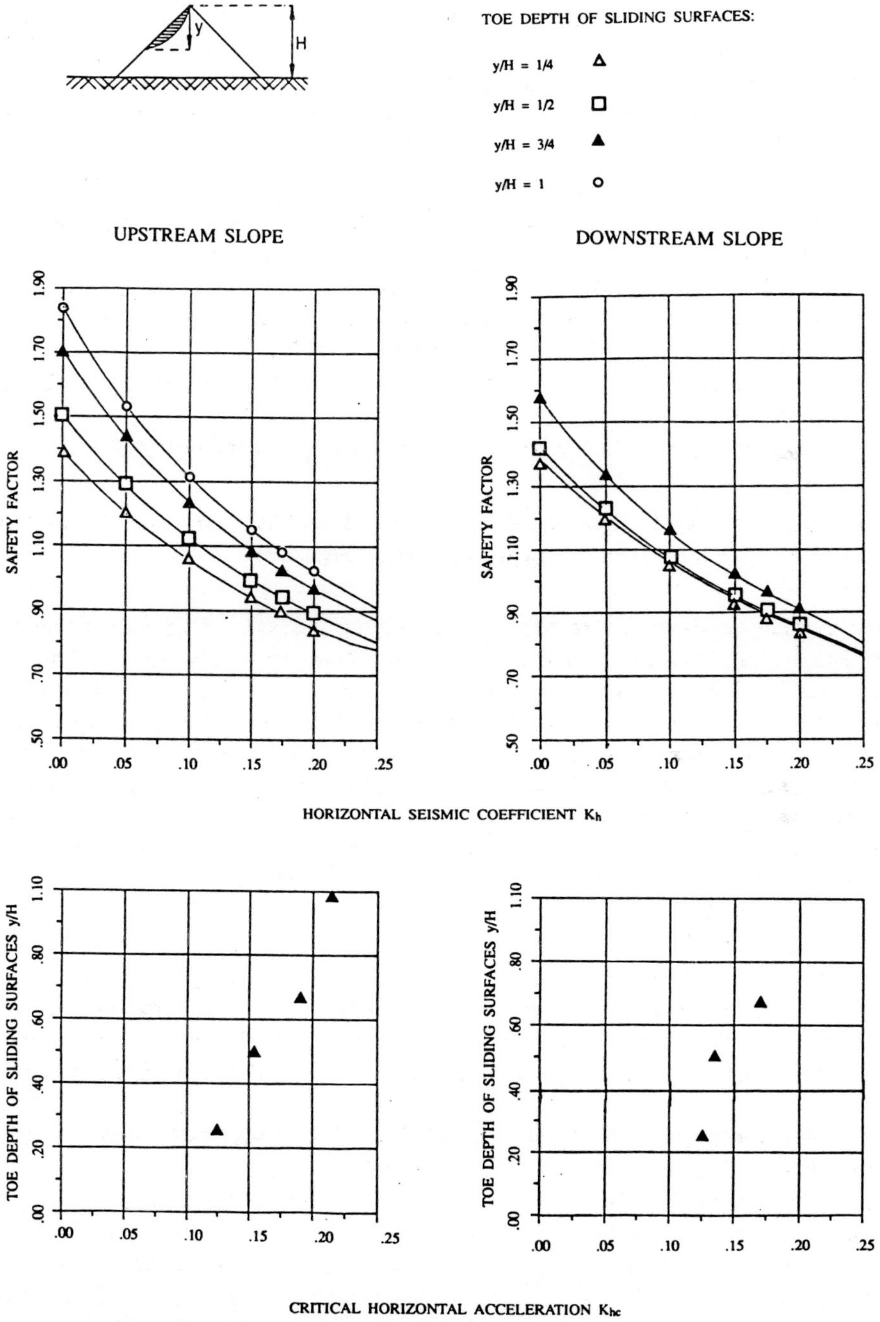

Fig. 10: Calculated critical horizontal accelerations

numerical computations have been carried out in an ALLIANT FX80 at the ISMES computer facilities in Bergamo.

3.1.1 Evaluation of the initial stress distribution

In an elasto-plastic effective stress approach, it is of crucial importance to evaluate the initial state variables, such as effective stress and plastic variables of the material, from which to start the dynamic analysis. The numerical evaluation of these quantities requires an accurate simulation of the loading history which the structure has been subjected to. Unfortunaly, the loading history of the Infernillo embankment dam is not reported in literature. Consequently, a drastically semplified analysis has been carried out assuming an elastic-perfectly plastic constitutive model; the yield (potential) function has been assumed to be of Mohr-coulomb type.

The analyses performed consist of:

1. Evaluation of the stress distribution at the end of the construction stage. In this analysis, only the body forces due to the wet weight of the materials have been considered. These body forces have been simply linearly increased up to reach the actual weight of the materials.
2. Evaluation of the stress distribution at the end of the water filling of the reservoir. In this analysis, only the buoyant forces due to the pore pressure distribution have been considered. Starting from the initial stress distribution determined by the first analysis, these buoyant forces have been simply linearly increased up to reach their actual value.

3.1.2 Seismic response

The dynamic response of El Infernillo embankment dam has been calculated for the earthquake accelerograms EQ1.DAT. For the numerical calculations the following assumptions have been made:

1. The effective stress obtained from the 2*nd* analysis reported in the previous Section 3.1.1, has been taken as the initial effective stress distribution.
2. The materials have been assumed to responde according to the ECAM model. [18]. The additional parameter values required by the model have been estimated in order to be as close as possible to the soil behaviour resulting from the eodometer tests reported in the literature.
3. The core and the filter (materials 1,2 and 3 in Fig. 6) have been analysed assuming undrained conditions. In other terms, it has been assumed that the permeability values of these materials are sufficiently low, so that the fluid stored in them has not the possibility to escape during the short time of the earthquake excitation.
4. The rockfill (materials 4,5 and 6 in Fig. 6) in the upstream side of the embankment, have been analysed assuming fully drained conditions. In othter terms, it has been assumed that the permeability values of these materials are sufficiently high, so that no significant excess pore pressure is built up in them during the earthquake excitation.

The accelerations, displacements and excess pore pressures response calculated during the earthquake excitations for the points shown in Fig. 7, are reported in Figs. 12, 13, 14 and 15.

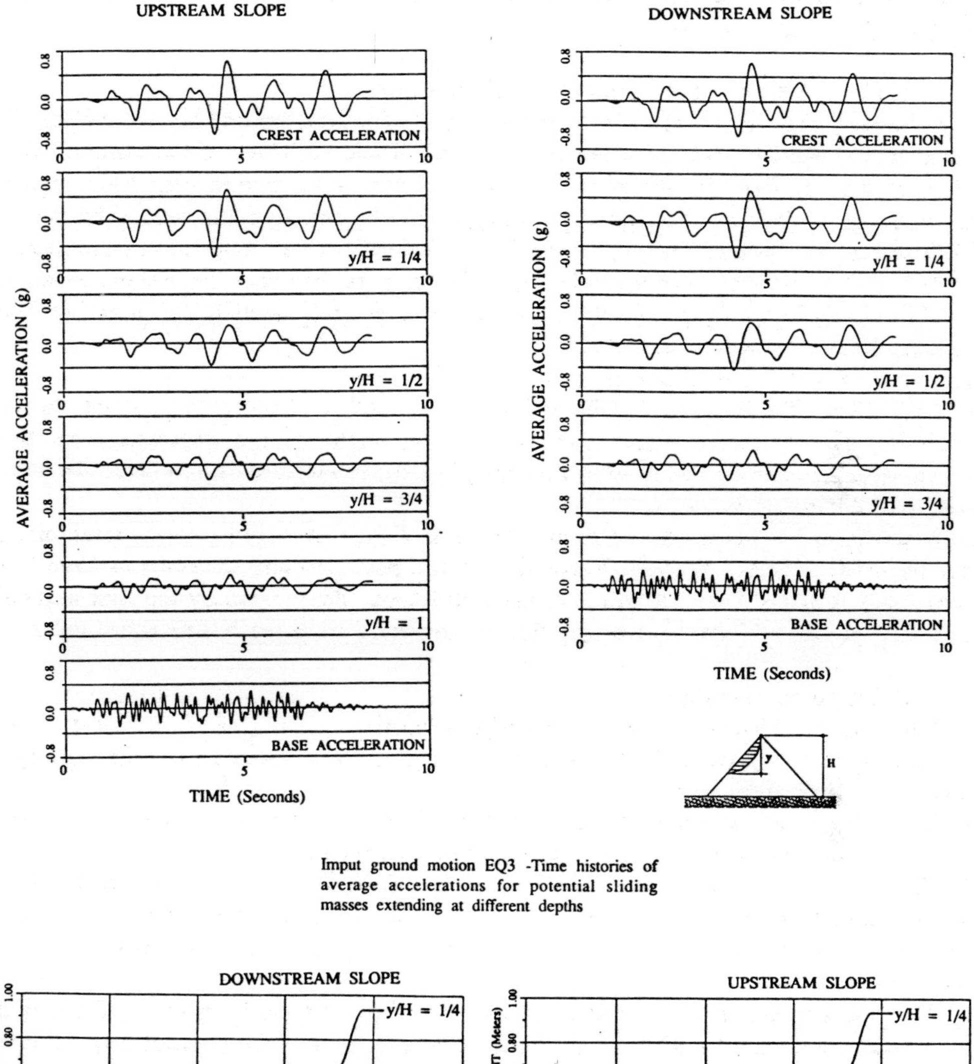

Imput ground motion EQ3 -Time histories of average accelerations for potential sliding masses extending at different depths

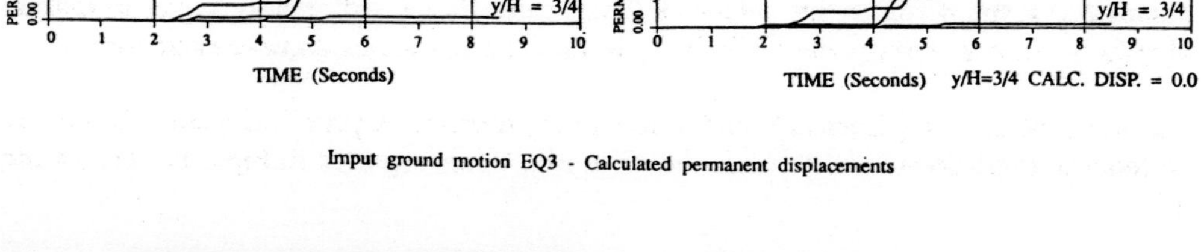

Imput ground motion EQ3 - Calculated permanent displacements

Fig. 11: Calculated permanent displacements

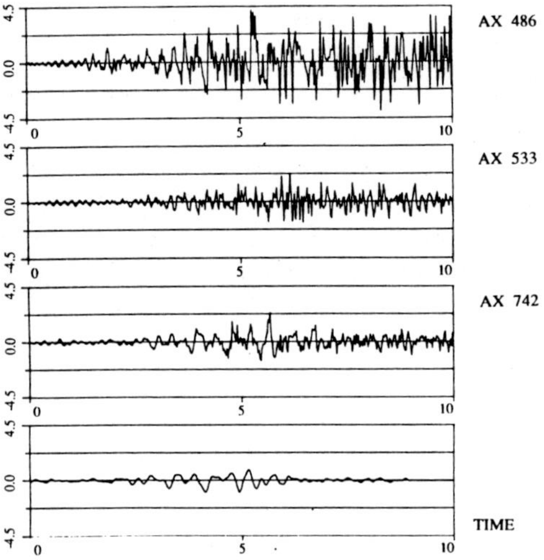

Fig. 12: Horizontal component of the total acceleration

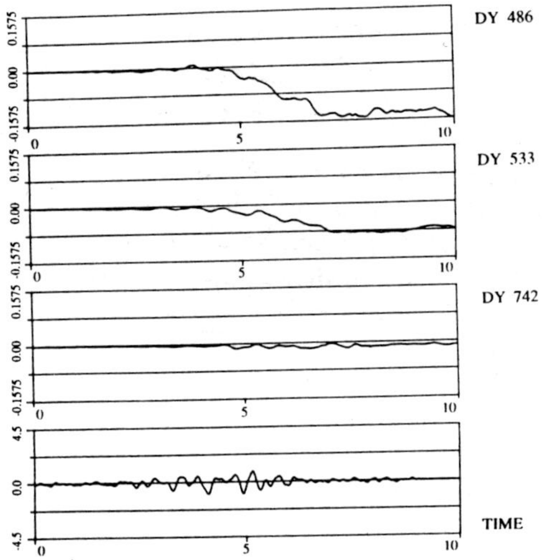

Fig. 14: Vertical component of the displacements

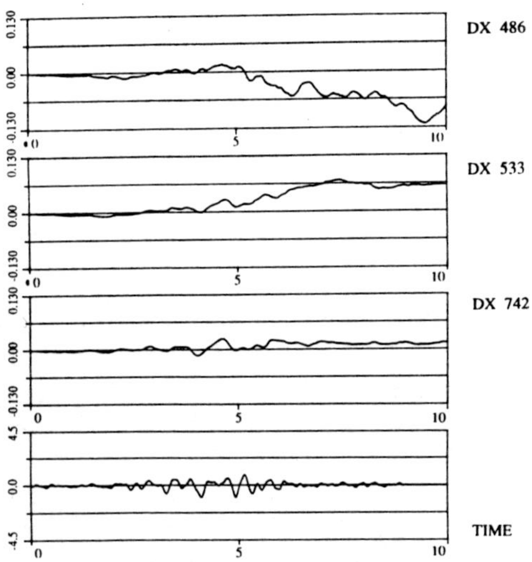

Fig. 13: Horizontal component of the displacements

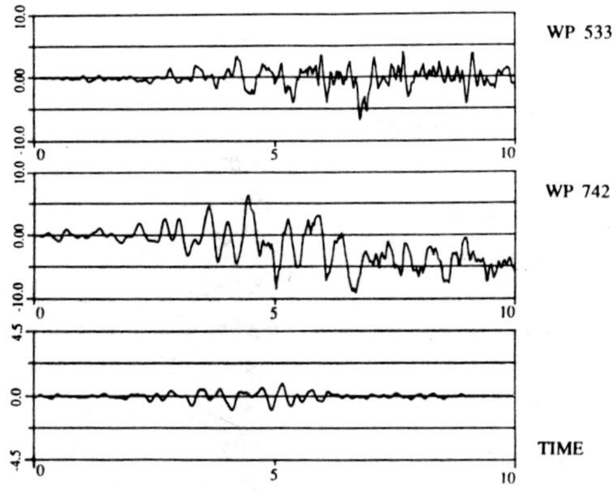

Fig. 15: Excess pore pressure

4.0 Conclusions

This paper briefly describes three different approaches to the dynamic analysis of earth dams. The safety evaluation of a dam carried out with the method of local safety factors has allowed to point out the most stressed areas of the embankment in seismic conditions and therefore with lower safety factors. However it is important to point out the approximation degree that characterizes the definition of the safety factor, which has been determined by assuming that the "effective" shearing stress has a constant effect, while it always has a instantaneous for very short times. Infiernillo dam has been studied both with Newmark method and with a coupled elastic-plastic effective stresses analysis. The permanent displacement computed with Newmark method can surely be considered as a stability index of the embankment under the action of different earthquakes. Anyhow it is difficult to give a precise physical significance to this parameter to be later compared with the same quantity measure in situ. However this quantity is computed considering the whole dynamic response of the structure, therefore it can be considered as a reliable safety index, above all in terms of comparison. The analyses have been carried out on DIGITAL VAX 8600 computer, the static analysis has taken a computation time of 50 minutes, the dynamic analyses for the two examined accelerograms have taken a computation time of 25 hours. The results obtained with the coupled effective stress dynamic analysis are certainly very promising, namely for the higher accuracy of this procedure that provides the time history of stresses, displacements and pore pressures inside the embankment. At present a limitation to the application of this procedure could depend on the shortage of data relating to mechanical properties of the materials characterizing the constitutive law under dynamic conditions. Another limitation could be represented by the engagement for the computation costs. The results reported in this note have been obtained by analyzing a time interval of 10 seconds with integration steps Δt of 2.5×10^{-4} sec, for a total of 40,000 steps. The computations have been carried out non parallel and vectorial ALLIANT FX80 computer and have taken a total computation time of 43 hours and 20 minutes.

Bibliography

[1] Seed H.B. (1979). *Considerations in the earthquake-resistant design of earth and rockfill dams.* Geotechnique 29, No.3.

[2] Seed H.B. (1973). Stability of earth and rockfill dams during earthquakes. Embankment-dam engineering. Hirchfeld R.C. and Poulos S.J. eds. New York; John Wiley and Sons, 239-270.

[3] Seed H.B. and Idriss I.M. (1970). *Soil moduli and damping factors for dynamic response analysis.* Earth. Eng. Res. Center, report n. EERC 70-10.

[4] Duncan J.M. and Chang C.Y. (1970). *Non linear analysis of stress and strain in soils.* J. SMFD-ASCE vol.96 SMS, pp. 1629-1653.

[5] Ministero LL.PP. *Norme tecniche relative alle costruzioni sismiche.* Legge n.64 del 2 Febbraio 1974 -DM del 19 giugno 1984, D.M. del 29 Gennaio 1985.

[6] Schnabel G., Lysmer J. and Seed H.B. (1972). *SHAKE: a computer program for earthquake response analysis of horizontally layered sites.* Report n.EERC 72-12, University of California, Berkeley.

[7] Idriss I.M., Lysmer J., Hwang R. and Seed H.B. (1973). *Quad4, a computer program for evaluating the seismic response of soil structures by variable damping finite element procedures.* Report n. EERC 73-16, University of California Berkeley.

[8] Newmark N.M. (1965). *Effects of earthquake on dams and embankments.* Geotechnique, 15, No. 2.

[9] Ambraseys M.N. and Sarma S.K. (1967). *The Response of earth dams to strong earthquakes.* Geotechnique, 17, No.3.

[10] Makdisi F.I. and Seed H.B. (1978). *Semplified procedure for estimating dam and embankment earthquake induced deformations.* J. Geotechique Engineering Division, vol.104, No.Gt7.

[11] Liam Finn W.D. (1990). *Seismic analysis of embankment dams.* Dam Engineering, vol. I, Issue I.

[12] Purdue University. *User guide for PC STABL 5M.* Joint Highway Research Project JHRP 88/19.

[13] Ruggeri G. et al. (1991). *The seismic Response of El Infiernillo embankment dam: equivalent linear approach and Newmark method for the evaluation of permanent displacements.* First ICOLD Benchmark Workshop, Bergamo (Italy), May 28-29.

[14] Biot M.A. (1941). *General theory of three dimensional consolidation.* Journal of Applied Physic, vol.12, pp.155-164.

[15] Sandhu R.S. and Wilson E.L. (1969). *Finite element analysis of seepage in elastic media.* Proceeding ASCE, Mechanics Division, vol.95, No.EM3.

[16] Zienkiewicz O.C. and Shiomi T. (1984). *Dynamic behaviour of saturated porous media: the generalized Biot formulation and its numerical solution.* International Journal of Numerical and Analytical Methods in Geomechanics, vol.8, pp.71-96.

[17] Fusco A. (1985). *Continuum mechanics and finite element numerical solutions in geomechanics.* Ph . D. Thesis, University of Ottawa.

[18]Fusco A. et al. (1991). *The seismic Response of the El Infiernillo Embankment dam: coupled effective stress elasto-plastic analysis.* First Benchmark Workshop - Bergamo (Italy), May 28-29.

NUMERICAL MODELLING OF DYNAMIC BEHAVIOUR OF NATURAL SLOPES IN PRESENCE OF NON-LINEAR SEDIMENTS AND PORE WATER

Hormoz MODARESSI* and Ikhlef BENZENATI**

Abstract: The authors have already developed a complete methodology to handle with the whole problem of the seismic response of inelastic saturated soils in a unified approach. In the present paper this technique is applied to a real case for which the existing *in situ* recordings permit the control of the quality of the results.

Résumé: Une méthodologie complète fondée sur l'approche directe est proposée et employée par les auteurs pour évaluer la réponse sismique des sols anélastiques saturés. Elle est appliquée ici à un cas réel pour lequel les enregistrements existant sur quelques points de contrôle permettent de contrôler la qualité des résultats.

1- Introduction

It is now well established that the geometrical irregularities modify the seismic ground motion. Many slope failures are reported as a result of dynamic loadings especially the earthquakes. For instance, serious damages were observed during the 1968 Tokachi-oki earthquake on the buildings situated near slopes.

Mainly two major factors seem to play an important role : The geometry of the slope and the mechanical behaviour of the soil. The geometry affects the nature of the incident seismic waves while the mechanical behaviour of the soil dictates the deformation as well as the stress generation in the soil.

In general, in Earthquake Engineering problems we are interested in a frequancy range between 0.1 and 20 Hz and the velocity of shear waves in upper layers varies from 0.1 to 3 km/s so that the wave lengths are from tens of meters to tens of kilometers. The geometrical irregularities with these dimensions will then induce considerable modifications on the seismic response of the ground. Moreover, the non-linear behaviour of sediments as well as the pore pressure generated during the seismic process can largely affect the stability of soil slopes.

In the methods commonly used in practice, the dynamic loadings are handled in a very simple way, usually by introducing an equivalent force without any regards to the geometry of the problem or the intrinsec behaviour of soils.

---

* Senior Engineer, Bureau de Recherches Géologiques et Minières, Orléans, France
** Graduate Student, Ecole Centrale de Paris, Châtenay Malabry, France

Numerous works have been dedicated to the study of topography and lateral inhomogeneity effects on the ground response at the surface (Trifunac (1971), Wong *et al.* (1974), Bard*et al.* (1980), Wong (1982)). In all these works a linear elastic behaviour of sediments is assumed. An interesting work carried-out by Ohtsuki *et al.* (1983) leaded to some very usefull conclusions.

As far as the topographical effects are concerned some general results can be given. They are summarized in figure 1. We note that for smaller ratios between the wave lenght and the height of the slope the horizontal displacements at the upper corner are higher than those of the bottom corner, while the vertical displacements at the upper corner are smaller compared to those of the bottom corner. The same is valid for higher values of the slope angle.

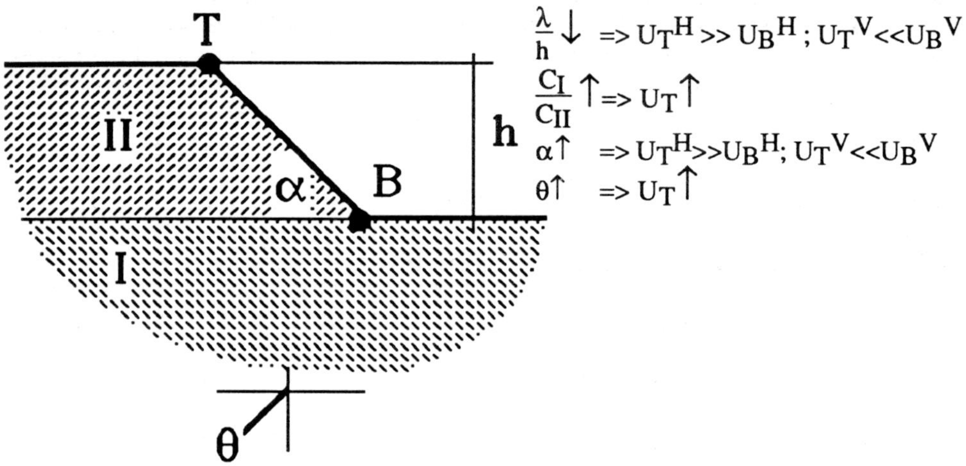

Fig. 1

The influence of non-linear and two-phase behaviour of sediments as well as the topography effects have been already studied by Modaressi *et al.* (1988) using a Finite Element approach. For a given slope, they studied the influence of the elastoplastic behaviour of sediments and compared the computed results to those obtained in the case of linear elasticity. They noted that while the horizontal acceleration does not seem to be affected by the inelastic behaviour of the soil the displacement field is significantly amplified. This observation confirms that the only assumption of an equivalent force determined from a hypothetical horizontal acceleration can not be considered as sufficient.

2- Illustrative Example

To assess the validity of an approach it is always necessary to compare the computed results to those obtained in real cases. Unfortunately, a complete set of results in the case of a natural slope submitted to a strong motion was not available to authors. For this reason we have decided to focus our attention to the case of Ashigara Valley which will be briefly presented in the following paragraph.

However, we have to underline that this chosen example can not be considered as a very significant example as firstly the studied slope is enough steep and secondly the seismic input motion of magnitude 5.1 does not induce very high distortions in the sediments (about 2.10^{-4}). In any case, the results obtained on a control point can be compared to recorded results and in this way ensures the quality of predictions. On the other hand, in a phenomenological point of view, the computed results are of some value for understanding the physical phenomenon occuring in a saturated soil submitted to seismic loading.

2.1 ASHIGARA VALLEY

Ashigara Valley is a sedimentary valley situated in south-west of Tokyo and has been chosen for an international workshop for the evaluation of numerical methods for the blind prediction of site effects during earthquakes. This valley has essentially a tri-dimensional aspect but a large majority of predictions have been carried out in two-dimensional geometry either on longitudinal or transversal cross sections. The longitudinal section considered here presents in average a smooth slope (≈1/20). However, in a small portion of it there is sharper slope of about 1/9 on which we will focus in this paper (fig. 2).

The sedimentary soil is very porous (porosity ≈ 0.7) saturated (saturation degree = 96%) and exhibits an inelastic behaviour observable through G-γ curves (fig. 3). The permeability of the soil has been fixed to 10^{-9} m/s. Two sets of recordings are available respectively for 2.9 and 5.1 magnitudes. In this study only the higher magnitude is considered as the non-linear effects are searched. The input acceleration and its Fourier Transform are presented in figures 4 and 5. Taking into account the shear wave velocity of the upper sediment layer (≈170 m/s), the wave lengths of some interest to the studied slope vary from 20 m to 80 m.

It is important to notice that the shear wave velocity in the surface sediment changes during the seismic loading due to the dependancy of the shear modulus to the distortion. We have remarked during the numerical simulations that this velocity has been decreased at the end of the motion.

The shear velocites in underlying layers are given below:

layer	domain	shear wave velocity
Alluvial sand	5	170 m/s
Gravel [Hp(g)]	4	700 m/s
Sand [Hp(s)]	3	400 m/s
Bedrock	1	800 m/s

2.2 COMPUTED RESULTS

The studied domain considered for the compuations and different subdomains are presented in figure 6. The Finite Element mesh size is chosen as to be able to represent fairly the frequencies up to 8 Hz. In order to limit the computational cost the whole longitudinal section is not considered in our analysis. In fact, only the first control point KS1 is included into the Finite Element analysis.

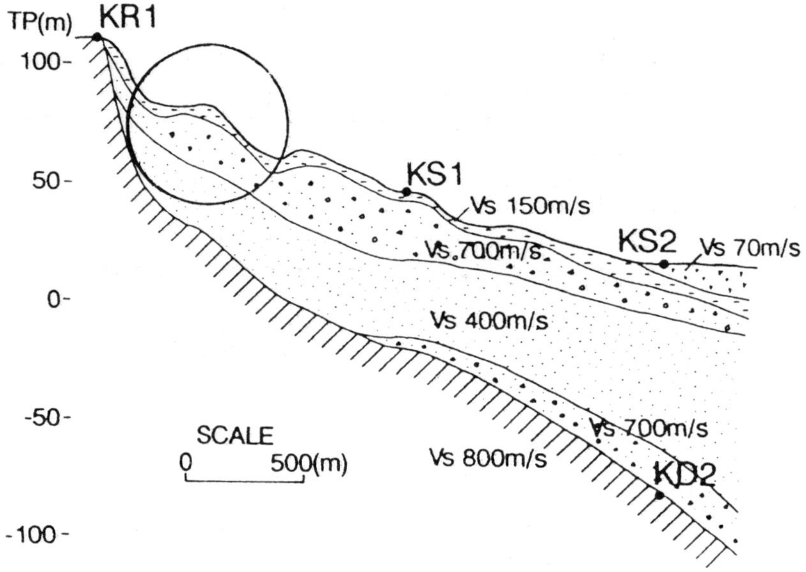

Fig 2 - Longitudinal cross section of Ashigara Valley and position of studied slope

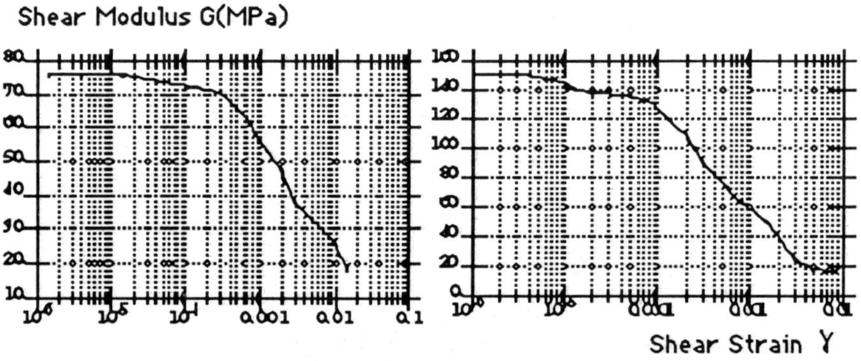

Alluvial Humus Soil (Hp) younger Loam Sediment (TMP)

Fig. 3 - G-γ curves for surface sediments (experimental results)

The surface sediments are considered as elastoplastic two-phase media. The G-γ curves obtained from numerical simulations are presented in figure 7. The lower layers are assumed to be elastic one-phase materials.

The seismic incident field is given on the rock site (KR1) and introduced by the paraxial elements which act also as dynamic absorbing boundaries.

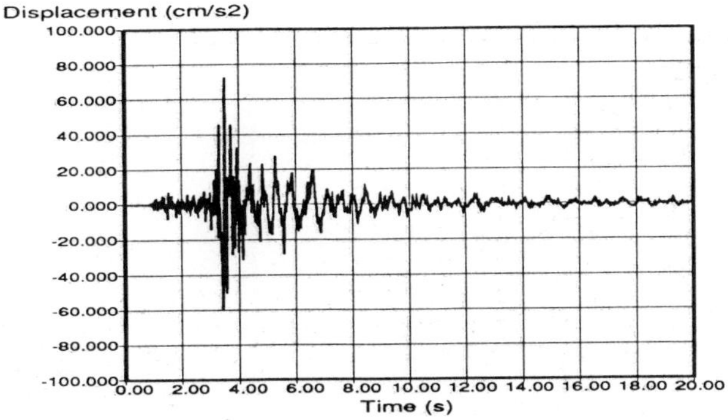

Fig. 4 - Input seismic acceleration

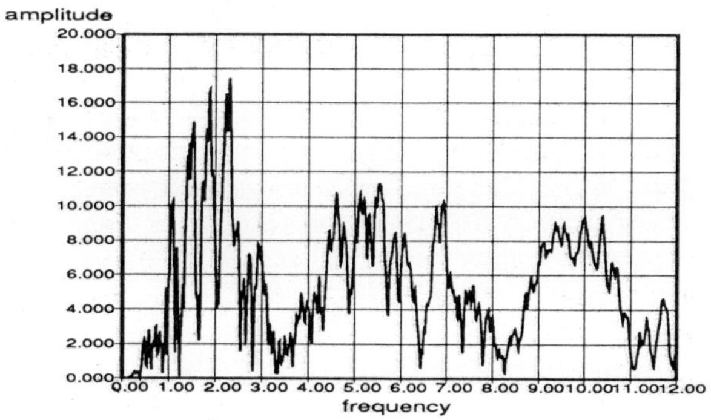

Fig. 5 - Frequency content of input seismic acceleration

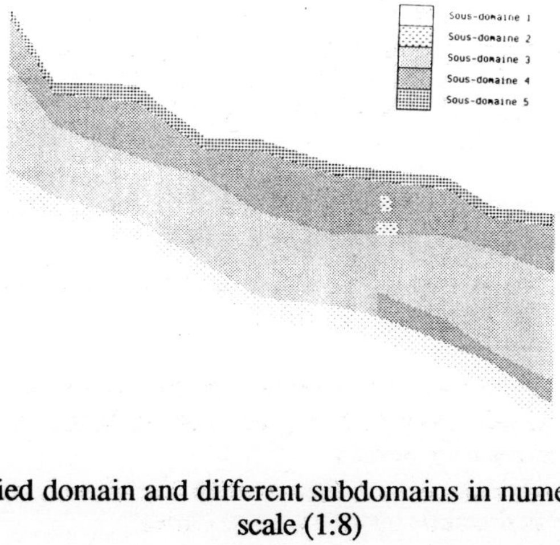

Fig. 6 - Studied domain and different subdomains in numerical simulation scale (1:8)

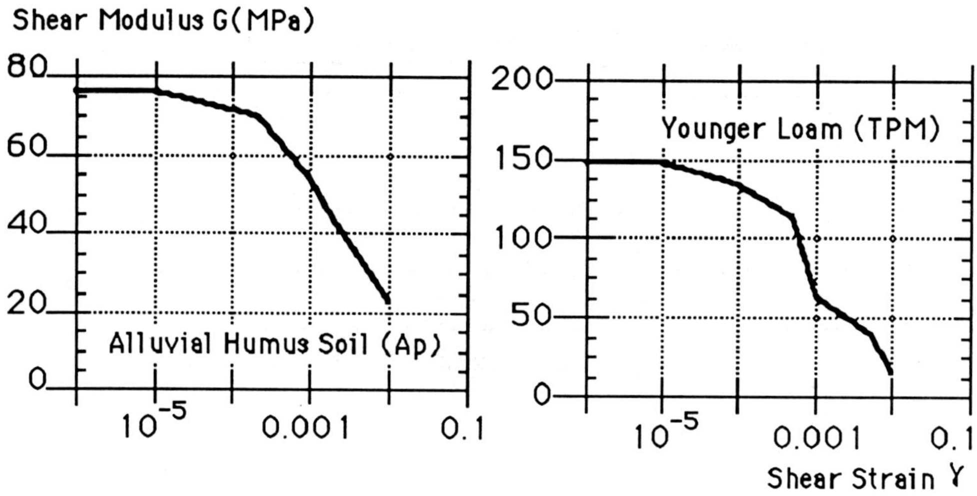

Fig. 7 - G-γ curves for surface sediments (numerical simulations)

The horizontal and vertical displacement time histories computed on the surface of the slope are presented in figures 8 and 9. As expected the horizontal displacements are higher on the upper corner of the slope. A slight settlement is also observed at the upper corner of the slope. The horizontal displacement fields are presented after 4 and 8 seconds (fig. 10 and 11). Although the values of the displacements are very small the expected tendancy of the motion is obtained.

The excess pore pressure field after 8 seconds is given in the figure 12. Its maximum value is 12 kPa and is obtained at 3.5 m below the surface at the vicinity of the upper corner of the slope. The mean effective pressure at this depth is small due to the high porosity of the soil and equals to about 18 kPa. Despite the low generated strain in the soil the liquefaction is not so far to happen.

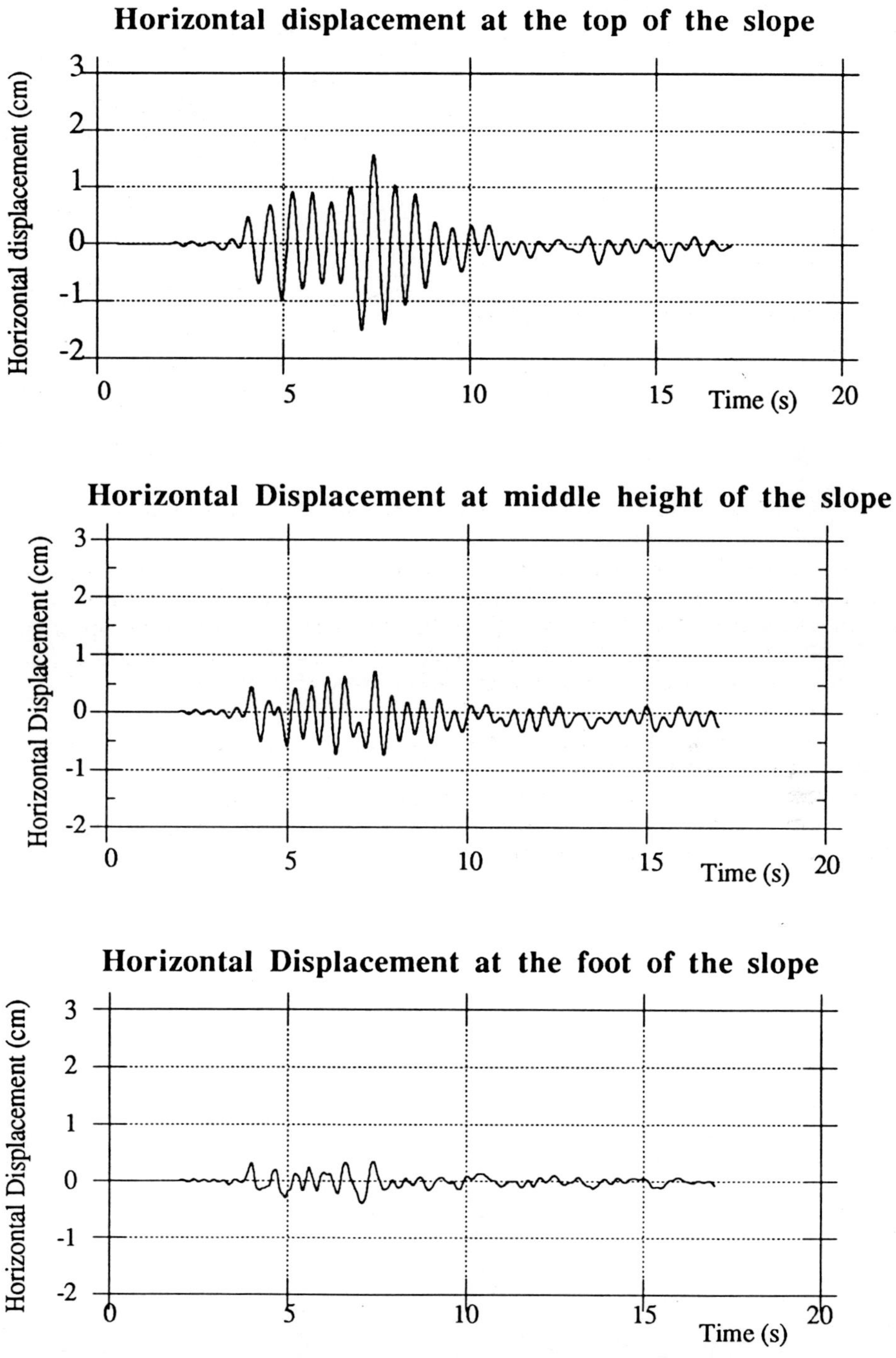

Figure 8 - Computed horizontal displacements on the slope

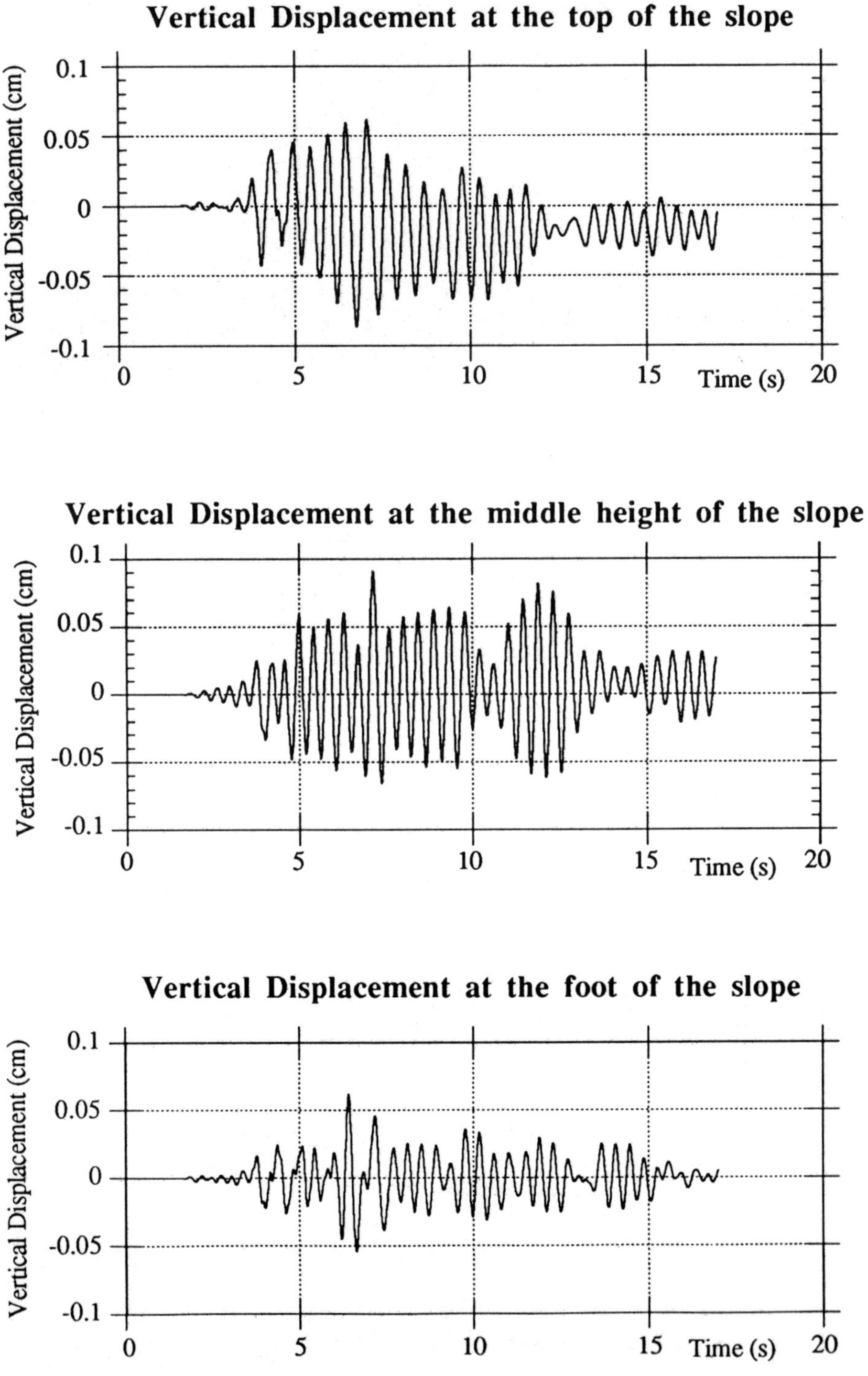

Figure 9 - Computed vertical displacements on the slope

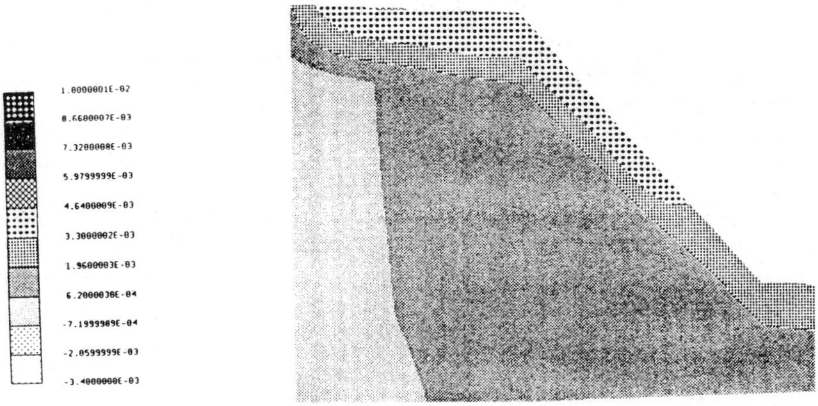

Fig. 10 - Horizontal displacement field after 4 seconds (scale 1:12)

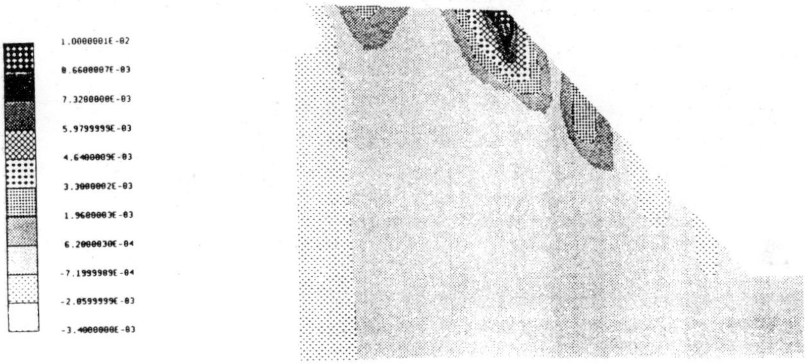

Fig. 11 - Horizontal displacement field after 8 seconds (scale 1:12)

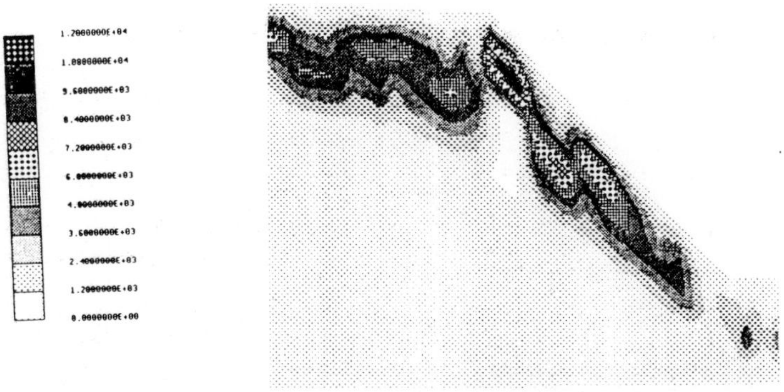

Fig. 12 - Pore pressure field after 8 seconds (scale 1:12)

3- Conclusions

The behaviour of a natural slope submitted to an earthquake of magnitude 5.1 is studied. The elastoplastic and two-phase behaviour of the upper layer is taken into account. The slope is very smooth and therefore stable and in addition, the seismic motion does not induce high distortions in the surface soil. However, some intereseting aspects as the generation of pore pressure and a slight irreversible settlement at the vicinity of the upper corner of the slope is observed. The generation of the pore pressure resulting into the decrease of mean effective pressure introduces some heterogeneities which affect the wave pattern in the studied domain. The horizontal displacements are amplified at the upper corner and deamplified at the bottom corner of the slope. Unfortuantely, the seismic intensity and the slight slope do not permit us to examine the failure mechanism as has been expected.

Bibliography

[1] Bard P.Y. & Bouchon M. (1980) The seismic response of sediment-filled valleys.Part 1:The case of incident SH waves. B.S.S.A., vol 70, n° 4, 1263-1286.

[2] Trifunac M.D. (1971) Surface motion of a semi cylindrical alluvial valley for incident plane SH waves. B.S.S.A., vol 61, 1755-1770.

[3] Modaressi H., Aubry D. & Mouroux P. (1988) Influence de la topographie et du comportement anélastique des sols sur les ondes sismiques SV, *afps*.

[4] Ohtsuki A. & Harumi K. (1983) Effect of Topography and Subsurface Inhomogeneities on Seismic SV waves, Earthq. Eng. and Struc. Dyn., vol 11, 441-462.

[5] Proceedings of the International Symposium on The Effects of Surface Geology on Seismic Motion, Odawara, Japon, 1992.

[6] Wong H.L. & Trifunac M.D. (1974) Scattering of plane SH waves by a semi elliptical canyon. Int. J. Earthq. Eng. Struc. Dyn. , vol 3 , 157-169.

[7] Wong H.L. (1982) Effect of surface topography on the diffraction of P, SV and Rayleigh waves. B.S.S.A., vol 72, n° 4, 1167-1183.

STABILITY OF OFFSHORE SLOPES IN SEISMIC REGIONS

Fabrizio PELLI*, Gian Paolo VASSALLO**
and Fabrizio LAGASCO***

Abstract: The stability of an offshore slope in a seismic area is evaluated using two different procedures: one based on a deterministic approach and the other on a probabilistic approach. From the results of a seismic response analysis, the two procedures are applied for the computation of the permanent displacements induced by the earthquake.

The deterministic procedure is based on a conceptual model consisting of a rigid block on a sloping plane and is developed through double integration of the accelerograms. The probabilistic procedure is based on stochastic models for the definition of peak acceleration and yield acceleration and allows the evaluation of the probabilistic distribution of the permanent displacements.

The results obtained with the two procedures are compared and discussed with reference to a case history.

Sommario: La stabilità di un pendio sottomarino in zona sismica viene valutata con riferimento a due procedure diverse, rispettivamente di tipo deterministico e di tipo probabilistico. A partire dall'analisi di risposta sismica, i due metodi vengono utilizzati per il calcolo delle deformazioni permanenti indotte nel terreno dal sisma.

La procedura di tipo deterministico è basata su un modello di blocco rigido su piano inclinato ed è sviluppata mediante doppia integrazione degli accelerogrammi. La procedura di tipo probabilistico è basata su modelli stocastici per la descrizione di accelerazione di picco e accelerazione di soglia e permette di valutare la distribuzione probabilistica delle deformazioni permanenti.

I risultati ottenuti con le due procedure sono confrontati e discussi con riferimento ad un caso reale.

---

* Project Manager, D'Appolonia S.p.A., Genoa, Italy.
** Project Manager, D'Appolonia S.p.A., Genoa, Italy.
*** Project Engineer, D'Appolonia S.p.A., Genoa, Italy.

1. INTRODUCTION

The assessment of slope stability under seismic conditions is a key design issue in areas characterized by relevant seismicity. This aspect is particularly relevant offshore, where even relatively gentle slopes may be affected by strong motion earthquakes, due to the poor mechanical properties often characterizing the marine sediments.

The stability assessment can be carried out by evaluating the permanent displacements associated with a seismic event by double integration of the acceleration time history, or according to an empirical approach such as the method developed by Lin and Whitman [1]. The empirical approach is suitable for use within a probabilistic framework where the key parameters influencing the slope behaviour are treated as random variables.

In the following discussion, the two methods are applied to the seismic stability analysis of a site located in the Mediterranean Sea, for which a detailed geotechnical characterization is available. The purpose of the study is to illustrate the use of the two methods for design purposes, evaluate the accuracy limitations of the simplified empirical procedure and assess the sensitivity of the basic random parameters influencing the earthquake induced displacements.

2. SLOPE STABILITY ASSESSMENTS IN SEISMIC CONDITIONS

The magnitude of induced slope displacements, rather than considerations on the possibility that the factor of safety drops to unity, should be the criterion for assessing the degree of stability or instability of a slope, during an earthquake [2]. These displacements can be computed by a procedure similar to that used for analyzing the movement of a sliding block on a sloping plane. Such principle has been applied in the past [3] to compute earthquake induced displacements in sand embankments by integrating twice the portion of the acceleration time history that exceeds the critical acceleration a_c (i.e., acceleration value producing a state of limit equilibrium when applied to the block). Sarma [4] presented a simplified procedure where each cycle of the acceleration time history exceeding a_c is given a rectangular, triangular or half sine shape. He found that for a_c/a_{max} greater than 0.5 the triangular pulse is generally appropriate, whereas for a_c/a_{max} values smaller than 0.5 a rectangular shape is preferable. More recently, based on integration of 86 recorded time histories, Yegian et al. [5] have observed that the triangular shape is generally suitable for a_c/a_{max} greater than 0.1, whereas for a_c/a_{max} smaller than 0.1 all three simple base motions tend to underestimate the displacements.

For seismic stability assessments the acceptable deformation (i.e., the limit between stable and unstable conditions) has to be defined depending on the specific design case. As an example, for a jacket offshore structure founded on sloping seabed the residual displacement must be small enough not to cause structure collapse, and, in particular, the integrity of the foundation piles has to be assured [6].

3. DESCRIPTION OF THE CASE STUDY

The site is located in the Mediterranean Sea and is characterized by a water depth of approximately 100 meters. The slope angle is slightly variable across the area investigated, and a value of 4.5 degrees was selected for the analysis.

The soil properties were defined on the basis of extensive field and laboratory programs. The soil profile consists primarily of normally consolidated clay down to a depth of at least 200 meters. [6]

The shear strength properties were assessed by means of triaxial (UU and CIU), simple shear, torvane and CPT testing. The soil behavior under cyclic loading was investigated by means of cyclic triaxial tests on both granular and cohesive samples.

4. RESPONSE TO SEISMIC EVENTS

A free-field seismic response analysis was carried out to evaluate accelerations, strains and stresses induced in the soil mass by the earthquake [6].

The analysis was conducted using the finite element code QUAD4 [7], where the non linearity of the shear modulus and damping is accounted for by the use of equivalent linear soil properties. Due to the gentle slope of the seabed a one-dimensional finite element model of a 200 meter soil column was used.

A horizontal acceleration time history, obtained from an actual acceleration record scaled to reproduce a peak ground acceleration of about 0.4 g, was applied at the base of the soil column. The peak ground acceleration value was computed through a seismic risk study of the area for the RIE (Rare Intense Earthquake).

Soil damping and deformability where defined based on available experimental information for sand and clay, and using the relationships developed by Hardin and Drnevich [8].

For the evaluation of permanent displacements, acceleration time histories at several depths were considered. Typical records computed at three significant depths are shown in Figures 1a, 1b and 1c.

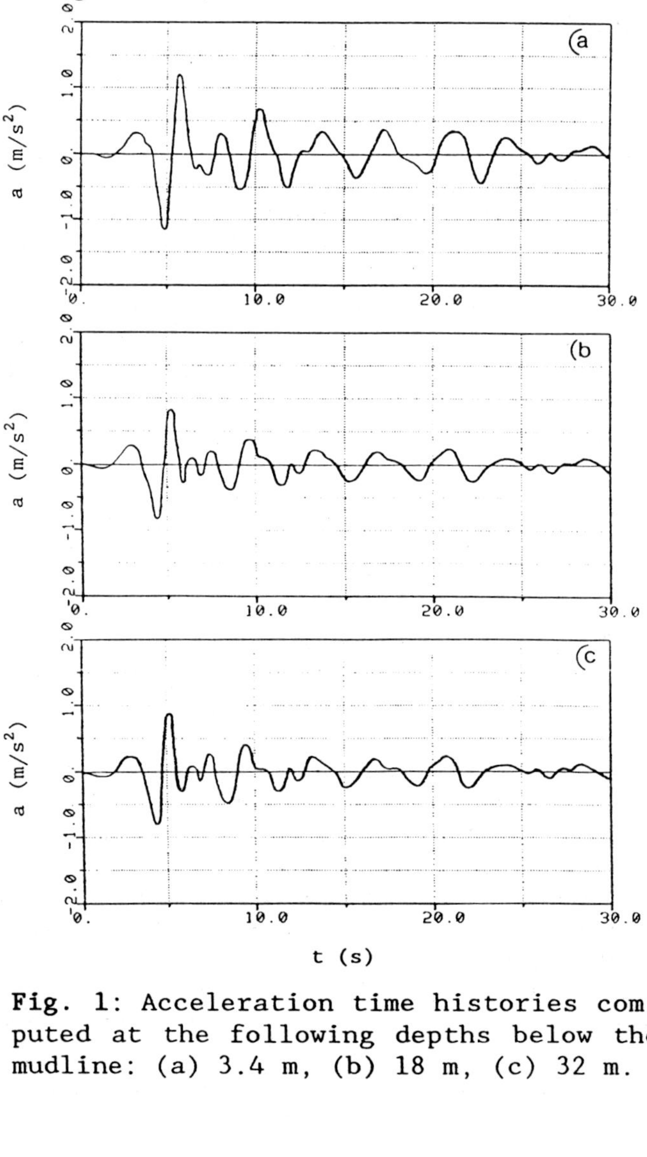

Fig. 1: Acceleration time histories computed at the following depths below the mudline: (a) 3.4 m, (b) 18 m, (c) 32 m.

In order to evaluate a proper peak ground acceleration to be used in conjunction with the empirical method, smoothing of the computed records was required to eliminate the effects of spurious high frequencies resulting from the numerical procedure. The response analysis showed that a significant deamplification of the base acceleration takes place within the soil column. The discussion presented herein will concentrate on the potential failure surface located at a depth of 32 meters, which resulted to be the most critical.

5. CRITICAL ACCELERATION

The sliding block model is depicted in Figure 2. The block, located on a sloping plane, tends to move downwards when the acceleration time history along the inclined plane exceeds the critical value $a_c = a_y - a_w$, where a_y is the yield acceleration, related to the shear strength of the soil, and a_w is the acceleration due to the self weight of the soil along the inclined plane.

For an infinite submerged slope in cohesive soil, the critical acceleration value can be inferred from the static factor of safety according to the following relationship:

$$a_c = \frac{\gamma}{\gamma'} \sin \alpha \; (FS - 1) \; g \qquad (1)$$

where γ' and γ are the buoyant and total soil unit weights, respectively, FS is the static factor of safety, α is the seafloor slope angle, and g is the acceleration due to gravity.

In defining the yield acceleration a_y the shear strength of the soil has to be assessed considering the high strain rate and the cyclic loading conditions associated with a seismic event.

The undrained shear strength of clayey soils tends to increase with the strain rate [9]. For the case study, the cyclic triaxial tests conducted in the laboratory revealed that for a frequency of one hertz the dynamic shear strength of the soil exceeds the static undrained shear strength S_u by a factor of about 1.3. On the other hand, the effects of cyclic loading may lead to strength degradation associated with pore pressure build up, if a limiting strain threshold value is exceeded [10]. Lefebvre et al. [11] presented stress threshold values frequently higher than 50 per cent of the undrained strength obtained at proper strain rates.

For the case study, based on cyclic triaxial tests, an a_y value consistent with the static S_u was selected as adequately conservative for the deterministic analysis. An undrained strength equal to 1.3 times the static S_u value is considered appropriate for the probabilistic approach.

6. DISPLACEMENT COMPUTATION BY NUMERICAL INTEGRATION

The deterministic analysis was carried out assuming the sliding block model depicted in Figure 2 and integrating twice that portion of the acceleration time history that exceeds the critical acceleration value a_c (Figure 3). It was found that the most critical slip plane was located at about 32 meters below mudline where the soil resistance was taken as S_u=50 kPa; for this plane a maximum residual displacement of four centimeters was obtained.

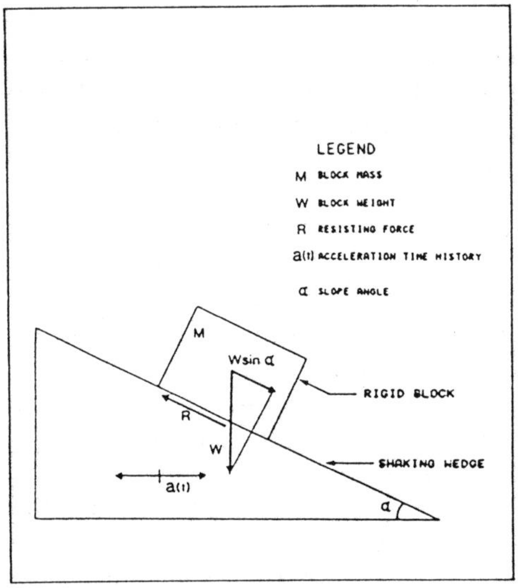

Fig. 2: Sliding block model.

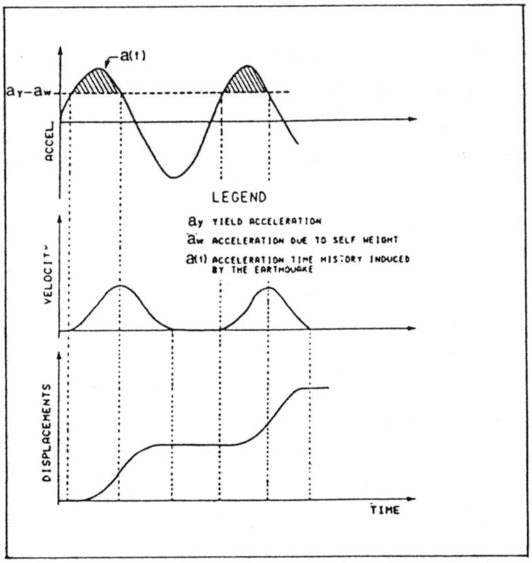

Fig. 3: Method to compute residual displacement (modified after [3]).

7. DISPLACEMENT PREDICTION BY A SIMPLIFIED MODEL

The procedure developed by Lin and Whitman [1] is based on the deterministic model proposed by Newmark [2], in combination with the ground motion model described by

Vanmarcke [12]. The model by Vanmarcke describes the damaging effect of a seismic excitation by means of four parameters, transforming a non-stationary motion into a normalized Gaussian equivalent stationary motion having a specified duration S:

$$S = \frac{\pi}{\Omega} \exp\left(\frac{a^2_{max}}{2\sigma^2}\right) \qquad (2)$$

where Ω is the dominant circular frequency of the ground acceleration, σ is the root mean square of the acceleration and a_{max} is the peak of the filtered acceleration at the location of the potential slippage.

Assuming that the excitation is narrow-banded, the expected value of the number of cycles E[N(S)] exceeding the critical acceleration is given by multiplying the duration S by the associated up-crossing rate for a Gaussian excitation.

Based on the assumptions presented above, a relationship providing the expected total displacement has been derived [1]. The formulation includes two empirical functions which were obtained carrying out simulated excitations using 140 acceleration records [1].

For the case study a value of 18 seconds was assigned to the duration of the equivalent stationary motion S, in consideration of the soil type, and of the typical magnitude and epicentral distance of a RIE event [13]. The value of the root mean square of the acceleration σ was then computed based on Eq. (2). The analysis of the response spectrum led to a dominant period of two seconds, and a value of the bandwidth δ equal to 0.5 was estimated.

The analysis was carried out for several values of the critical acceleration and the results are presented in Figure 4. It can be noted that for a_c consistent with the static S_u (50 kPa) the displacement predicted is larger than the value computed by means of the numerical integration. The difference is partly due to the inherent approximation of the simplified method. Furthermore, examining the acceleration time history resulting from the seismic response analysis, it is found that only one peak exceeds the a_c level, as opposed to 3.6 peaks predicted by the model based on the Gaussian assumption. This reveals that the assumed Gaussian ground motion [12] does not fit very well this specific time history.

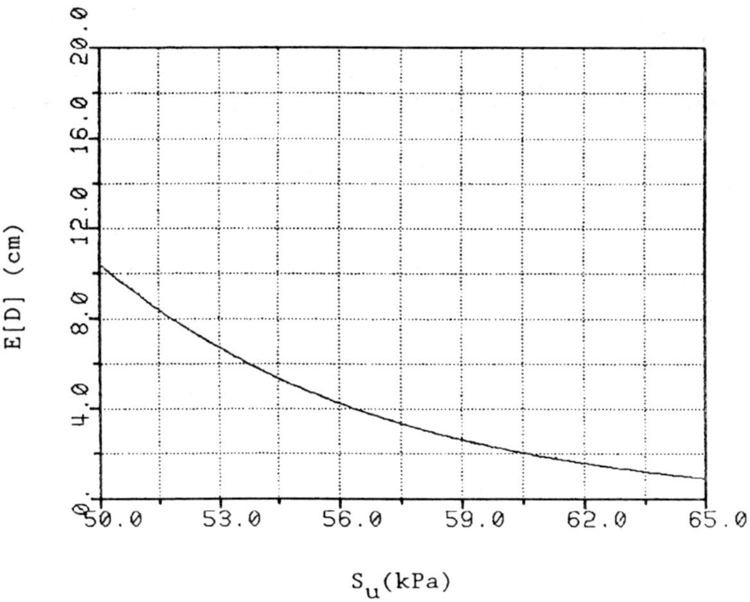

Fig. 4: Expected displacement vs. undrained shear strength.

Regardless of the accuracy limitations discussed above, the sensitivity analysis presented in Figure 4 indicates the strong influence of S_u, and consequently of a_c, on the expected earthquake induced displacement. As an example, increasing the undrained strength value by 30%, the displacement decreases by almost 90%. Emphasis must be placed on the importance of proper field and laboratory programmes, including static and cyclic soil testing, in order to estimate as accurately as possible the expected value of the critical acceleration.

8. PROBABILITY DISTRIBUTION OF THE ANNUAL EXPECTED PERMANENT DISPLACEMENT

Based on the assumptions associated with the simplified model presented above, the probability distribution of the permanent displacement can be assessed. In general, the following relationship can be written:

$$P(D) = \int P(D|a_c, a_{max}, \sigma, \Omega, \delta) \; dF(a_c, a_{max}, \sigma, \omega_g, \xi_g) \qquad (3)$$

where: $P(D|a_c, a_{max}, \sigma, \Omega, \delta)$ is the conditional probability distribution of the permanent displacement, ω_g is the ground natural frequency and ξ_g is the ground damping. Lin and Whitman [1] describe a procedure to evaluate the conditional probability $P(D|a_c, a_{max}, \sigma, \Omega, \delta)$. The joint distribution F $(a_c, a_{max}, \sigma, \omega_g, \xi_g)$ of Eq. (3) can be computed by defining the individual probability distributions of the parameters a_c, a_{max}, σ, ω_g, ξ_g and by considering possible correlations among these parameters. However, the peak ground acceleration and the critical acceleration can be considered as the most influencing random variables.

The statistical distribution of the peak ground acceleration has been assessed through a previous seismic hazard analysis.

The conditional probabilities of exceeding specified displacement values are shown, for displacements up to one meter, in Figure 5.

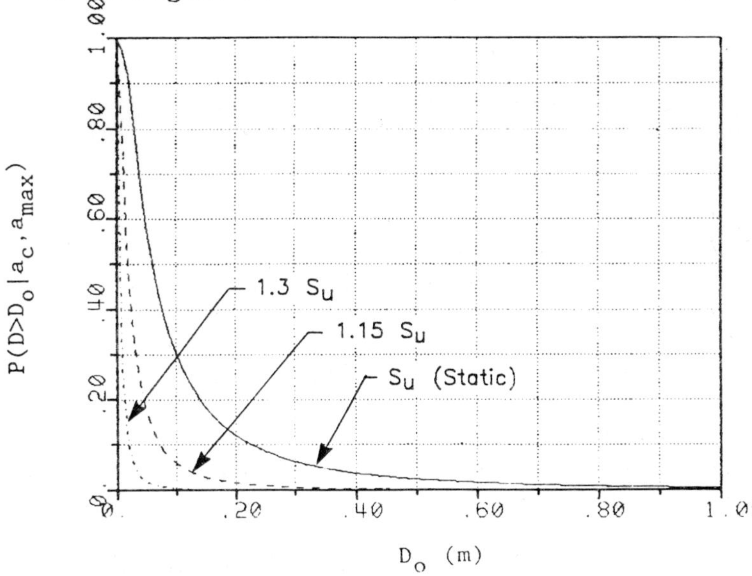

Fig. 5: Conditional probabilities of exceeding specified displacement values D_o.

The three curves, which correspond to expected values of a_c consistent with S_u (static), 1.15 S_u and 1.3 S_u, respectively, were obtained on the basis of the conditional probability function presented by Lin and Whitman [1]. The large scatter associated with variable a_c values is clearly visible; as an example, the conditional probability of exceeding a displacement of 20 cm varies from less than 1‰ (1.3 S_u) to more than 12% (S_u).

The annual probability of exceeding displacement values up to one meter are shown in Figure 6. The plot was obtained considering the return period of the RIE event. The random variables a_c and a_{max} were assumed to follow lognormal distributions [1]. For a_c an expected value consistent with a shear strength equal to 1.3 S_u and a coefficient of variation equal to 30% were assumed. Based on Figure 6, it can be observed that the predicted annual probability of exceeding the displacement of four centimeters computed by numerical integration is close to $3x10^{-6}$.

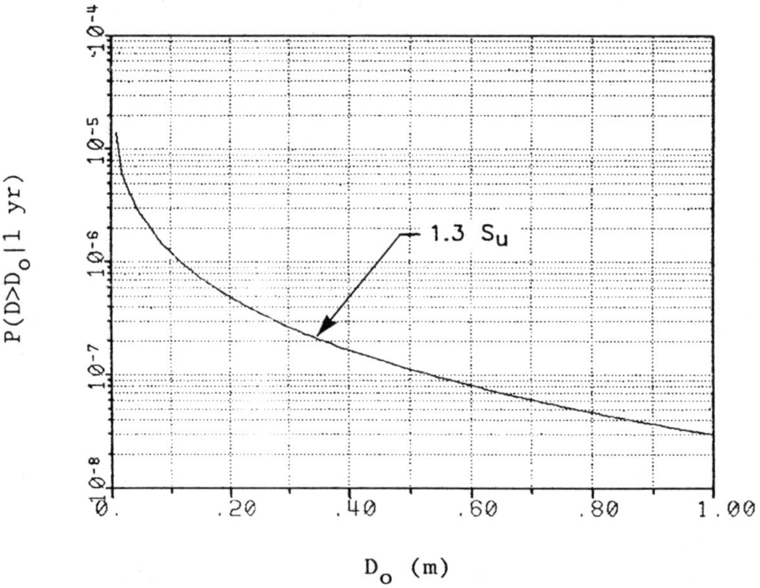

Fig. 6: Annual probability of exceeding specified displacement values at a depth of 32 m.

9. CONCLUSIONS

Two different methodologies for the computation of the residual displacements induced in an offshore slope by a seismic event are compared. The first methodology is based on a rigorous, double integration analysis of the acceleration time history. The second relies upon a simplified description of the earthquake, where the ground motion is assumed consistent with a Gaussian distribution.

In light of the discussion presented above, the following conclusions can be drawn:

1) The characteristics of the ground motion in proximity of the potential failure plane were found to influence considerably the earthquake induced displacement in the slope; therefore a proper seismic response analysis, providing a realistic acceleration time history, is crucial.
2) The sensitivity analysis carried out by using the simplified method indicated the key role played by the critical acceleration a_c in controlling the earthquake induced displacement. For the case study an increase of a_c associated with a 30% variation of the soil strength resulted in a reduction of the expected displacement in excess of ten times. Great care should be taken in defining a_c with sufficient accuracy, by means of proper field and laboratory testing programmes.
3) The simplified empirical method proved to be simple to use and suitable for preliminary assessments. However, its inherent inaccuracies and poor fitting of the record considered herein resulted in displacement values larger than computed by double integration of the acceleration time history.
4) One advantage of the simplified model is the possibility of using it in the framework of a probabilistic approach without major difficulties. An application of the method to determine the annual probability of exceeding the specified displacement value is illustrated herein, with reference to a case study.

References

[1] Lin J.S. and Whitman R.V. (1986). Earthquake-induced displacements of sliding blocks. *Journal of Geotechnical Engineering (ASCE)*, vol. 112, No. 1, p. 44-59.

[2] Newmark N.M. (1965). Effects of earthquakes on dams and embankments. *Géotecnique*, vol. 15, No. 2, p. 139-160.

[3] Goodman R.E. and Seed H.B. (1966). Earthquake-induced displacements in sand embankments. *Journal of Soil Mechanics & Foundation Division (ASCE)*, vol. 92, No. SM2, p. 125-146.

[4] Sarma S.K. (1975). Seismic stability of earth dams and embankments. *Géotecnique*. vol. 25, No. 4, p. 743-761.

[5] Yegian M.K., Marciano E.A. and Ghahraman V.G. (1991). Earthquake-induced permanent deformations: probabilistic approach. *Journal of Geotechnical Engineering (ASCE)*, vol. 117, No. 1, p. 35-50.

[6] Pelli F. and Vanzini R. (1991). Stabilità di strutture offshore fondate su pendio e soggette a sisma. (in Italian with summary in English). *Rivista Italiana di Geotecnica*, No. 4.

[7] Idriss I.M., Lysmer J., Hwang R. and Seed H.B. (1973). QUAD4 - A computer program for evaluating the seismic response of soil structures by variable damping finite element procedures. *University of California, Report No. EERC 73-16*, Berkeley.

[8] Hardin B.O. and Drnevich V.P. (1972). Shear modulus and damping in soils: design equations and curves. *Journal of Soil Mechanics & Foundation Division (ASCE)*, vol. 98, No. SM7, p. 667-692.

[9] Bjerrum L. (1973). *Problems of soil mechanics and construction on soft clay*. State-of-the-art report, Proceedings 8th International Conference on Soil Mechanics and Foundation Engineering, Moscow, vol. 3, p. 109-159.

[10] Sagaseta C., Cuellar V. and Pastor M. (1991). *Cyclic loading*. Proceedings 10th European Conference on Soil Mechanics and Foundation Engineering, vol. 3, p. 981-999.

[11] Lefevbre G., LeBoeuf D. and Demers B. (1989). Stability threshold for cyclic loading of saturated clay. *Canadian Geotechnical Journal*, vol. 26, No. 1, p. 122-131.

[12] Vanmarcke E.H. (1977). Structural response to earthquake. *Seismic Risk and Engineering Decisions*, edited by C. Lomnitz and E. Resenblueth, Elsevier Publishing Co..

[13] Theofanopoulos N. and Drakopoulos J. (1986). A Study of strong ground motion duration of earthquake in Greece. *Proceeding of the 8th European Conference on Earthquake Engineering*, Lisbon, vol. 1, p. 3.2/65-3.2/72.

DESIGN CHARTS FOR THE SEISMIC BEARING CAPACITY OF STRIP FOOTINGS ON SLOPES.

Abdul-Hamid SOUBRA* and Franck REYNOLDS**

Abstract: The seismic bearing capacity of a strip footing located on the top of a slope is calculated. The approach used is pseudo-static. The calculation scheme is based on the upper-bound method in limit analysis for a rotational log-spiral mechanism. Some charts are presented and discussed.

Résumé: On présente dans cet article le calcul de la capacité portante d'une fondation superficielle en tête de talus en présence de séisme. L'approche proposée est de type pseudo-statique et le schéma de calcul est basé sur la méthode de la borne supérieure en analyse limite pour un mécanisme rotationnel en spirale logarithmique. Des abaques sont présentés et commentés.

1 Introduction

A major consideration in the design of a strip footing in seismic areas is its stability against catastrophic failure. This problem of the ultimate bearing capacity of shallow strip footings under earthquake loadings has not been studied in detail in the past.

The very little studies available in literature describing the seismic effect on the bearing capacity of foundations concern those of Meyerhof [1] and Shinohara et al [2]. Both the approaches are pseudo-static where horizontal and vertical accelerations are applied to the centre of gravity of the structure and the problem is reduced to a static case of bearing capacity with inclined eccentric loads. However, in these solutions, the inertia of the soil mass is not included.

Recently, Sarma and Iossifelis [3] have considered this problem of the bearing capacity of strip footings in seismic areas by considering the inertial forces on all parts of the soil-foundation system. However, their approach is based on a limit equilibrium method making assumptions concerning the slip surface and the distribution of the inter-slice forces.

In this paper, the inertial forces are considered on all parts of the soil-foundation system and the method used is a variational limit equilibrium method: This method does not make assumptions concerning the slip surface and the normal stress distribution along this surface. It is also an upper-bound method in limit analysis

* Enseignant - Chercheur, E.N.S.A.I. Strasbourg
** Ingénieur, E.N.S.A.I. Strasbourg

for a rotational log-spiral mechanism. Hence, the solution obtained is an upper-bound one for an associated flow rule material obeying Hill's maximal work principle.

2 Theory

The following assumptions have been made in the analysis:

a. As was mentioned before, all inertias of the soil-structure system are considered in this analysis.

b. The earthquake acceleration for both the soil and the structure is assumed to be the same: Only the horizontal seismic coefficient k_h is considered in this analysis, the vertical seismic coefficient is often disregarded.

c. The earthquake load on the structure is represented by the base shear load acting at the foundation level and an eccentricity for the vertical foundation load.

d. A one sided failure is assumed to occur along the surface y(x) (fig. 1).

e. An infinite slope is considered in this analysis where the slip surface y(x) intersects the slope at point C. However, for a relatively small fill, the base failure might be predominant and little modification in calculation is necessarily.

f. Only the reduction of the bearing capacity due to the increase in driving forces is investigated under seismic loading conditions. The shear strength of the soil is assumed to remain unaffected by the seismic loading.

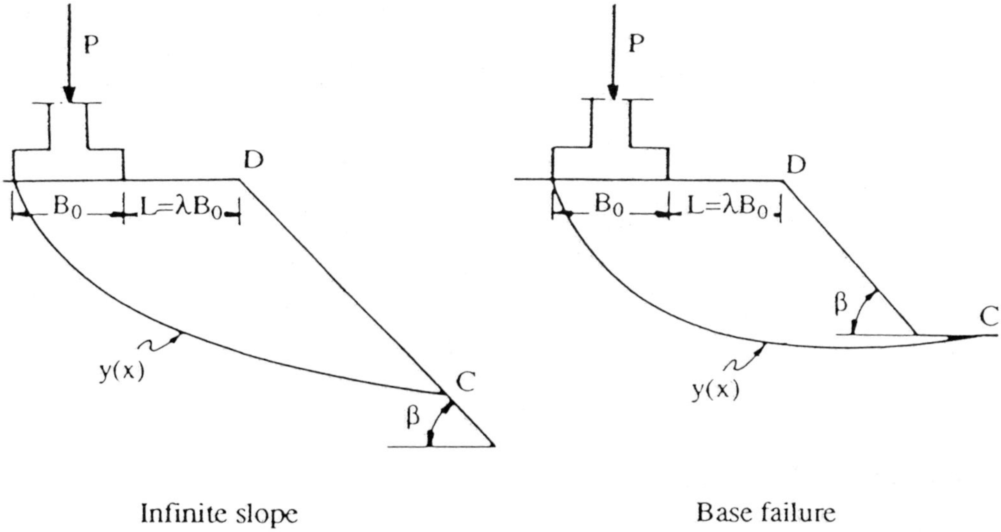

Fig. 1 : Slip surface for seismic bearing capacity analysis

As it is generally known, the traditional limit equilibrium methods give approximate solutions for the stability problems in geotechnical engineering. This fact is due to the a priori hypothesis concerning the shape of the slip surface and the normal stress distribution along this surface. In this paper we briefly describe a variational limit equilibrium method. This method allows us to determine the shape of the mechanism giving the minimum value of the seismic bearing capacity and for which the three limiting equilibrium equations are satisfied. For further details concerning this approach, refer to previous works of Soubra [4].

2.1 LIMITING EQUILIBRIUM EQUATIONS

As shown (Fig. 1), the strip footing rests on the top of a slope of inclination β, apart from the slope shoulder D by a distance $L=\lambda.B_0$ where B_0 is the breadth of the footing. The analysis considers the case of a surface footing without any surcharge pressure.

Writing the three equations of the static equilibrium for the soil mass (Fig. 2) and combining these equations with the Mohr-Coulomb criterion, one obtains the three limiting equilibrium equations as follows

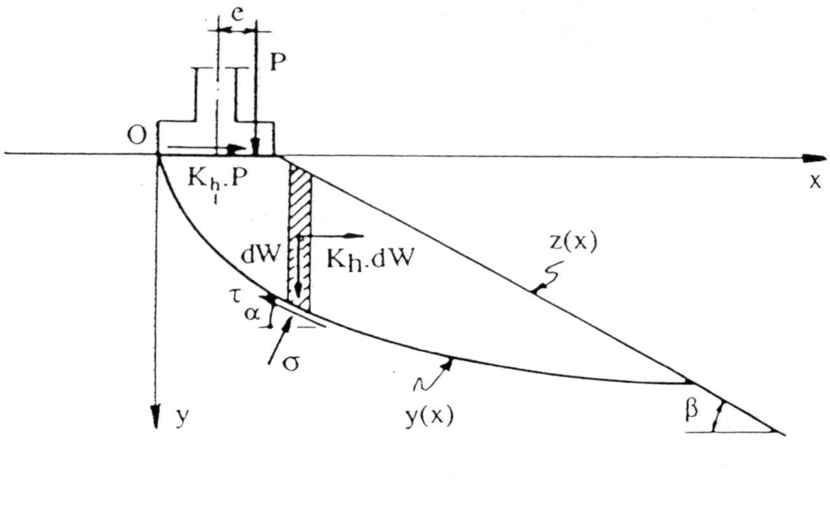

Fig. 2 : Free body diagram

$$P-\int_0^{x_1}[(c+\sigma tg\phi)y'+\sigma-\gamma(y-z)]dx=0 \qquad (1)$$

$$\int_0^{x_1}[\sigma y'-(c+\sigma tg\phi)+k_h\gamma(y-z)]dx \ +k_hP=0 \qquad (2)$$

$$\int_0^{x_1}[c+\sigma(tg\phi-y')]y-[(c+\sigma tg\phi)y'+\sigma]x-\frac{k_h}{2}\gamma(y^2-z^2)+\gamma x(y-z)dx$$

$$+P(\frac{B_0}{2}+e)=0 \qquad (3)$$

where equations (1) and (2) represent the equilibrium equations in the y and x directions respectively and equation (3) represent the moment equation around point O.

From these equations, it is easy to see that the foundation load P is a functional of two functions y(x) and σ(x). Thus, the rigourous bearing capacity problem is a variational one of the isoperimetric type as follows

$$\text{Min } P=\int_0^{x1}F(x,\ y,\ y',\sigma)dx$$

subject to

$$\int_0^{x_1}G_i(x,\ y,\ y',\sigma)dx=a_i \qquad (i=1,2)$$

where F(x, y, y', σ) is simply obtained through one of the equations (1, 2, 3). G_i(x, y, y', σ) and a_i can be obtained from the two remaining equilibrium equations. It was shown by Petrov [5] that the solution of such a problem is obtained using the Euler equations as follows

$$\frac{\partial H}{\partial\sigma}=\frac{d}{dx}\frac{\partial H}{\partial\sigma'} \qquad (4)$$

$$\frac{\partial H}{\partial y}=\frac{d}{dx}\frac{\partial H}{\partial y'} \qquad (5)$$

where H is an intermediate functional given by

$$H=F+\lambda_i.G_i \qquad (6)$$

Notice that H is linear in σ, hence equation (4) is equivalent to $f(x, y, y')=0$. Solving this equation, one obtains the equation of the slip surface which is a log-spiral in the case of a constant ϕ. Replacing this equation into equation (6), one can see that H becomes independent of the normal stress distribution. This result is a direct consequence of the shape of the slip surface. It was shown by Soubra [4] that any equation of the normal stress distribution having at least two degrees of freedom will satisfy the three equations of static equilibrium and that only the equation of moments around the centre of the log-spiral is sufficient to calculate the foundation load P. It is easy to see that the moment equation of the rotational log-spiral mechanism around the centre is identical to the work equation for the same mechanism in the upper-bound method in limit analysis. Thus, solving the bearing capacity problem by the upper-bound method in limit analysis for a rotational log-spiral mechanism will give a rigourous solution in regard to the limit equilibrium method. This method is detailed in the following section

2.2 UPPER-BOUND METHOD

According to the upper-bound theorem in limit analysis, for a kinematically admissible mechanism, the rate of external work exceeds the internal rate of dissipation of energy along the plastically deformed region. Thus, equating the external rate of work of all external forces to the internal rate of dissipation of energy gives an upper-bound of the exact solution for an associated flow rule material.

As was mentioned in the preceeding section, the mechanism considered in this paper consists of a rotational log-spiral mechanism whose equation is given as

$$r=r_0.\exp(\theta-\theta_0)\tan\phi$$

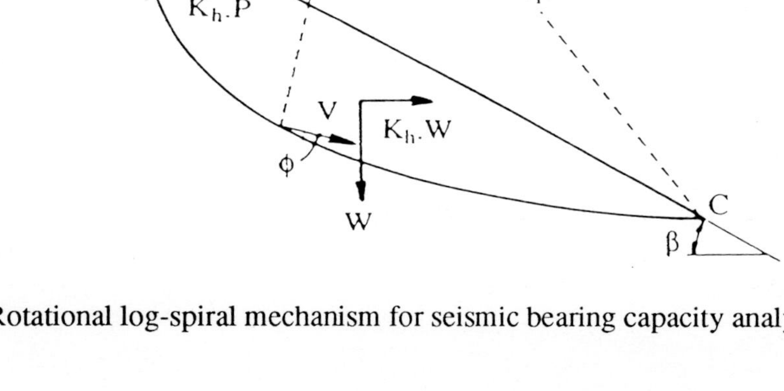

Fig. 3 : Rotational log-spiral mechanism for seismic bearing capacity analysis

For a rigid body rotation, this mechanism is kinematically admissible since the velocity V along the plastically deformed surface BC makes an angle ϕ with the transition layer according to the normality condition for an associated flow rule material.

2.2.1 Calculations of incremental external work

The incremental external work due to an external force is the external force multiplied by the corresponding incremental displacement or velocity. As shown (Fig. 3), the external forces contributing in the incremental external work consist of:

-The foundation load
-The weight of the soil mass
-The inertial forces on:
a.The soil-structure interface $K_h.P$
b.The soil mass $K_h.W$

The currently used values of K_h ly between 0.05 and 0.15 in the United States and between 0.15 and 0.25 in Japan. Notice that the choice of the seismic coefficient is completely emperical (Seed [6,7,8]).

The incremental external work for the different external forces can be easly obtained as follows:

a. Incremental external work due to self weight and inertia of the soil mass ABC.

$$\Delta W_{ABC}=\gamma.A\,(X_1 + K_h.Y_1).\Omega \tag{7}$$

b. Incremental external work due to the foundation load and the corresponding shear load:

$$\Delta W_P = P\,(X_2 + K_h Y_2).\Omega \tag{8}$$

The total incremental external work is the summation of these two parts of contributions, equation (7) and (8). That is:

$$\Sigma[\Delta W]_{ext}=\Delta W_{ABC} + \Delta W_P \tag{9}$$

2.2.2 Calculations of incremental energy dissipation

According to Finn [9], the incremental energy dissipation per unit length along a velocity discontinuity or a narrow transition zone can be expressed as follows

$$\Delta D_L=c.\Delta V.\cos\phi$$

It is noted that the internal dissipation is totally independent of the seismic coefficient. The energy dissipation takes place along surface BC, therefore;

Along BC, there is:

$$\Sigma[\Delta D]=\Delta D_{BC}=\int_{\theta_0}^{\theta_1} c.\frac{r.d\theta}{\cos\phi}.V\cos\phi \quad =c.r_0^2.f_1(\theta_0,\theta_1).\Omega \qquad (10)$$

where $f_1(\theta_0,\theta_1)$ is given by Soubra [4]. For a cohesionless soil, this dissipation is vanishing.

By equating $\Sigma[\Delta W]_{ext}$ in equation (9) to ΔD_{BC} in equation (10), we have:

$$q_c=\frac{P}{B_0}=\gamma\frac{B_0}{2}.N_\gamma+c.N_c \qquad (11)$$

where N_γ and N_c are the seismic bearing capacity factors. The most critical q_c-value can be obtained by minimization with respect to θ_0 and θ_1 angles [Fig. 3].

A pascal computer program for assessing seismic bearing capacity coefficients has been developed with equation (11) as a basis. The program gives the critical angles corresponding to the critical bearing capacity .

3 Numerical results and discussions

As mentioned previously, the present limit analysis solutions are valid when there is no reduction in soil strength due to an earthquake. This is the case only when the seismic acceleration a<=0.3g since most soils will either liquify or seriously weaken if a>0.3g (Okamoto [10]).

3.1 BEARING CAPACITY FACTORS

The bearing capacity factors N_γ and N_c are presented (Fig. 4). It is evident from these figures that these factors follows the same trend with respect to ϕ and β.These factors increase with an increase in ϕ and decrease with an increase in β.

N_γ	Saran & Agarwal	Present solution	
	β=0	β=0	β=5
φ=20	6,36	7,33	4,91
φ=25	–	16,52	9,88
φ=30	29,45	38,07	18,76

N_C	Saran & Agarwal	Present solution	
	β=0	β=0	β=5
φ=20	17,71	17,39	15,96
φ=25	–	25,02	22,58
φ=30	37,24	37,60	33,30

Figure 4 : Bearing capacity coefficients N_γ and N_c

3.2 COMPARISON OF BEARING CAPACITY FACTORS

Recently, Saran and Agarwal [11] have proposed a non-symetrical mechanism to study the bearing capacity of eccentrically obliquely loaded footing. Their mechanism consists of a triangular wedge under the footing, a radial shear zone and a passive Rankine zone. The approach used in their analysis is a limit equilibrium method. Figure 4 compare the N_γ and N_c values obtained from the proposed analysis and those obtained from the available Saran and Agarwal's solution concerning the case (β=0°).

These values concern the static case (k_h=0) and a footing subject to central vertical loads. Figure 4 indicates that the proposed N_γ-values compare reasonably with Saran and Agarwal's solution, however; the values of N_c obtained from the present analysis are very close to those obtained by Saran and Agarwal. Notice that the numerical results presented in this paper do not consider the moment due to the seismic loading on the structure (i.e. e=0), and the λ-value is taken equal to zero.

Kusakabe et al [12] have presented an upper-bound solution for the bearing capacity of foundations on slopes by introducing a failure mechanism which consists of a triangular active wedge, a log-spiral radial shear zone and a quadrilateral passive zone. For (φ=30°, $c/\gamma B_0$=1, β=15°), the value of $q/\gamma B_0$ as obtained by the present approach is equal to 25.05. The corresponding value given by Kusakabe et al and Bishop are respectively 28.9 and 29. This means that our upper-bound method gives interesting solution in this case. Notice that Bishop's method is a limit equilibrium method. His approach is based on a priori hypothesis. Hence, his solution is an approximate one and can not be classified as an upper-bound solution. Our method is more rigourous than the Bishop one since it is not based on a priori hypothesis and it verifies all the kinematical conditions of the problem.

Notice however that for large β-values, the coefficient N_γ decreases very rapidly and becomes negatif due to the unfavorable effect of the soil mass. Notice also that the determination of the coefficients N_γ and N_c requires the minimization of each factors separetly: The rigourous solution consists in determining the total load P causing failure due to both weight and cohesion by unique minimization of the foundation load. For the two reasons mentioned above, we use in the following section the dimensionless parameters $q/\gamma B_0$ and $c/\gamma B_0$.

3.3 SOME DIMENSIONLESS CHARTS

To enable a general representation of results, the nondimensional parameters $q/\gamma B_0$ and $G=c/\gamma B_0$ are employed. Figures 5 indicate the variation of the $q/\gamma B_0$ values with the $c/\gamma B_0$ values for $\phi=30°$, ($k_h=0$, 0.15) and for different values of slope angles β.

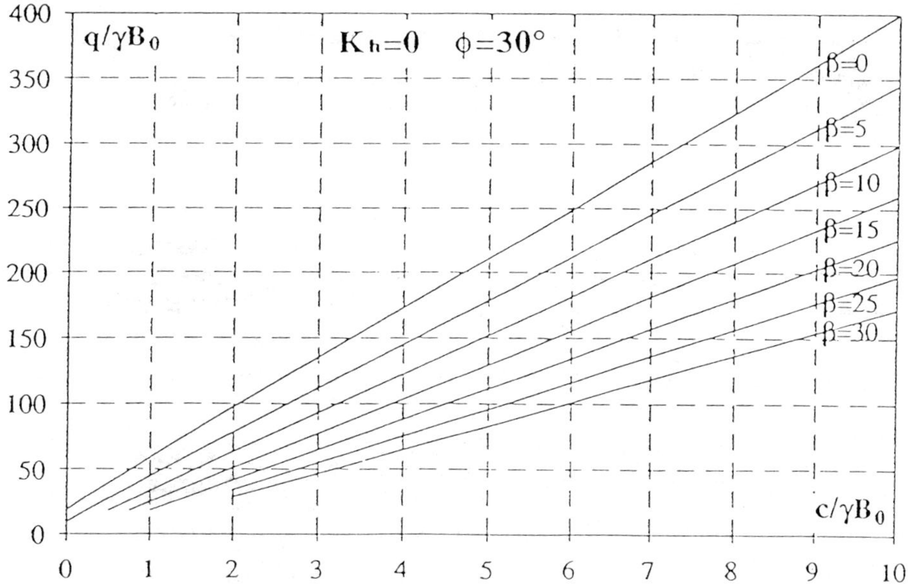

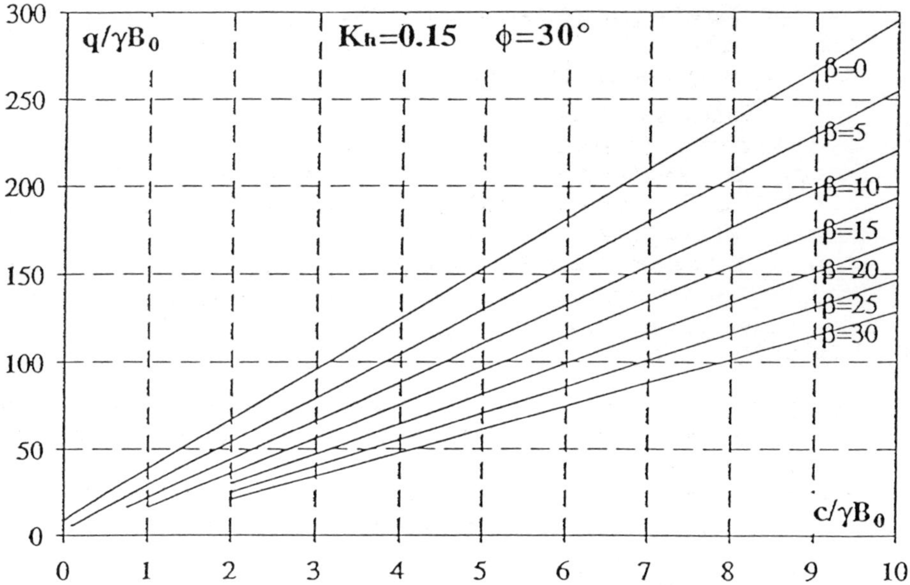

Fig. 5 : Variation of $q/\gamma B_0$ with $c/\gamma B_0$

Due to (Fig. 5), it is easy to see that for ϕ=30°, $c/\gamma B_0$=10, β=30°, the reduction in the $q/\gamma B_0$ value is about 26% when the horizontal seismic coefficient increases form zero to 0.15. Thus, the calculation of the nondimensionalized bearing capacity $q/\gamma B_0$ taking into account the earthquake forces is of great interest in areas of high earthquake risks.

4 Conclusion

The upper-bound method in limit analysis for a rotational log-spiral mechanism is applied to the seismic bearing capacity problem for a strip footing resting on the top of a slope in a quasi-static manner. The approach presented is interesting since the solutions so obtained are in good agreement with the theoretical and experimental results of Saran and Agarwal in the case (K_h = 0) available in literature.

Notations

k_h	horizontal seismic coefficient
$\sigma(x)$	normal stress distribution along y(x)
z(x)	equation of the ground surface
x_1	abscissa of end point C
r	ray of the log-spiral mechanism

r_0, θ_0	initial ray and angle of the log-spiral mechanism
V	velocity
W	weight of the soil mass
Ω	angular velocity
A	area of the soil mass ABC
X_1, Y_1, X_2, Y_2	distances from the point O to the line of action of forces W, K_h.W, P and K_h.P respectively.
q_c	ultimate bearing pressure
a	seismic acceleration

Bibliography

[1] Meyerhoff G.G. (1953). The bearing capacity of foundations under eccentric and inclined loads. Proceedings of the 3rd international conference in soil mechanics and foundation engineering, Vol. 1, p. 440-445.

[2] Shinohara T., Tetsuro Tateishi and Koichi Kulso (1960). Bearing capacity of sandy soils for eccentric and inclined loads and lateral resistance of single piles embedded in sandy soil. Proceedings of the 2nd world conference on earthquake engineering, Vol. I, Japan, p. 265-280

.[3] Sarma S.K. and Iossifelis I.S. (1990). Seismic bearing capacity factors of shallow strip footings. Géotechnique, 40, N° 2, p. 265-273.

[4] Soubra A.H. (1989). Application de la méthode variationnelle au problème de détermination des pressions passives des terres. Influence des forces d'écoulement. Theisis, INSA Lyon, 200 p.

[5] Petrov I (1968). Variational methods in optimum control theory, academic press, New-York and London, 216p.

[6] Seed H.B. (1966). A method for earthquake resistant design of earth dams, Journal of the soil mechanics and foundation division, ASCE, Vol XCII, N° SM3.

[7] Seed H.B. and Martin G.R. (1966). The seismic coefficient in earth dam design, Journal of the soil mechanics and foundation division, ASCE, Vol XCII, N° SM3,.

[8] Seed H.B., Lee K.L. and Idriss I.M. (1969). Analysis of the Sheffield dam failure, Journal of the soil mechanics and foundations division, ASCE, Vol XCV, N° SM6.

[9] Finn W.D.L. (1967). Applications of limit plasticity in soil mechanics, Journal of the soil mechanics and foundations division, ASCE, Vol 93 (SM5) p. 101-119.

[10] Okamoto S. (1956). Bearing capacity of sandy soil and lateral earth pressure during earthquake, Proceedings 1st World conference on earthquake engineering, Tokyo, p. 231-247.

[11] Saran S. and Agarwal R.K. (1991). Bearing capacity of eccentrically obliquelly loaded footing, Journal of geotechnical engineering, ASCE, Vol 117, N° 11, p. 1669-1690.

[12] Kusakabe O., Kimura T. and Yamaguchi H. (1981). Bearing capacity of slopes under strip loads on the top surfaces, Soils and foundations, Vol. 21, N° 4, p. 29-40.

STABILITY ASSESSMENT UNDER DYNAMIC LOADING OF THE RIGHT BANK OF THE SIR DAM (TURKIYE)

Luc VAULOUP *

Abstract: A non-linear dynamic analysis allowed to understand how the Sir Dam right bank, despite its very stable appearance, had been altered by earthquakes, to ascertain that the designed excavation would not appreciably change its behaviour, and to design a light rock support insuring its perenniality.

Résumé: Une analyse dynamique non-linéaire a permis de comprendre que la rive droite du barrage de Sir, malgé son apparence de grande stabilité, avait été dans le passé affectée par les tremblements de terre, que les excavations qui y étaient prévues ne changeraient pas sensiblement son comportement , et qu'un soutènement léger permettait d'assurer sa pérennité.

1. Introduction

Series of shear planes were discovered during the excavations of the spillway and dam's right abutment of the SIR Dam in Turkiye. These joints, having a gouge infilling of up to several tens of centimetres, were not critical for the slope stability under static conditions; the infilling shear strength was however not sufficient to prevent movements during earthquakes. The approach to the problem and to the necessity and design of a rock support has been to appraise to what extent a dynamic loading could jeopardize the bank, both in terms of global stability and of irreversible movements.

A non linear analysis of the bank was carried out in this respect, both on the natural ground's shape of the bank and after the proposed excavation, in order to better understand the mechanism of formation of the bank as well as to design a rock support insuring the perenniality of the works.

---

* Chief Engineer, Coyne et Bellier Consulting Engineers, Paris.

2. Information and data

2.1 BRIEF DESCRIPTION OF THE PROJECT

The Sir project consists of a 120 m high double curvature arch dam located in the south-centre of Turkiye, feeding a 315 MVA power plant located at its downstream toe. The spillway, designed for a peak outflow of 7 500 m³/s, is located on the right bank and has required considerable rock excavations: 450 000 m³ out of the 850 000 m³ of excavation for the whole of the project. The other main features of the scheme will be found on Fig. No 1.

The main part of the civil works were commenced in 1987. The scheme is now under operation, with the last unit commissioned in July 1991.

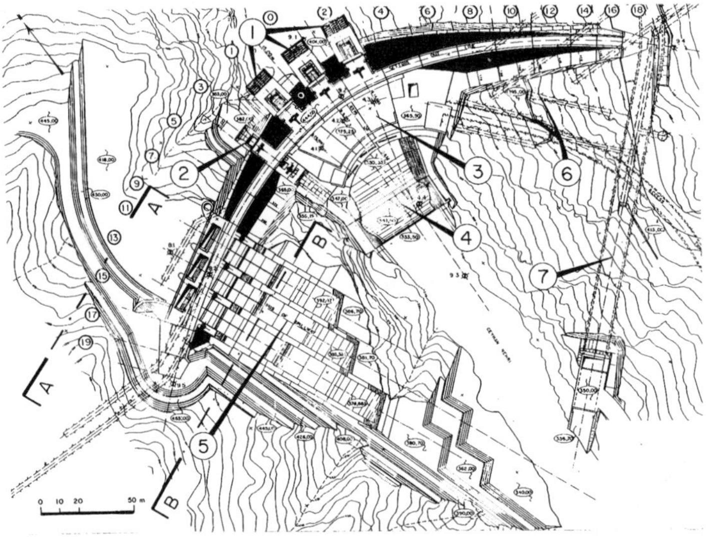

Fig No 1 - Plan View

1 - Power Intakes

2 - Bottom Outlet

3 - Powerhouse

4 - Tailrace

5 - Spillway

6 - Anchor Galleries

7 - Diversion Tunnel

2.2 TOPOGRAPHY

Over the dam's height, the site topography is asymetrical, with a milder general slope on the left bank (35° to 40°) than on the right bank (55° in average). Above the dam's crest, the slope flattens on both sides; on the right side, the natural ground slopes at approximately 30° over a great distance, up to the foot of a mountain culminating at an altitude of 1 500 m.

While it appears at the first glance that the left bank is entirely modelled by the major rock discontinuities (bedding faults) with large surface areas of rock slab exposed being the trace of large slip failure the steep, almost vertical, right bank was giving an impression of great stability, with absolutely no sign of recent failure. Before the start of the works, it appeared as a succession of high (10 to 30 m) vertical faces, interrupted by 10 to 15 m wide ledges on which footpaths had been established for access to the investigation works (boreholes and exploratory adits).

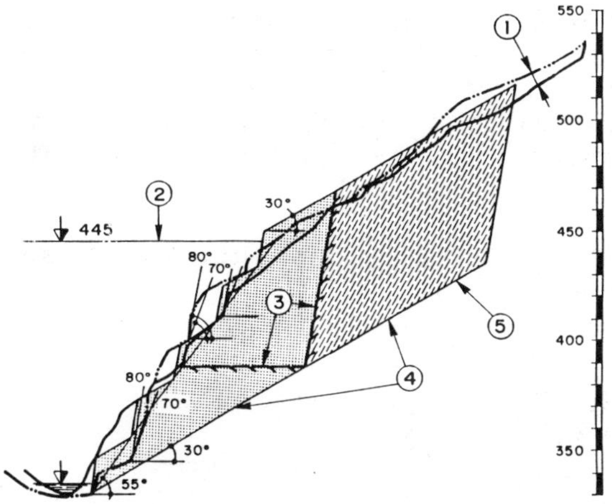

Fig No 2- Cross sections of the Right Bank

1 - Sections A & B of Fig.No 1
2 - Dam Crest Elevation

3 - Spillway Excavations

4 & 5 - Finite Element Mesh Limits
4 - Original
5 - After Spillway Excavations

2.3 GEOLOGY - GENERAL

The project site is located in the Paleozoïc formations, composed here of (hard) quartzites and (soft) phyllites alternations, at the upstream end of a 20 km long gorge through which the Ceyhan river crosses the Amanos mountains. The works are founded on quartzites with a few phyllite intercalations, whereas the reservoir lies on formations where phyllites are predominant. The borderline between these two facies is a major reverse fault diagonally crossing the valley 100 to 200 m upstream of the dam.

2.4 SEISMICITY

Three continental plates meet in the area: the Anatolian, Arabian and African plates; two major active faults meet at approximately 40 km of the site: the Dead Sea fault and the East Anatolian fault. The seismic activity is therefore appreciable in the area, with the occurrence of more than 15 earthquakes with magnitudes exceeding 5 in the last 50 years.

All the works, including the slope stabilities and relevant rock support, are therefore designed to resist such dynamic loadings. The seismic studies have led to consider for the Design Basic Earthquake a peak acceleration of 0.2g corresponding to earthquakes of frequent period of return (100 years),of magnitudes of 6.0 to 6.5, with epicentres at 15 km of the site. The Maximum Design Earthquake, which would correspond to seismic events on the East Anatolian Fault, with estimated period of return of 1000 years and magnitudes at the site of 7.5 to 8.0, would result in peak ground accelerations of 0.3g.

2.5 ROCK DISCONTINUITIES

The main families of discontinuities affecting to some extent the banks' stability are summarized herebelow.

(i) The weakest and most concerning discontinuities are the bedding faults, which are tectonically sheared bedding planes generally corresponding to phyllite intercalations; strike: N50°E, dip: 37° NW in average. They generally present gouge infilling which greatly vary in thickness (a few millimitres to 30 to 40 cm) with spacing of the order of 5 to 15 m. They do not affect the right bank's stability.

(ii) Two families of subvertical joints, striking roughly N-S, dip 80°E, dip direction N110°, and E-W, dip 90°. Their spacing is in the range 5-20 cm, with a greater density on the right bank. They are generally unweathered and without infilling, and closed on the left bank.

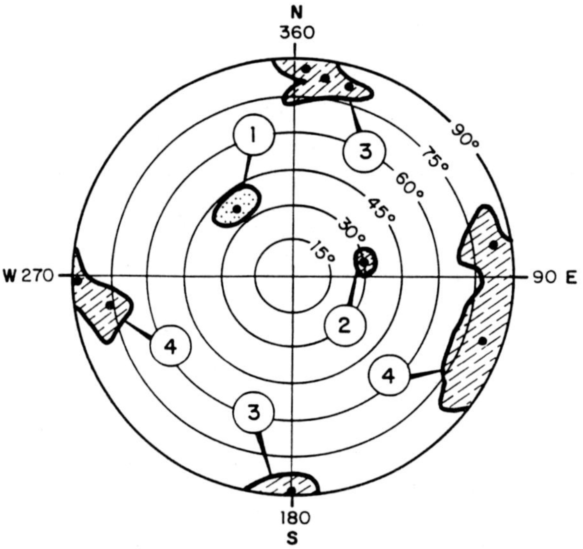

Fig No 3- Diagram of the Discontinuities (Kwaraiski, Upper Hemisphere)

1 - Bedding Faults

2 - Shear Planes

3 - E-W Joints

4 - N-S Joints

(iii) A series of shear planes, striking N20°E and dipping 30°E in average, with a 15 m spacing. These planes are specially observed in the right bank, where they form, with the above N110° subvertical joints, wedges the stability of which entirely relies on the shear strength available on these planes. They are generally filled with 5 to 40 cm gouge very similar to that of the bedding faults. Their geomechanical characteristics will be given in the next paragraph.

A special mention must be made here to the fact thet in the right bank, a systematic survey of the exposed rock surface has shown that some of the subvertical joints of the N110°E family were appreciably opened (1 to 5 cm) at the ground surface. A survey carried out in the exploratory adits at depth shows that such openings also exist, but are millimetric. The spacing of these open joints is typically of 10 to 20 m at the ground surface.

The poles of these main families of discontinuities are shown on Fig. No 3.

2.6 GEOMECHANICAL CHARACTERISTICS OF THE MATERIALS

The quartzite rock matrix is not questioned here, having an unconfined compressive strength of more than 100 MPa in average, and the stability entirely relies on the shear strength of the shear planes which in turn is governed by that of their infilling. They can schematicaly be divided into two parts: a coarse fraction on both sides close to the fault walls, mainly composed of crushed quartzites, and a finer fraction at the centre; its effect is predominant on the total shear strength;

This finer part mainly consists of non-plastic silt and clay, composed of quartz (51%), illite (36%) and kaolinite (4%), with variable and much smaller proportions of calcite and feldspar. These mineralogical analyses, their thickness and material grading confirm the similarity with the bedding fault infilling, and allow for the use of mechanical tests results made on these discontinuities. Laboratory direct shear tests, and above all, direct in situ large scale shear tests had revealed a residual shear strength equivalent to a friction angle of **31°**.

This value, obtained on thick infillings, is considered as the lower limit of the total available strength: it takes into account neither the variability of the material over great surface areas, nor the undulations of the fault walls, which provides extra strength by dilation.

Direct observation of the rock wedges during the course of the excavations brought additional valuable information: while no instability was noticed on planes dipping at 30°, or less, local instabilities were observed on the left bank bedding faults dipping at 37° or more, and a complete failure occured on the right bank on one of these shear planes locally sloping at 40°.

From these tests and observations, the estimate of the average global shear strength was in the range of **35°** to **37°**, with the gouge's own shear strength in the range of **30°** to **35°**.

An important question was also to appreciate whether the original shear strength may be altered by movements on these planes; it is clear that the gouge's strength is purely residual, and the answer is therefore directly related to the extra strength provided by dilation. The plane walls' undulations directly observed on large surface areas have a wave length of several meters, and it was clear that their contribution to the stability would only be written off by displacements of the same order of magnitude, which would not be acceptable; the shear strength can therefore be considered as unaffected by displacements acceptable for the works.

3. The stability problem

Under static conditions, and in absence of uplift water pressures, the bank is clearly stable. The best evidence is its natural appearance.

On the other hand, it is also obvious that movements cannot be avoided during major eathquakes: indeed, even if one assumes that the shear strength may be as high as 40°, it would entirely be mobilized by a relative horizontal acceleration of:

$$\gamma_h = \tan(\Phi - D) \qquad = 0.18 \text{ g}$$

with: D = dip of the planes = 30°
Φ = friction angle = 40°

For friction angles of 35° and 37°, the corresponding limit acceleration are 0.09g and 0.12g respectively.

The above mentioned design peak accelerations (0.2 and 0.3g) exceed these values, and an amplification with the altitude, resulting from the bank geometry, has to be expected. This means that, regardless of the actual shear strength available on these planes, it would be exceeded for short periods of time even during the so-called Design Basic Earthquake, and the rock masses would therefore experience displacements.

However, this does not mean that a major slope failure would occur: indeed, if a rock mass is put into movement by a peak acceleration exceeding the available strength, this only lasts a few hundreaths of seconds, and as soon as the acceleration reduces below the critical value, the rock mass slows down and stops after a while; this phenomenon is even quicker if a pulse in the opposite direction occurs.

The concept of stability expressed in static or pseudo static terms is therefore of little use in such a case, as the rock masses will in any case experience displacements. The real questions are therefore:

(i) does the shear strength dramatically decrease during such movements, leading to large irreversible displacements and a large scale failure, and, if not,

(ii) are the expectable movements acceptable for the bank itself and the structures built at its toe?

The answer to Question (i) has already been dealt with in the preceding paragraph, and no appreciable drop of the shear strength is feared in the present case; the real question is therefore Point (ii).

4. The Non-linear Modelization of the Bank

The idea is therefore to estimate the magnitude and pattern of such displacements in order to firstly understand why the bank presents this appearance of great stability, and secondly to appraise its expectable future behaviour during earthquakes, after excavation. A two-dimensional non-linear dynamic finite element analysis of the bank was carried out in this respect.

A simplified geometry was chosen, made of elastic rock blocks separated by shear planes and subvertical joints both modelized by non linear joint elements having no tensile strength and a limit shear strength defined by a friction angle and a cohesion; when this shear strength is exceeded, differential movements occur between both sides of the joints, calculated by Newmark's method of integration. The code used is the GEFDYN program, and more detailed information on the joints' law of behaviour is given in Appendix.

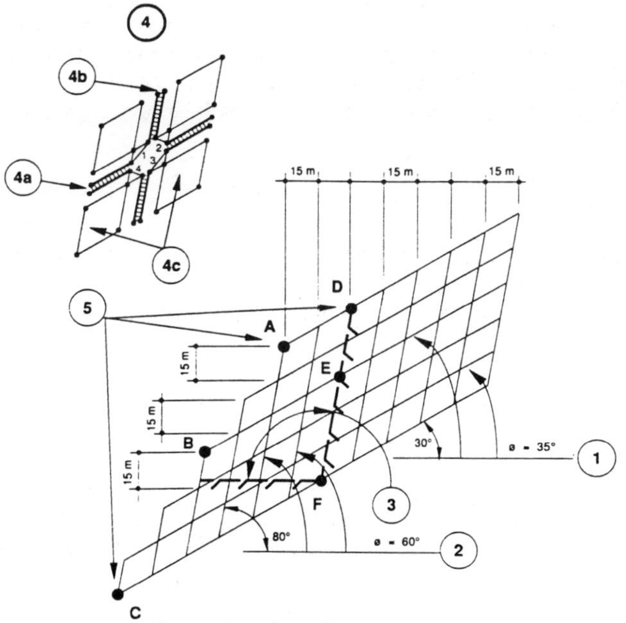

Fig No 4 -
Finite Element Mesh

1 - Shear Planes

2 - Subvertical Joints

3 - Trace of Spillway Excavation

4 - Typical Arrangement of the Volume and Joints Elements
4a & 4b : Non-linear Joints
4c : 2D Rock Matrix Elements

5 - Points referred to in the result plots

The boundaries of the model are shown on Fig. No 2, superimposed to the bank cross sections, the mesh and detail of the joint arrangement on Fig.No 4. The idealized cross section was chosen to simplify the calculations, the purpose being here to illustrate a mechanism and derive orders of magnitude, not to obtain accurate values of displacements: the modelized spacing of both the shear planes and the open subvertical joints is the average observed spacing of these discontinuities i.e. 15 m in both directions.

With the same purpose of illustration, a voluntarily simplified dynamic loading is applied to the model under the form of volumic inertia forces: a sinusoïdal acceleration with a 5 Hz frequency (around the maximum response to the ATHEDF accelerogram), and two maximum amplitudes: 0.1 and 0.2 g, is used. The response to this loading is calculated step by step at a time interval of 0.01 s over a period of 2 seconds.

Such a loading, well suited to understand the bank's behaviour, is however pessimistic as regards the displacements: indeed, it delivers the peak acceleration 5 times per second over the 2 second period, while this peak is only reached once during the total duration of the reference accelerogram.

The geomechanical data used in the calculations are detailed in Appendix. One will just mention here a 35° friction angle and a 0.2 MPa cohesion are taken on the shear planes.

5. The Behaviour of the Bank in its Original State

The irreversible displacements and joint openings at the end of the 2 second loading are shown on Fig. No 5. The displacement history of selected points of the mesh are shown on Fig. No 6.

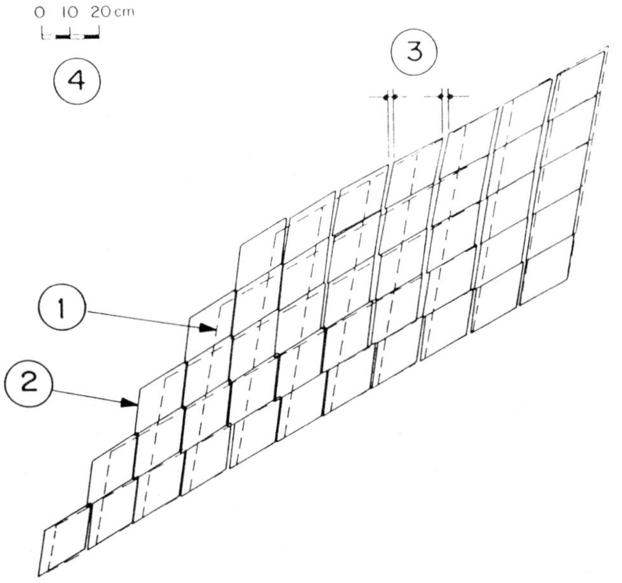

Fig No 5 - Displacements & Joint Openings (t=2s, 0.2g)

1 - Original Mesh

2 - Deformed Mesh

3 - Joint Openings

4 - Displacement & Joint Opening Scale

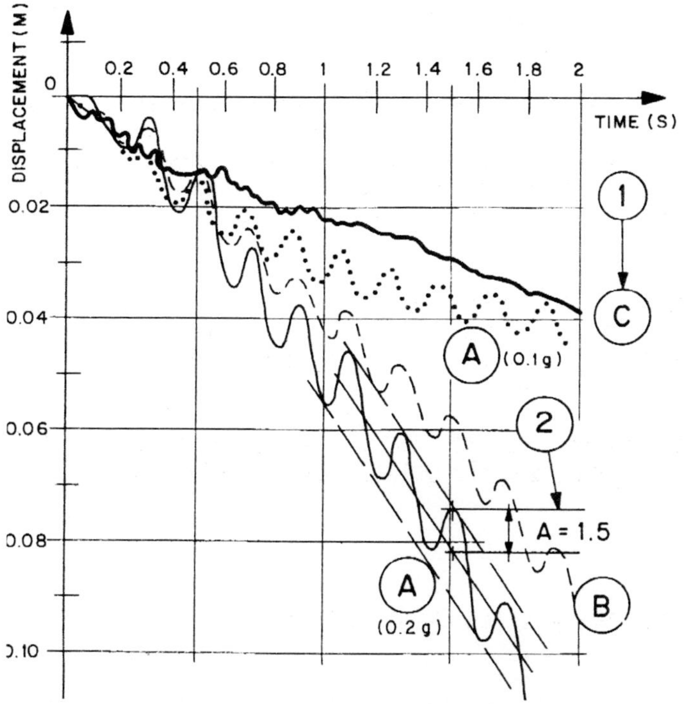

Fig No 6 - Displacement History of Points shown on Fig. No 4

1 - Displacements of Points A (at 0.1 & 0.2g), B & C; Original Bank

2 - Amplification of Relative Acceleration at Point A

Several observations can be derived from these plots:

- The non-linearity of the phenomenon is illustrated by the comparison of the total displacements for 0.1 and 0.2 g. This simply results from the fact that only relative accelerations above 0.09 g (= tan (35°-30°)) produce irreversible displacements.

- The time-displacement curves show the expectable amplification with height on the bank face. At Point A which experiences the larger displacements, the maximum relative acceleration is approximately 1.5 times the peak ground acceleration. One will note at this occasion the two components of the movement: reversible, almost sinusoïdal, and irreversible, almost linear.

- Under this loading, the total irreversible displacements are expressed in centimetres (12 cm at Point A). They decrease towards the hill and towards the bank's foot (4 cm at the toe).

- The joint opening pattern with 0.2 g acceleration is not uniform, with larger openings towards the bank; towards the river they do not increase with time and fluctuate with values not exceeding 1 cm: after an initial period during which a steady movement establishes, the further riverside displacements increase the opening on the hillside, and the riverside columns of blocks move in parallel. One will also note the progressive closing of the joints with depth.

- The differential horizontal displacements are well distributed over the height and never exceed 2 cm, to be compared to the aggregate 12 cm.

6. Mechanism of Formation of the Bank

These results suggest a possible explanation of the bank's present shape, as well as of the appearance of stability and of the mechanism which led to it. The external agents which may have been involved are the river erosion, the earthquakes and the water uplift pressures.

In the absence of earthquakes, the discontinuities which can have modelled the bank are the subvertical joints, by buckling and toppling of the upper blocks progressively falling down as the erosion progresses. One can therefore imagine an original much steeper slope (around 80°) progressively shaped to the present 55° by surface run-off seeping into these joints producing local failures.

But this does not explain the two other main features of the bank, i.e. the opening of some of the vertical joints at ground level, and the flatter slope above the crest elevation, sloping at 30°, i.e. the average slope of the shear planes.

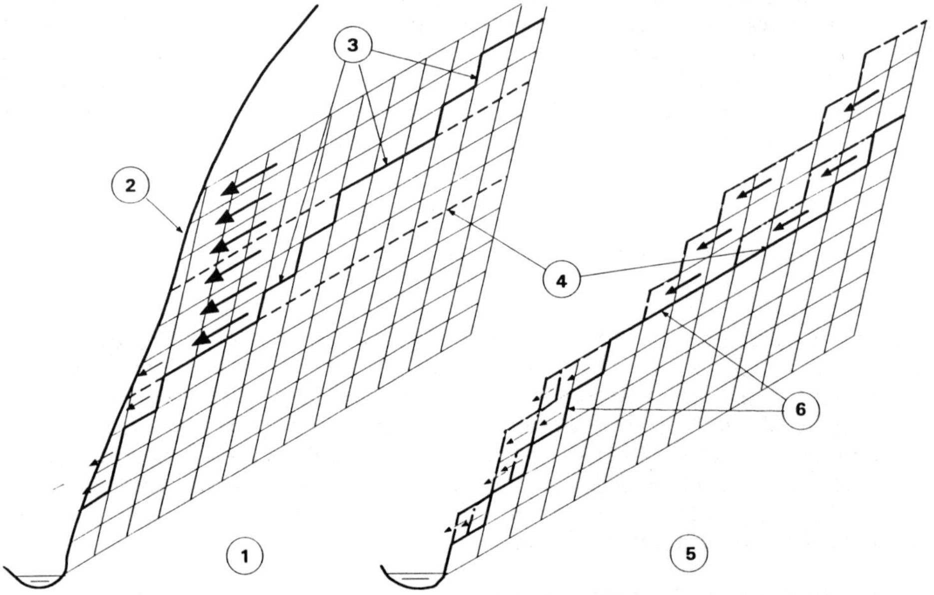

Fig. No 7 - Mechanism of Formation of the Right Bank - (1) Original Stages and effects of major earthquakes - (2) Assumed Original Shape - (3) Shape after major earthquakes - (4) Possibly thicker shear planes - (5) Possible further stages to the present shape - (6) Present shape.

One may obviously wonder whether these open joints may be the result of stress relief following the river erosion, and to what extent. An indication is here given by the first step of the calculation which consists in finding the original stress distribution in the bank, starting with a condition of nil displacement on the river side, condition which is progressively released: the displacements and joint openings are obtained after this operation, which simulates the stress relief, are smaller by one power of ten than those obtained by the dynamic calculation, and much smaller than those observed in the field: the maximum total displacement is 1 cm, with a maximum joint opening of 1 mm.

As the bank certainly experienced many earthquakes in the past, it is therefore more likely that the observed openings, which fit well with the calculated ones, are the result of such dynamic loading. The top slope strongly suggests that it could be the trace of an ancient shear plane, and a tentative explanation of the mechanism of successive failures which led to the present shape is sketched on Fig. No 7.

Due to the amplified relative acceleration at height, possibly combined with a thicker shear plane, major slides could have taken place on the top planes resulting in the almost complete disappearance of the top quartzites.

During the same events, in the lower part of the bank, the surface rock submitted to smaller accelerations experienced smaller movements on the shear planes, which, combined with loosening of the unconfined faces, progressively led to the stepped section and shaped the observed ledges. This phenomenon occurred during the more ancient major earthquakes. The more recent ones did not produce appreciable extra joint opening and the bank globally moved parallel to itself, without appreciable failures, giving the impression that the bank had not recently been affected by the earthquakes. The conclusion is therefore that a state of pseudo equilibrium has been reached, in the sense that the bank is and will be affected by future seismic events, with almost no visible effect at the rock surface.

6. The Behaviour of the Excavated Bank

Fig No 8 directly compares the displacements and joint opening of the original and excavated banks after the same loading. No appreciable difference is highlighted as regards the order of magnitude of the total maximum displacements and the general joint opening pattern. However the now unconfined toe of the excavation moves more than the original one. The relative accelerations applied to the top part are similar, but larger accelerations are applied to the lower blocks. This results in a greater irreversible displacement on the lower shear plane, which account for half of the total displacement.

The behaviour of the bank after excavation is globally similar to that of the original one, and no large instabilities are expectable even during major earthquakes. The main difference is the absence of confinement of the freshly excavated face, with local centimetric overhangs resulting from movements on the shear planes, doubtlessly accompanied by a general loosening of the rock close to the exposed face. Without support, local rock falls and failures are expectable, until the bank has recovered a slope similar to the present one, which provides the above-mentioned pseudo equilibrium.

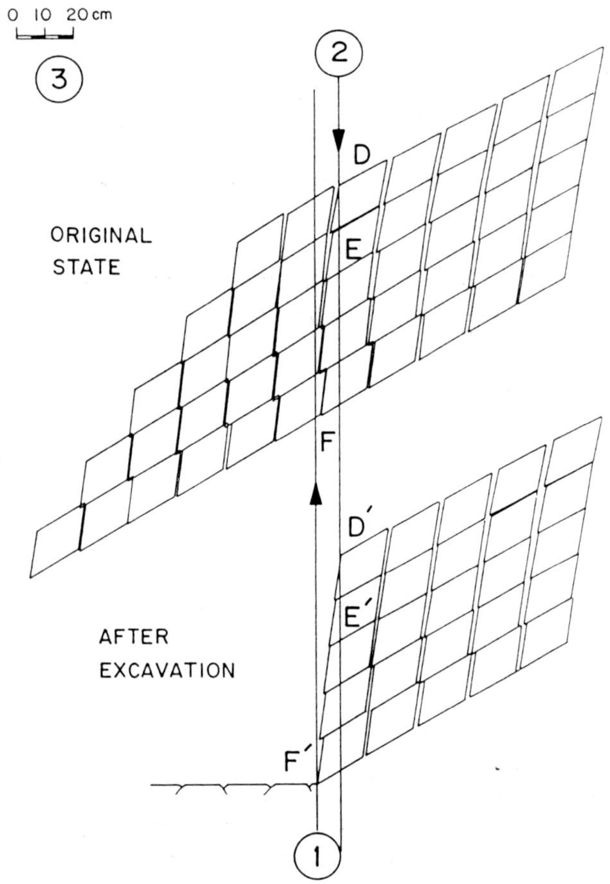

Fig No 8 - Comparison of Displacements before & after Excavation.

1 - Excavation toe

2 - Excavation top corner

3 - Displacement scale

7. Approach to the Bank's Rock Support

Are the expectable displacements acceptable or not for the works? If not the inertia forces in excess of the available shear strength should be counterbalanced by a suitable rock support which in fact would be aimed at improving the safety factor of the mountain: the required supporting measures would be in this case quite considerable.

Except at limited locations of the structures, where special measures have been applied (high capacity tendons), displacements of the calculated order of magnitude are not detrimental, provided they do not induce rock failures of several cubic metres, which would fall on the spillway channels and damage the concrete works.

The purpose of the rock support was therefore to **prevent loosening** of the freshly excavated face by re-establishing **a confinement under the form of a flexible but monolithic facing** able to cope with these displacements without appreciable disorder.

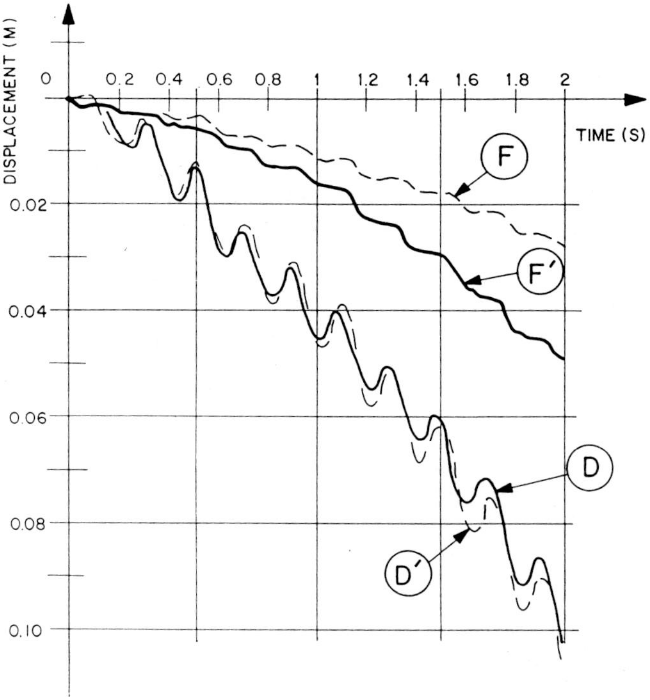

Fig No 9 - Comparison of Displacement History before & after Excavation.

Displacement versus time of Points D, D', F & F' shown on Fig. No 8.

The volume of rock to be so confined was the most delicate item to appraise; from the observed width of the ledges and the average spacing of the open joints (15 m), it was concluded that the result should be achieved by reinforcing a 15 m (horizontally) deep slice of rock from the excavation face by means of distributed passive anchors between the shear planes, tied to a wire mesh and a general shotcrete, and an increased anchoring through these planes for appropriate transmission of the inertia forces applied to the upper blocks, down to the unaffected rock.

A special mention must be made to the dam's abutment itself, for which a special calculation (not presented here) shows that the stressing of the bank by the arch thrust would prevent any movement of this abutment.

8. Conclusions

The stability of large rock slopes in areas of high seismic activity is often questioned: such slopes can be stable under static conditions, and may become unstable when submitted to a dynamic loading. In the present example, an aspect of apparent stability was hiding features resulting from ancient earthquakes.

Both the understanding and treatment of such problems can hardly be dealt with by means of a static or pseudo-static approach, the determining parameters being the irreversible displacements: indeed, the size of the rock mass may be such as to render impossible or uneconomical the support required to prevent movements.

The use of a non-elastic, dynamic method allows for estimating the expectable magnitude of such displacements, and for stating on the two fundamental questions:

(i) has a major failure (resulting from a drop of the shear strength) to be feared?

(ii) if not, are the expectable displacements acceptable for the designed works?

The correct understanding of the displacement pattern, and a reliable estimate of their order of magnitude can be derived of such an analysis, and allow for rehabilitating a solution which would be condemned by a pseudo-static approach, and design of cheaper rock support; this is at the cost of a permanent monitoring of the movements during earthquakes, cost which is in general largely covered by the savings made on this support.

Bibliography

Hendron A.J. and al. (1971). *Analytical and graphical Methods for the Analysis of Slopes in Rock Masses*. NCG Report No 36. Department of Civil Engineering, University of Illinois.

Appendix

GEFDYN non-linear joints

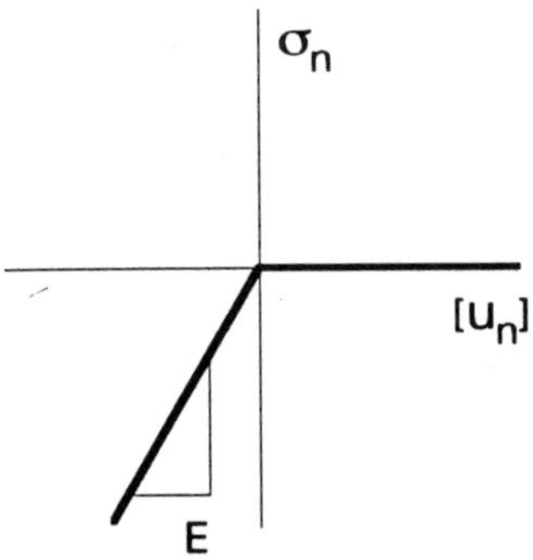

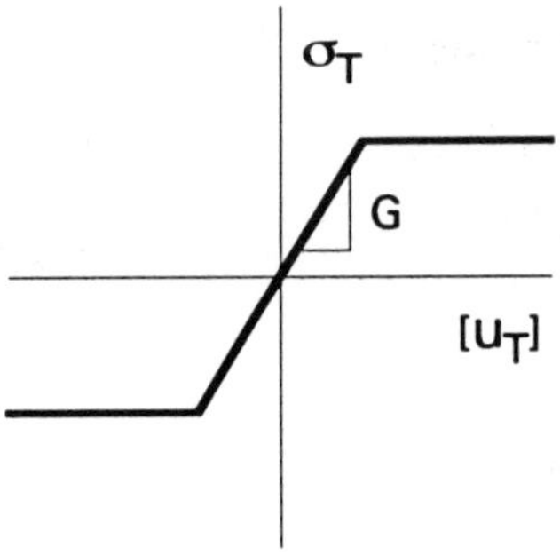

(i) Non penetration of the two volumes:

$[u_n] \geqslant 0$

(ii) Opening:

If: $[u_n] > 0$, then $\sigma.n = 0$

(iii) Contact with compressive strength:

If: $[u_n] = 0$, then $\sigma_n < 0$, and $/\sigma_T/ < /\sigma_n/\tan\Phi$

(iii.a) No plastic shear strain:

If: $/\sigma_T/ < /\sigma_n/ \tan\Phi$, then: $[\delta_t u_T{}^p] = 0$, and $[\delta_t u_n{}^p] = 0$

(iii.b) Plastic shear strain possible:

If: $/\sigma_T/ = /\sigma_n/ \tan\Phi$, then: $\exists\lambda > 0$

$[\delta_t u_T{}^p] = \lambda\, \sigma_T / /\sigma_T/$,

$[\delta_t u_n{}^p] = \lambda \tan\Psi$

with Ψ = dilation angle.

Numerical values of the parameters in the model

Rock matrix:

Elasticity Modulus:	$E=10^4$ MPa
Poisson's coefficient:	$\nu=0.2$

Joints representing the shear planes:

Elasticity coefficients:	$E' = 5\ 10^3$ MN/m^2, $G' = 2.1\ 10^3$ MN/m^2
Friction angle:	$\Phi = 35°$
Cohesion:	c= 0.2 MPa

Joints representing the subvertical joints:

Elasticity coefficients:	same as above,
Friction angle:	$\Phi = 60°$
Cohesion:	c= 0.2 MPa.

THREE-DIMENSIONAL NUMERICAL SIMULATION OF ROCKFALLS

Thomas ZIMMERMANN* and François DESCOEUDRES**

Abstract : Existing numerical simulation models usually consider the problem to be two-dimensional and follow the block motion along the line of maximum slope. The study of actual rockfalls shows that this can be totally inadequate.A fully three-dimensional model for single block dynamics is developed in this article. It considers the motion of a rigid block with elasto-plastic and frictional contact, impact, rolling and sliding on the ground.

Résumé : Les modèles de simulation existants considèrent en général une situation bidimensionnelle, l'étude de situations réelles montre que cette hypothèse peut être tout à fait inappropriée. Un modèle tridimensionnel général pour la dynamique de blocs isolés est développé ici. On considère le mouvement d'un bloc rigide avec possibilités de contact élastoplastique avec frottement, impact, roulement et glissement sur le sol.

1. Introduction

Near rock cliffs in mountainous regions, falling rocks and rockslides make dynamic studies of possible block trajectories necessary for the evaluation of the extent of the threatened zones as well as for the definition of adequate measures of protection.

Until now, existing models of computation are essentially two-dimensional. They simply apply to a ground profile selected in a vertical plane or along the line of the deepest slope, without being able to take care of the lateral dispersion of the block trajectories. In many cases of complex topography, with corridors, funnels, conic slopes or any other shape, the modelling of a single profile can be very misleading.

Faced with a real rockfall of considerable extent, in which a million cubic meters are involved, the authors have developed a numerical model for the study of spatial movement of blocks striking a three-dimensional topography.

The model is first described, followed by benchmark problems. It is then applied to the case of a major rockfall which occured in the swiss mountains in August 1985. For this simulation, the topography was captured by aerial photogrammetry and the response characteristics of the ground were identified through video observations taken during the main event. The simulation lead to delimitation of the risk areas with respect to additional rockfalls. The results show a good correlation with site observations.

---

* Chargé de cours, adjoint scientifique au Laboratoire de mécanique des structures et milieux continus, Ecole Polytechnique Fédérale de Lausanne

** Professeur, directeur du Laboratoire de mécanique des roches, Ecole Polytechnique Fédérale de Lausanne

2. Description of the numerical model

The numerical model applies rigid body dynamics to simulate the motion of ellipsoïdal or polygonal blocks. The driving force is gravity. During its motion the block encounters successive impacts on the topography. These impacts involve frictional contact and elasto-plastic response.

Solution algorithm : The main steps of the solution algorithm are : computation of the motion, block localization and contact-impact management. The algorithm is summarized in table 1 and the key steps are described next.

Motion : Compute the acceleration, velocity, displacement and position. Euler's equations apply, i.e :

$$m \frac{d\mathbf{v}}{dt} = \mathbf{F}$$

$$\mathbf{J} \frac{d\omega}{dt} = \mathbf{M} + \mathbf{M}(\omega)$$

where m is the mass, **v** the translation velocity of the center of gravity **F** the applied force (- mg for gravity), **J** the rotary inertia, ω the rotation velocity and **M** the applied moment, **M**(ω) is a nonlinear function of ω.
The trapezoïdal algorithm is used for time discretization of the equations of the motion.

Block localization : Localize the block with respect to the discretized topography.

The topography has to be discretized first. This is illustrated in Figure 1 for the specific case of les Créteaux, the event which is described later on. The topography was captured by aerial photogrammetry. A horizontal map was generated, and discretization into quadrilateral elements was performed next, in a way which approximates the topography as closely as possible. Bilinear interpolation is performed, within each element, to localize points on the surface with respect to the coordinates of element nodal points.

Figure 1a) illustrates a restitution of the topography 1b), the discretized topography and 1c) the interpolation scheme, within each element of surface.
Once a block is located in space, its projection on the topography can be calculated and, by comparizon, it can be determined whether impact occurs or not.

Contact-impact : if impact occurs, balance laws must be applied to determine the reflected motion (essentially $\mathbf{v}^+$, the velocity after rebound) from the incident motion (essentially $\mathbf{v}^-$).

Assuming that the velocity of the contact point is zero at some time $\bar{t}$, during the impact, the following coupled equations apply, in a body fixed referential; the incident impulse $\mathbf{I}^-$ results :

$$m\left(\bar{\mathbf{v}} - \mathbf{v}^-\right) = \mathbf{I}^-$$

$$\mathbf{J}\left(\bar{\omega} - \omega^-\right) = \mathbf{r} \times \mathbf{I}^- \qquad (1)$$

$$\bar{\mathbf{v}} + \bar{\omega} \times \mathbf{r} = \mathbf{0}$$

For each time step,

1. Solution of the equations of motion

   There are 3 possible cases :

   a. Free body motion

   b. Rolling motion

   c. Sliding motion

   Set : MOTION = "**FREE**" ; free body motion is assumed until evidence of the contrary

2. Localization of the block with respect to the topography

3. Test if the block is in possible impact position

   d < TOL ; TOL = tolerance

   d = distance to ground

   yes : set : MOTION = "**IMPACT**", go to 4

   no : go to 10

For each impact,

4. Computation of the incident impulse ($\underset{\sim}{I}^-$)

4.1 Test sign of incident normal impulse ($\underset{\sim}{I}_n^-$)

(+) : impact is realized, go to 4.2
(-) : impacting edge moves away, go to 10

4.2 Compute $\underset{\sim}{I}$; total impulse assuming sticking

4.3 if $|I_t| > \mu |I_n|$, instantaneous sliding occurs; set

$\underset{\sim}{I}_t^{(i+1)} = - \mu |I_n|^{(i+1)} \; \underset{\sim}{i}_t$; the tangential impulse is proportional to the friction coefficient and the normal impulse

$(I_t^-)^{(i+1)} = \frac{1}{2} \underset{\sim}{I}_t^{(i+1)}$, go to 4 and iterate until

$|\underset{\sim}{I}_n^{(i+1)} - \underset{\sim}{I}_n^{(i)}| / |\underset{\sim}{I}_n^{(i+1)}| <$ TOL; assume symmetry and iterate

Table 1 : Global transient algorithm

5. Compute reflected motion

6. Test current type of motion :

   MOTION = "IMPACT", go to 7; contact point velocity = 0

   MOTION = "ROLLING", go to 8; contact point velocity = 0

   MOTION = "SLIDING", go to 9; normal contact point velocity = 0

7. Impact motion, test for permanent contact : if the reflected normal velocity is insufficient for a rebound; set MOTION = "ROLLING", set the coefficients of restitution e_n, e_t to zero and recompute the impulse : go to 4

   Else : instantaneous contact occurs : go to 10

   For permanent contact ,

8. MOTION = "ROLLING", test for sliding motion :

   if $|I_t| > \mu |I_n|$, then sliding should occur; set MOTION = "SLIDING", and recompute the impulse; go to 4

   Else : sticking contact occurs, go to 10

9. Compute ground reaction

10. Compute energy balance for control purposes

11. Perform next time step : go to 1

Table 1 : Global transient algorithm (continued)

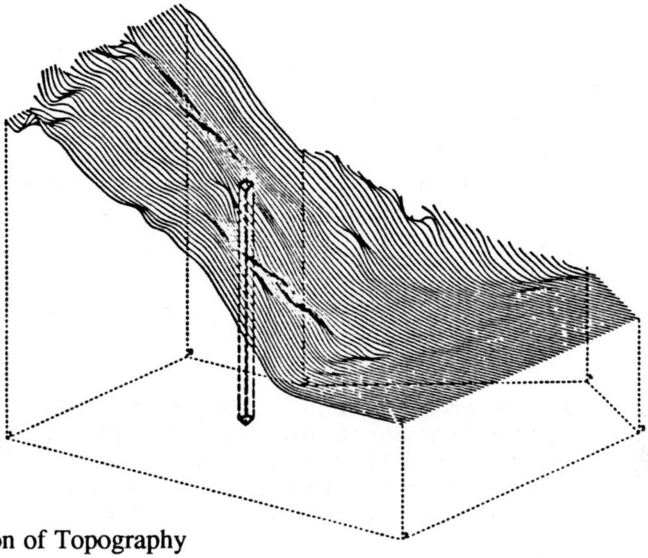

Fig 1 a) : Restitution of Topography

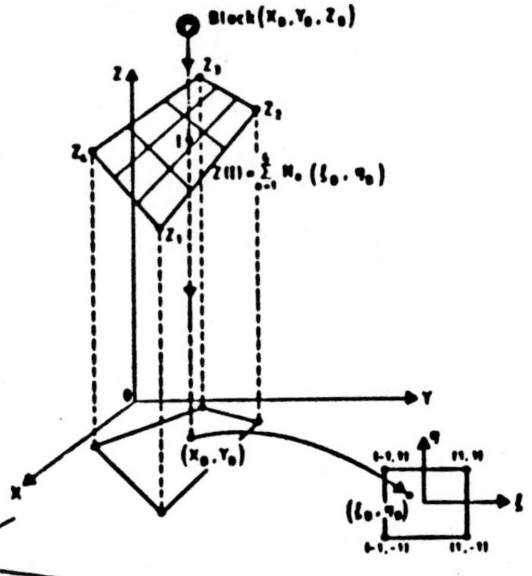

Fig. 1 c

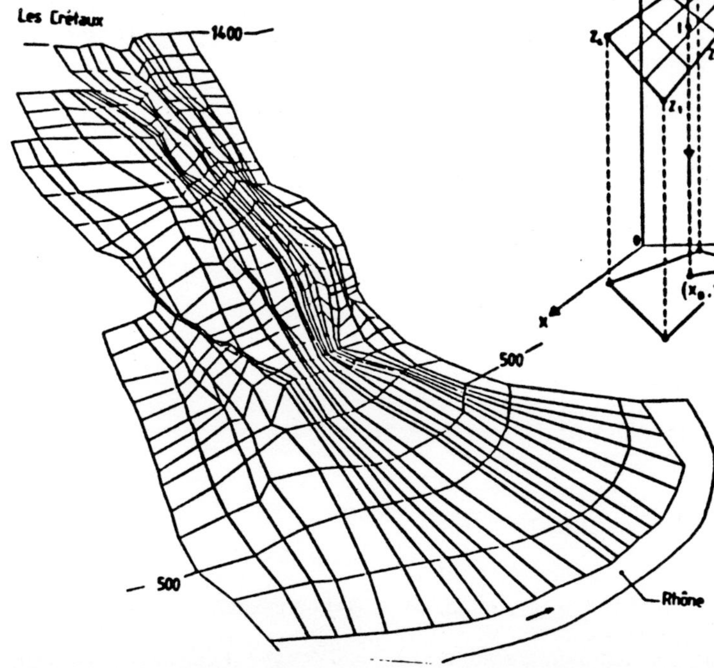

Fig. 1 b) : Les Créteaux - Perspective view of the discretized topography

Impact conditions : A coefficient of restitution, e_n and a coefficient of friction are used to describe nonconservative impact conditions.

Let the total impulse be $\mathbf{I} = \mathbf{I}^- + \mathbf{I}^+$ and split $\mathbf{I}^-$ into normal and tangential components. The reflected impulse, $\mathbf{I}^+$, is defined by

$$I_n^+ = e_n\, I_n^-$$

$$I_t = 2\, I_t^- \leq \mu\, I_n \quad \text{(see note)} \qquad (2)$$

where

$$\mathbf{I}^+ = m\left(\mathbf{v}^+ - \overline{\mathbf{v}}\right)$$

by analogy with $\mathbf{I}^-$.

The quantities e_n and μ can sometimes by assumed to be constants. Here, only μ is assumed constant and

$$e_n = \min\left(e_{ref},\, m^{0.5} * c/I_n^-\right)$$

where e_{ref} is the maximum value e_n can take and a function of the material involved. c is a site-dependent constant proportional to the maximum elastic energy of deformation of the ground (per unit surface), considered to be elastic-perfectly plastic. c can be derived from theory or tuned until the maximum rebound duration corresponds to the one observed on the site. Obviously, alternative definitions are possible.

Reflected motion : Once the total impulse is known the reflected motion results from

$$m\,(\mathbf{v}^+ - \mathbf{v}^-) = \mathbf{I}$$

$$\mathbf{J}\,(\omega^+ - \omega^-) = \mathbf{r} \times \mathbf{I} \qquad (3)$$

Rolling and sliding (Fig. 2) : if $\mathbf{v}_n^+$ is zero the block remains on the ground and the following equations apply.

$$m\,\mathbf{a} = m\,\mathbf{a} + \mathbf{R}$$

$$\mathbf{J}\,\dot{\omega} = \mathbf{r} \times \mathbf{R} + \mathbf{M}\,(\omega) \qquad (4)$$

$$\mathbf{a} + \dot{\omega} \times \mathbf{r} = -\,\omega \times \dot{\mathbf{r}}$$

The first two repeat the equations of motion with the ground reaction **R** added. The third maintains rolling contact. Again, the last equation is only valid if the normal and tangential components of the reactions, R_n and R_t, satisfy

$$R_T \leq \mu\, R_n \qquad (5)$$

If this is not true, an iterative solution is required.

Note : If equation (2) is not satisfied by the initial solution then system (1) must be solved iteratively with the 3rd equation replaced by $(I^-)_t^i. = 0.5 * \mu * (I_n)^i$

Arrest : Friction progressively diminishes the velocity of the block. However, the dominant contribution to arrest is given by the shape of the blocks. The assumption is made that contact is always located at a single point and this point will always be at an edge if the block is polygonal. Therefore, the reaction **R** alternatively opposes and activates the motion and the necessary condition for a block to be able to stop is that its shape allows a stable static equilibrium (Fig. 3).

An illustration of three-dimensional block-motions is shown in Figure 4.

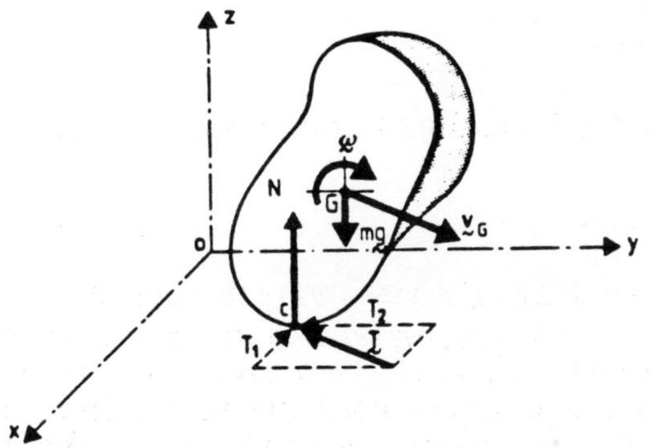

Fig. 2 : Equilibrium of block in contact

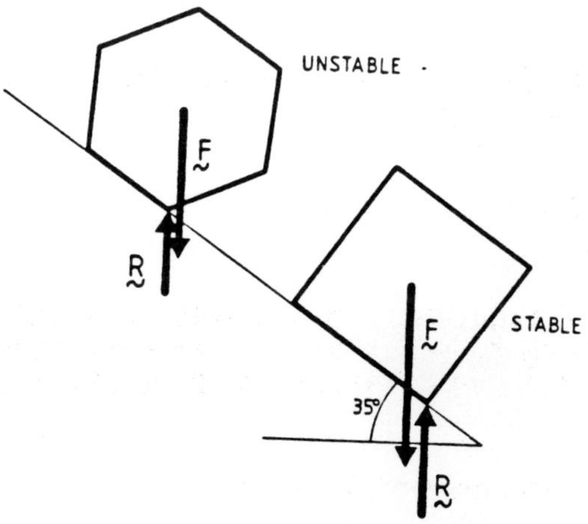

Fig. 3 : Influence of block shape

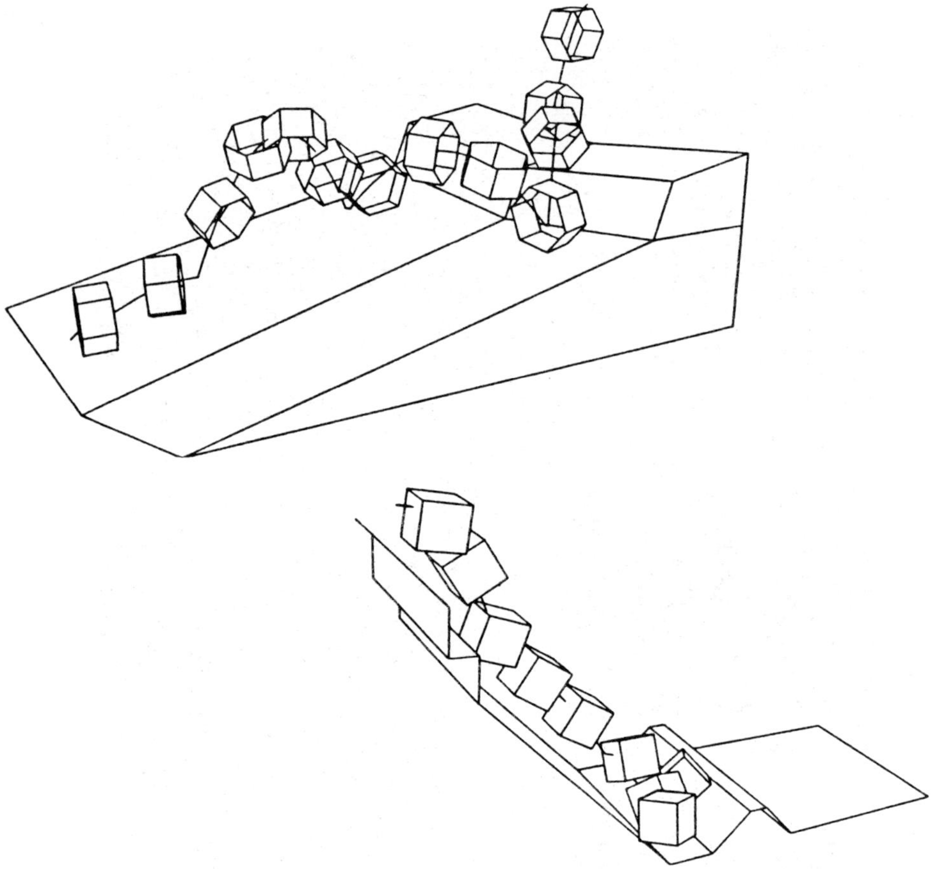

Fig. **4** : Typical block motions

3. Application to the rockfalls of les Créteaux in Vallis, Switzerland

A few months after the important rockfall of les Créteaux, which occured in August 1985, an analysis was performed, using the numerical model described above, to delimit the risk zones and to design remedial measures.

A view of the site is shown on figure 5; the element mesh of the site topography is shown in Figure 1. After some preliminary tests for tuning the model with the observed block rebounds and velocities (estimated from cine-camera takes), a series of runs was simulated.

Fig. 5 : View of site

The basic data is composed of the following informations :

- Starting zone : 10 different block positions, spaced every 20 m, in front of the rock cliff at the top of the corridor (Fig. 5)
- Geometry of the block : hexagonal prisms best fitting the shape of the fallen blocks
- Coefficient of restitution : $e_n = 0.85$ (corridor), $e_n = 0.40$ (vineyard)
- Size of the blocks : 4 m^3 (average observed size) or 10 m^3 (max.)
- Initial falling height : 2 or 20 m above topography
- Coefficient of friction : $\mu = 0.50$ (probable value) or $\mu = 0.30$ (unfavourable value)
- Coefficient of plastic shock (site tuning) : 331 $kg^{1/2} * m * s^{-1}$ (or 2 c or c/2).

From the large quantity of graphical output obtained, results for selected runs are shown in Figure 6 and 7. Figure 6 illustrates a typical simulation. Figure 6a shows a perspective view of the block trajectory with successive impacts. Figure 6b shows the same trajectories from an aerial point of view.

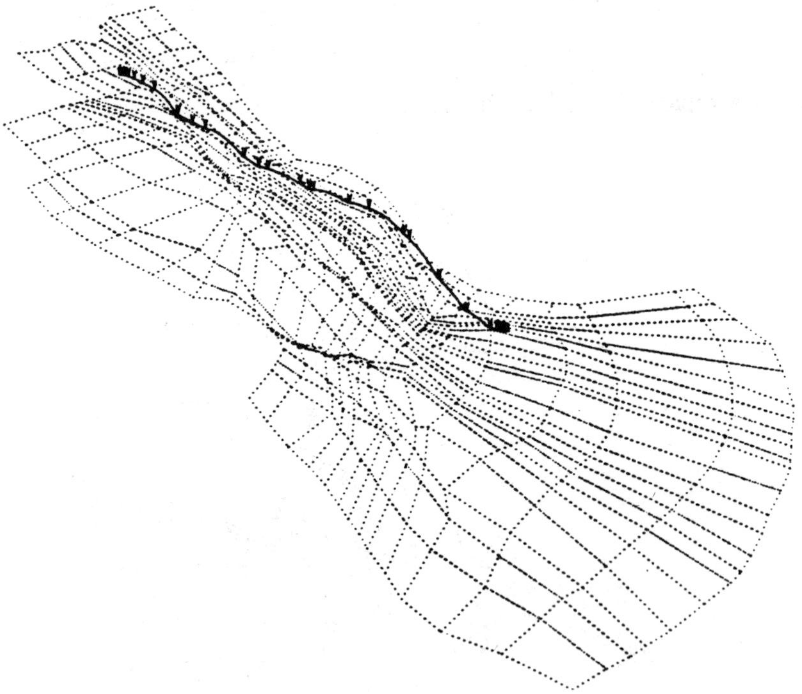

Fig. **6** : A typical simulation - **a**) Perspective view

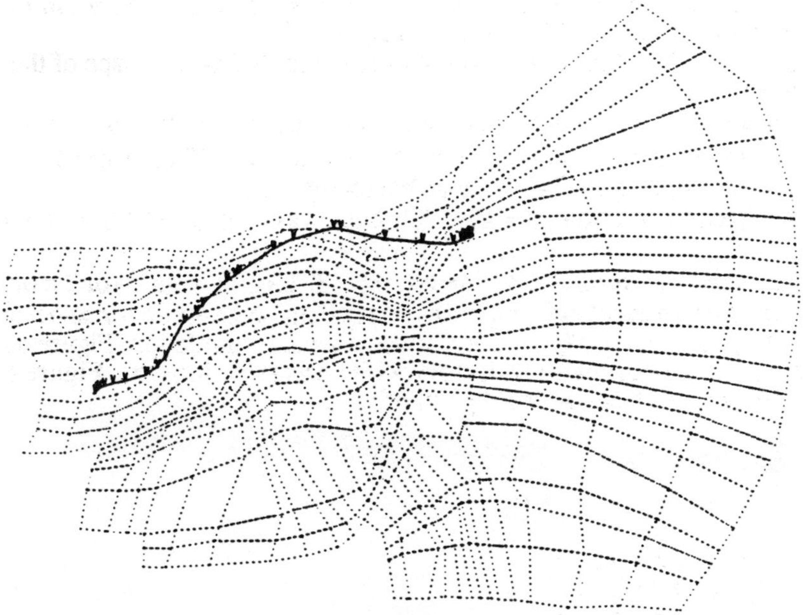

Fig. **6** :. A typical simulation - **b**) Aerial view

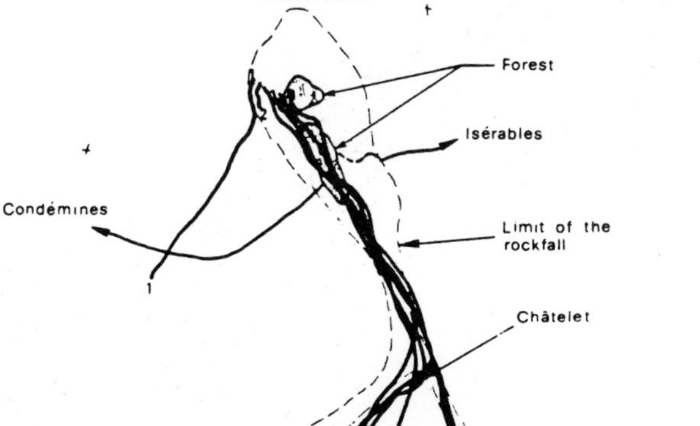

Fig. **7a**) : Les Créteaux - Trajectories of 8 (10 m^3) blocks (horizontal projection)

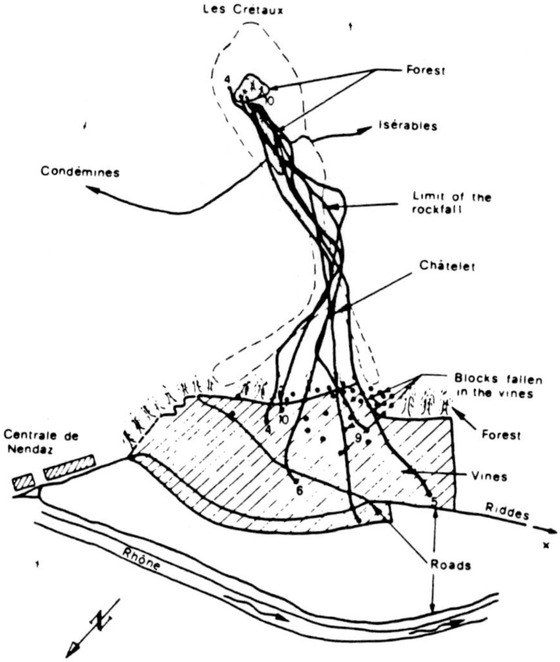

Fig. 7b) : Les Créteaux - Trajectories of 8 (4 m^3) blocks (horizontal projection)

4. Conclusions

The proposed numerical model's current stage of development is sufficient to allow for its use as a valuable tool for simulations of rock dynamics. Its usefulness for the prediction of the extension of threatened zones has been established. The calculated trajectories and the arrest zone agree well with site observations.

Acknowledgements

The contribution of E. Davalle and B. Rebora in the computer implementation of the model and the financial support of the canton of Vallis are gratefully acknowledged.

Reference

[1] Zimmermann Th., Rebora B., Davalle E, Descoeudres F., (1989). A three-dimensional numerical simulation model for rockfalls. *IREM internal report 89/1, EPF-Lausanne, Switzerland.*